Geometric Approaches to Differential Equations

AUSTRALIAN MATHEMATICAL SOCIETY LECTURE SERIES

Editor-in-Chief: Professor J.H. Loxton, School of Mathematics, Physics, Computing and Electronics, Macquarie University, NSW 2109, Australia

Editors:
Professor C.C. Heyde, School of Mathematical Sciences, Australian National University, Canberra, ACT 0200, Australia

Associate Professor W.D. Neumann, Department of Mathematics, University of Melbourne, Parkville, Victoria 3052, Australia

Associate Professor C.E.M Pearce, Department of Applied Mathematics, University of Adelaide, SA 5005, Australia

1   Introduction to Linear and Convex Programming, N. CAMERON
2   Manifolds and Mechanics, A. JONES, A GRAY & R. HUTTON
3   Introduction to the Analysis of Metric Spaces, J. R. GILES
4   An Introduction to Mathematical Physiology and Biology, J. MAZUMDAR
5   2-Knots and their Groups, J. HILLMAN
6   The Mathematics of Projectiles in Sport, N. DE MESTRE
7   The Petersen Graph, D. A. HOLTON & J. SHEEHAN
8   Low Rank Representations and Graphs for Sporadic Groups,
                        C. PRAEGER & L. SOICHER
9   Algebraic Groups and Lie Groups, G. LEHRER (ed)
10   Modelling with Differential and Difference Equations,
                        G. FULFORD, P. FORRESTER & A. JONES
11   Geometric Analysis and Lie Theory in Mathematics and Physics,
                        A. L. CAREY & M. K. MURRAY (eds)
12   Foundations of Convex Geometry, W. A. COPPEL
13   Introduction to the Analysis of Normed Linear Spaces, J. R. GILES
14   The Integral: An Easy Approach after Kurzweil and Henstock,
                        L. P. YEE & R. VYBORNY

Australian Mathematical Society Lecture Series. 15

Geometric Approaches to Differential Equations

Edited by

Peter J. Vassiliou & Ian G. Lisle

School of Mathematics and Statistics,
University of Canberra

CAMBRIDGE
UNIVERSITY PRESS

PUBLISHED BY THE PRESS SYNDICATE OF THE UNIVERSITY OF CAMBRIDGE
The Pitt Building, Trumpington Street, Cambridge, United Kingdom

CAMBRIDGE UNIVERSITY PRESS
The Edinburgh Building, Cambridge, CB2 2RU, UK   www.cup.cam.ac.uk
40 West 20th Street, New York, NY 10011-4211, USA   www.cup.org
10 Stamford Road, Oakleigh, Melbourne 3166, Australia
Ruiz de Alarcón 13, 28014 Madrid, Spain

First published 2000

Printed in the United Kingdom at the University Press, Cambridge

A catalogue record for this book is available from the British Library

Library of Congress Cataloging-in-Publication Data
Geometric approaches to differential equations / Peter J. Vassiliou & Ian G. Lisle.
  p. cm. – (Australian mathematical Society Lecture Note: vol. 15)
ISBN 0 521 77598 1 (pbk.)
1. Differential Equations. 2. Geometry, Differential.
I. Vassiliou, Peter J., 1953– . II. Lisle, Ian G., 1959– .
III. Series: Australian Mathematical Society Lecture Series ; 15.
QA372.G46 2000
515'.35 – dc21   99–36403 CIP

ISBN 0 521 77598 1 paperback

Contents

List of Contributors vii

Preface ix

Introduction: Geometric Approaches to Differential Equations
    Peter J. Vassiliou 1

Bäcklund and his Works: Applications in Soliton Theory
    Colin Rogers, Wolfgang K. Schief and Mark E. Johnston . 16

Recent Developments in Integrable Curve Dynamics
    Annalisa Calini 56

An Elementary Introduction to Exterior Differential Systems
    Niky Kamran 100

Cartan Structure of Infinite Lie Pseudogroups
    Ian G. Lisle and Gregory J. Reid 116

Cartan's Method of Equivalence
    David H. Hartley 146

The Inverse Problem in the Calculus of Variations and its
Ramifications
    Geoff E. Prince 171

Twistor Theory
    Michael K. Murray 201

Index 224

Contributors

Annalisa Calini
  University of Charleston, Dept. of Mathematics,
  66 George St., Charleston SC 29424, USA.
  `calini@stelvio.cofc.edu`

David H. Hartley
  Physics and Mathematical Physics,
  University of Adelaide, 5005, Australia.
  `DHartley@physics.adelaide.edu.au`

Mark E. Johnston
  Department of Applied Mathematics and Theoretical Physics,
  University of Cambridge, Silver Street, Cambridge, UK.
  `M.E.Johnston@damtp.cam.ac.uk`

Niky Kamran
  Department of Mathematics and Statistics, McGill University,
  Montreal, Québec, H3A 2K6, Canada.
  `nkamran@gauss.math.mcgill.ca`

Ian G. Lisle
  School of Mathematics and Statistics, University of Canberra,
  Australian Capital Territory, 2601, Australia.
  `lisle@ise.canberra.edu.au`

Michael K. Murray
  Department of Pure Mathematics,
  University of Adelaide, 5005, Australia.
  `mmurray@maths.adelaide.edu.au`

Geoff E. Prince
  Department of Mathematics, La Trobe University,
  Bundoora, Victoria, 3083, Australia.
  `G.Prince@latrobe.edu.au`

Gregory J. Reid
  Centre for Experimental and Constructive Mathematics,
  Simon Fraser University,
  Burnaby BC, V5A 1S6, Canada.
  `reid@cecm.sfu.ca`

Colin Rogers, Wolfgang K. Schief
  School of Mathematics, The University of New South Wales,
  Sydney, NSW, 2052, Australia.
  `schief@solution.maths.unsw.edu.au`

Peter J. Vassiliou
  School of Mathematics and Statistics, University of Canberra,
  Australian Capital Territory, 2601, Australia.
  `pierre@ise.canberra.edu.au`

Preface

This book began as a series of lectures given by various speakers during the "*Miniworkshop in Geometry and Differential Equations*", held at the University of Canberra in May, 1995. The purpose of the workshop was to bring together people interested in the interplay between geometry and differential equations; a subject that has experienced a revival, particularly over the last 30 years.

Rather than having many short and highly specialised contributions, it was decided that a smaller number of more broadly based expository talks would better serve the purposes of the workshop. To that end, a small number of speakers were invited to give two hour long "microcourses" each of which would delineate an interesting area of current research in a clear, concise and elementary way. The lectures were to be accessible to graduate students as well as to the more experienced who may be from other fields. At the conclusion of the workshop, it was realised that the lectures might be of interest to a wider audience and workshop speakers were invited to prepare their lectures for publication. In order to give the ultimate publication more currency and breadth, a number of other experts who did not speak at the workshop were invited to contribute to the book as well.

In preparing this volume, every effort has been made to avoid the production of a "conference proceedings". Such collections, of course, have their place but it isn't what we were trying to create. We would like to thank the authors of the individual chapters most sincerely for rising to this challenge. Apart from the aforementioned attempt to keep the exposition as crisp and accessible as possible, authors have tried, where appropriate, to cross reference thereby lending the collection greater coherence than is usual for a typical conference proceedings. This end is hopefully further assisted by the inclusion of an introductory chapter which, among other things, aims to set each chapter in context and to provide even more motivation for the ideas presented, especially historical motivation. In addition, the introductory chapter aims to give a very brief and elementary introduction to some of the basic ideas in the subject as a whole.

In this respect, two basic historical threads within the subject can be discerned. Firstly, an approach to differential equations arising from the classical geometry of surfaces in three-dimensional Euclidean space. This work is associated with nineteenth century geometers such as G. Darboux, L. Bianchi and A.V. Backlund. Secondly, the introduction of the notion of "transformation group" by S. Lie and F. Klein in the late nineteenth century and profoundly developed by É. Cartan, E. Vessiot and others in this century leading to the emergence of a "geometry of differential equations". These two threads are far from independent. In the main, the contributions of the various authors represented in this volume reflect this historical development. A particular focus concerns the application of geometric ideas to the study of completely integrable systems both finite and infinite dimensional. Indeed, the study of infinite dimensional completely integrable systems is one of the major new ideas introduced into the subject in this century.

It should be emphasised that no attempt has been made to offer a comprehensive account of the field – this would be impossible within the compass of a single volume. Neither do we claim that it covers the most interesting aspects.

We have instead tried to offer the reader an introduction to some of the "basic topics" such as *exterior differential systems* as well as a taste of some of the more "exotic" aspects such as *twistor theory*. In this way, the book provides some foundational material as well as perspectives within the subject. In addition, numerous references have been included allowing the reader to explore further.

This volume is offered in the belief that geometric approaches to differential equations represent a rich and rewarding field of mathematical endeavour with many beautiful ideas and intriguing open problems. We hope you find the presentation informative and stimulating.

Peter Vassiliou, Ian Lisle,
Editors.

Acknowledgements

The "*Miniworkshop in Geometry and Differential Equations*", from which this volume took its inspiration, was supported by a grant from the Australian Research Council. Additional support was provided by the School of Mathematics and Statistics, the Faculty of Information Sciences and Engineering and the Advanced Telecommunications Research Centre, all at the University of Canberra. All this support is gratefully acknowledged. The editors would like to thank all those who participated in the *Miniworkshop*.

Introduction: Geometric Approaches to Differential Equations*

PETER J. VASSILIOU

Keywords: Bäcklund transformations, integrable systems, inverse scattering, differential systems, Cartan equivalence, Lie groups, Lagrangians, twistors

The year 1997 marked the 30th anniversary of the discovery of the inverse scattering transform, a technique for solving the initial value problem for certain nonlinear partial differential equations of mathematical physics [GGKM]. This event opened the flood gate for a large amount of intriguing research and has had a powerful effect in large parts of mathematics and physics ranging from algebraic geometry to optical fibres (see, for example, [AC]; [AN]; [DEGM]; [Pal]). One of these effects has been to revive the study of the *geometric* aspects differential equations, owing to the discovery that the inverse scattering transform has significant geometric properties.

The study of the geometric properties of differential equations goes back at least to the work of Monge on the minimal surface equation. Monge's investigations were greatly extended and codified in one of the great works of late 19th century mathematics, Darboux's *"Leçons sur la Théorie de Surfaces"* [Da] which is an extensive study of the geometry of surfaces in Euclidean space. However, as Robert Hermann has remarked, much of the four-volume work is devoted to various topics in partial differential equations, such as Laplace transformations which occupies much of volume 2.

Thus, from the very beginning there has been a fruitful correspondence between differential geometry on the one hand and differential equations on the other. Questions in differential geometry could be solved, or at least illuminated, by studying the differential equations that were implied by the geometry. Conversely, questions in differential equations benefit by a study of the accompanying geometric setting.

To give an illustrative example, we briefly describe the classical Laplace transformation and some of its more recent generalisations. The Laplace transformation (unrelated to the Laplace transform of harmonic analysis) is a transformation for surfaces in $\mathbb{E}^3$, three-dimensional Euclidean space. Suppose a surface S in $\mathbb{E}^3$ has a net of curves conjugate for the second fundamental form. Then a parametrisation $X(u,v)$ of S satisfies a second-order hyperbolic equation

$$X_{uv} = \Gamma_{12}^1 X_u + \Gamma_{12}^2 X_v \tag{1}$$

where $\Gamma_{12}^i, i = 1, 2$, are the Christoffel symbols for the surface S. Fix the coordinate v, say v_0, and consider the ruled surface $Y(t, u; v_0) = X(u, v_0) + tX_v(u, v_0)$, where subscript v denotes partial differentiation. The surface Y determines a curve, the edge of regression (see [Sp], p.208). It follows that, for each u, there

---
*I would like to thank Dianne Williams for reading a draft of this chapter and making valuable suggestions.

is a parameter value $t = t_0$ such that $Y(t_0, u; v_0)$ is on the edge of regression (see [KT95]); this, in turn, defines a point in $\mathbb{E}^3$. As v_0 varies, we obtain, in general, a new surface parametrised by $X_1(u, v)$ with the remarkable property that the net of curves u, v are still conjugate for the second fundamental form and hence X_1 satisfies a partial differential equation of the same form as equation (1). The new surface S_1 parametrised by X_1 is one of the two possible Laplace transformations of S. The other one is obtained by reversing the role of u and v in the above construction.

Now, it can happen that the image of a Laplace transformation, say S_1, can be degenerate forming a curve rather than a surface, as, for example, when S is a surface of rotation (see [KT95]). In this case, it turns out that the corresponding partial differential equation is solvable by quadrature. This fact, together with the general Laplace transformations described above, leads to an integration procedure and general "transformation theory" for linear hyperbolic equations of the form

$$u_{xy} + a(x,y)u_x + b(x,y)u_y + c(x,y)u = 0.$$

This classical construction has been extended in a number of ways in the more recent literature. In [Ch], Chern has given a geometric description of a generalisation of the Laplace transformation to a class of n-dimensional submanifolds of projective space which had previously been studied by Cartan. These submanifolds, which Chern calls Cartan submanifolds, admit a parametrisation by a conjugate net. For each n-dimensional Cartan submanifold, Chern constructs $n(n-1)$ generalised Laplace transformations which, generically, define n-dimensional Cartan submanifolds. In [KT96], Kamran and Tenenblat consider Cartan submanifolds in Euclidean space. They give a generalisation of the classical Laplace transformations and the classical integration procedure alluded to above. It transpires that the systems of partial differential equations that Kamran and Tenenblat thereby construct find application in the study of the conserved quantities for semi-Hamiltonian, strongly hyperbolic systems of hydrodynamic type [Ts]. For the reader wishing to pursue these intriguing matters further, the papers by Kamran and Tenenblat referenced above are highly recommended as is the book [Ten] which may be regarded as a companion to the present volume. Finally, in [Va], it is shown how a consideration of the theory of Laplace transformations leads to an integration procedure for a generalised Toda field theory.

Thus, the example of the classical Laplace transformation together with its recent generalisations and applications provide a nice illustration of the above mentioned correspondence between geometry and differential equations. This interplay is one of the chief themes of this book.

Indeed, it is amply illustrated in the contribution to this volume by Colin Rogers, Wolfgang Schief and Mark Johnston who present an account of the role of geometry to the theory of completely integrable systems. In the classical literature, a *completely integrable system* is a system of ordinary differential equations for an even number of dependent variables, say $2n$, which has a Hamiltonian structure and possesses n constants of the motion which pairwise commute with respect to the natural Poisson bracket. For more details on this, the reader is referred to Geoff Prince's contribution to this volume and to the references therein. The discovery of the inverse scattering transform in 1967 introduced the notion of an *infinite-dimensional* Hamiltonian system with in-

finitely many "constants of the motion" which pairwise commute with respect
to an appropriate Poisson bracket. Thus, a theory has developed generalising
the classical theory of completely integrable systems such as may be found, for
example, in classical texts such as Whittacker [Wh]. The many beautiful and
intricate properties of infinite dimensional completely integrable systems are far
too numerous to be reviewed here. For an account of many of these matters,
the reader is directed to the contribution of Annalisa Calini to this volume. In
addition, the recent paper [Pal] by R. S. Palais and the book [AN] by A. C.
Newell are also highly recommended.

Nevertheless, without going into all the details here, let us observe that if an
equation is solvable by the inverse scattering transform, such as, for example,
the celebrated Korteweg–de Vries equation (KdV)

$$u_t + uu_x + u_{xxx} = 0, \tag{2}$$

then fortuitously the equation has many other properties which are what Newell
refers to as the "miracles of soliton mathematics". One of these miracles is a
"transformation" discovered by the Swedish mathematical physicist and geometer A.V. Bäcklund (1845–1922) and these days referred to as a *Bäcklund transformation*. The reason for the inverted commas is that one does not really have
a transformation in the sense of a locally defined diffeomorphism on a manifold but rather a "correspondence" among the solutions of a given equation or
between the solutions of two distinct equations. Bäcklund's original discovery
concerned another famous equation from the theory of solitons, the sine-Gordon
equation

$$u_{xy} = \sin u, \tag{3}$$

which arises naturally within the theory of surfaces in three-dimensional Euclidean space. Bäcklund discovered the remarkable fact that if u is any solution
of equation (3) then u' defined by the first order system

$$(\frac{u' - u}{2})_x = \lambda \sin(\frac{u' + u}{2})$$
$$(\frac{u' + u}{2})_y = \frac{1}{\lambda} \sin(\frac{u' - u}{2}) \tag{4}$$

is also a solution of equation (3). Here λ is an arbitrary real parameter. Similarly, if u' is a solution of (3) then system (4) provides another solution of (3).
For instance, if one takes for u the zero solution of the sine-Gordon equation, one
may easily solve the resulting equations (4) to obtain the so-called 1-soliton solution, that is a "stable" travelling wave solution which has remarkable "elastic"
properties under interaction with other solitons.

A powerful consequence of this *Bäcklund correspondence* is the so-called
Bianchi Permutability Theorem whereby, given three distinct solutions of the
sine-Gordon equation (3), a fourth solution may be constructed by purely algebraic means. This phenomenon, referred to as a nonlinear superposition formula, and the associated Bäcklund transformation, appears only to occur for
a very privileged class of nonlinear partial differential equations. Nevertheless,
it is observed that this phenomenon always accompanies any equation that is
solvable by the inverse scattering transform. In this regard we remark that the

Bäcklund transformation for the sine-Gordon equation has been known for over 100 years and hence its discovery considerably predates the inverse scattering transform. However, the Bäcklund transformation for the KdV equation was discovered by Wahlquist and Estabrook in 1973 [WE], approximately five years after the discovery of the inverse scattering transform. Since this time, Bäcklund transformations have been discovered for all the equations known to be solvable by the inverse scattering transform. The deeper understanding of Bäcklund transformations and similar phenomena within the theory of completely integrable systems are outstanding problems of great importance.

The chapter by Rogers, Schief and Johnston explores and reviews many of these ideas giving a clear introduction to a number of recent results which illustrate the role played by geometry in differential equations. By way of introducing the subject, the authors begin by studying the Bäcklund transformation for the sine-Gordon equation. Their derivation is particularly interesting in that it is shown to arise explicitly from a simple geometric construction. Beginning with the Gauss equations for the parametrisation of a surface in $\mathbb{E}^3$ and their compatibility conditions, the Mainardi–Codazzi equations and the Gauss formula, they show that making a geometrically motivated choice of new dependent and independent variables leads to the so-called *Bianchi system* for surfaces of negative Gauss curvature. In the special case of constant negative Gauss curvature, the Bianchi system reduces to the sine-Gordon equation. It follows that each solution of the sine-Gordon equation then leads, via the Gauss equations, to a surface of constant negative Gauss curvature (a pseudospherical surface). Given a pseudospherical surface in $\mathbb{E}^3$, a simple geometric construction enables the authors to construct another pseudospherical surface. The analytic counterpart of this geometric construction, via the Gauss-Weingarten equations, leads to the sine-Gordon Bäcklund transformation.

Here again, as in the work of Chern, and of Kamran and Tenenblat mentioned above, we see that a nontrivial fact within theory of differential equations has a simple counterpart in differential geometry. Rogers, Schief and Johnston go on to discuss a number of other interesting topics within the general theme of Bäcklund transformations. For instance, they produce graphs of pseudospherical surfaces arising from "breather solutions" of the sine-Gordon equation. As a natural development of these ideas the *motion* of pseudospherical surfaces is considered. A host of topics are then discussed ending with a striking connection with the work of C. Loewner on the equations of compressible gas flow and the so-called "infinitesimal" Bäcklund transformations leading to new completely integrable systems in more than two independent variables. It is clear from the chapter by Rogers, Schief and Johnston that Bäcklund transformations form a deep and compelling topic within both differential geometry and differential equations.

The theme of completely integrable systems and differential geometry continues in the contribution of Annalisa Calini. Here, the simplest of geometric objects, curves, play a fundamental role. The objects of interest are *vortex filaments*, approximately one-dimensional regions in a fluid where the velocity distribution has a rotational component. Smoke rings are a common example of this phenomenon. Modeling vortex filaments as closed curves in $\mathbb{R}^3$ leads to the vortex filament equation which is shown to have a Hamiltonian formulation on an infinite-dimensional phase space (the loop space). Remarkably, it transpires that if a closed curve evolves according to the vortex filament equation then a function expressed in terms of the curvature and torsion of the curve evolves in accordance with the cubic nonlinear Schrodinger equation (NLS), a well -known soliton equation. It follows that the vortex filament equation is completely integrable. A further link is made with a differentiated version of the vortex filament equation leading to an evolution equation for the tangent vector to the closed curve identified as the continuous Heisenberg model (HM). It has been known for some time that the HM and NLS equations are related by a gauge transformation of associated "linear problems", commonly referred to as Lax pairs within the theory of completely integrable systems. Among the beautiful results of this chapter is a geometric interpretation of this gauge transformation in terms of a canonical connection on the circle bundle of S^2. In addition, use is made of algebraic geometry to obtain solutions of the HM and the construction of associated closed curves in $\mathbb{R}^3$. The paper concludes with a study of Bäcklund transformations and immersed knots and a discussion of future directions in the field.

Another thread that runs through the geometric aspects of differential equations is the notion of symmetry. Here the contribution of Sophus Lie is decisive and all pervading. A *symmetry* of a differential equation is a transformation that maps solutions of the equation to other solutions. Lie discovered that the collection of all such transformations form what is now called a Lie transformation group. Gradually, the notion of a transformation group was refined and it became clear, with the work of Élie Cartan and others that there is an abstract structure underlying transformation groups, namely Lie groups. From the very beginning, a distinction arose between the finite groups (those whose elements are parametrised by finitely many real or complex numbers) and the infinite groups whose elements depend upon arbitrary functions. The abstract structure for Lie groups mentioned above applies to the finite case. The construction of an abstract structure in the infinite case is still largely an open problem. In the infinite case, these transformation groups have come to be called *infinite Lie pseudogroups*. The qualifier 'pseudo' comes from the fact that the elements of the 'group' do not form a group in the usual sense since composition between two arbitrary elements may not be defined.

Many differential equations of interest and indeed many other 'geometric objects' have symmetries which are precisely infinite Lie pseudogroups. Hence, the structure of these groups is of great practical and theoretical importance. For example, one of Lie's results states that if an ordinary differential equation of order k admits a symmetry Lie group which is *solvable* and has dimension k, then the construction of solutions of the equation may be reduced to quadrature (see Bluman and Kumei [BK]; Olver [O]). Lie's result underscores one of the most important aspects of the geometry of differential equations, namely, structural information on admitted symmetry groups yields significant results about the differential equations themselves. This may be compared with the role played by the Galois group in the study of polynomial equations.

Thus, historically, we see two geometric influences impinging upon differential equations. On the one hand, the role of the classical differential geometry of surfaces, mentioned above, and on the other, the development of the notion of symmetry group leading to a *geometry* for differential equations. But what does it mean to talk about a "geometry" for differential equations? In the current state of development a definitive answer to this question is still under construction. In their expository article "Towards a geometry of differential equations",

Bryant, Griffiths and Hsu [BGH] have been content to confine themselves more or less to the description of significant examples within the field. This is surely the path of wisdom and their article is highly recommended for anyone interested in the subject. However, a few broad features can be discerned which may help the reader interpret these words. Firstly, let's turn to a setting in which the word "geometry" has a well-understood meaning. A Riemannian manifold is a differentiable manifold M together with a smoothly varying inner product g on each tangent space. Two Riemannian manifolds (M_1, g_1) and (M_2, g_2) are *locally equivalent* if and only if there is a local diffeomorphism $\phi : M_1 \rightarrow M_2$ which identifies their Riemannian structures: $\phi^* g_2 = g_1$, where superscript $*$ denotes pullback.

Riemannian manifolds M_1 and M_2 are then said to be *locally isometric*. One knows, for example, that a necessary condition for a pair of two-dimensional Riemannian manifolds (surfaces) to be locally isometric is that their Gauss curvatures be equal: $\phi^* K_2 = K_1$. One can say that the broad goal of Riemannian geometry is to describe all Riemannian manifolds up to local isometry (the natural equivalence for this class of geometric objects). In the course of trying to answer this question, one discovers the Riemann curvature tensor and other invariants that label the equivalence classes within this set of objects. Of course, the problem as stated is impossible to solve in any meaningful way but it sets down the tone of the subject. It turns out that one can study differential equations in much the same spirit as one studies Riemannian manifolds. The basic venue is the kth order jet bundle of maps of $\mathbb{R}^m \rightarrow \mathbb{R}^n$, $J^k(\mathbb{R}^m, \mathbb{R}^n)$. In order to describe this, we will pause briefly to set up the basic structures of the subject. So as to make rapid progress, a somewhat elementary description of jet bundles and the associated contact structure will be given. The reader may consult [BC3G] and [KV] for more details. The last reference is a noteworthy attempt to present the modern geometry of differential equations in a concise and accessible form. Since most of the considerations of this book shall be local, no generality is lost for the present purposes by taking our manifolds to be open subsets of $\mathbb{R}^n$. We wish ultimately to study systems of differential equations in p independent variables $x_1, x_2, \ldots, x_p$ and q dependent variables $u^1, u^2, \ldots, u^q$. Denote by X the space of independent variables and by U the space of dependent variables and consider the trivial bundle $\pi : X \times U \rightarrow X$. Two smooth sections f and g of π are said to be equivalent to order k at x if and only if

$$\partial_{x_I}^{|I|} f(x) = \partial_{x_I}^{|I|} g(x), \ x_I = x_1^{\alpha_1} x_2^{\alpha_2} \ldots x_p^{\alpha_p}$$

where I is a multi-index of order less than or equal to k, that is, $|I| = \alpha_1 + \alpha_2 + \cdots + \alpha_p \leq k$. In other words, two smooth sections are equivalent to order k at x if their Taylor coefficients agree up to order k at x. Lie called the equivalence class of a smooth section f at a point x a contact element of order k. The modern terminology for this is the *k-jet of f at x* and is denoted $j_x^k f$. The set of all k-jets of smooth sections of π constitutes the bundle of k-jets of maps $X \rightarrow U$ and denoted $J^k(X, U)$. The set $J^k(X, U)$ can be given the structure of a differentiable manifold. We introduce local coordinates $(x_i, u^\alpha, u_{i_1}^\alpha, u_{i_1 i_2}^\alpha, \ldots, u_{i_1 i_2 \ldots i_k}^\alpha)$, where the $u_{i_1 i_2 \ldots}^\alpha$ are symmetric in the lower indices and $i_1, i_2, \ldots = 1, 2, \ldots, p$ and such that

$$x_i(j_x^k f) = x_i, \quad u^\alpha(j_x^k f) = f^\alpha(x),$$
$$u_I^\alpha(j_x^k f) = \partial_{x_I}^{|I|} f^\alpha(x),$$

The zeroth order jet bundle, $J^0(X, U)$, is identified with $X \times U$. The k-graph of a smooth section f of π is the map

$$j^k f : X \to J^k(X, U)$$

defined by

$$x \mapsto j_x^k f.$$

Thus, the image of the k-graph of a section is a p-dimensional immersed submanifold of $J^k(X, U)$. Each jet bundle $J^k(X, U)$ comes equipped with a module of differential 1-forms spanned by

$$\theta_I^\alpha = du_I^\alpha - \sum_{i=1}^{p} u_{I,i}^\alpha dx_i$$

where I is a multi-index of order less than or equal to $k - 1$. This module is denoted $\Omega^k(X, U)$ or simply Ω^k and called the *contact system* on J^k.

Example 1. $\Omega^1(\mathbb{R}, \mathbb{R}) = \{du - u_1 dx_1\}$, $\Omega^2(\mathbb{R}^2, \mathbb{R}) = \{du - u_1 dx_1 - u_2 dx_2, du_1 - u_{11} dx_1 - u_{12} dx_2, du_2 - u_{21} dx_1 - u_{22} dx_2\}$.

The contact system on J^k is an example of an *exterior differential system*, generating an ideal in the ring of differential forms on a differentiable manifold. Since this particular differential system is generated by 1-forms, it is also called a *Pfaff system*. As Niky Kamran describes in greater detail in his contribution to this volume, for a given differential system E on a manifold M, generated by 1-forms, 2-forms, ... there may be submanifolds $N \subset M$ where all the forms in E are zero when pulled back to N. That is, if $i : N \to M$ is the natural inclusion, then $i^* E = 0$. Such a submanifold is called an *integral submanifold*. Another way to view this is that the differential system annihilates each tangent space to the submanifold N. Thus, N is an integral submanifold of E if and only if $v_p \in \ker(E_p), \forall v_p \in T_p N, p \in N$. Note that the image of the k-graph of any function $f : X \to U$ is an integral submanifold of the kth-order contact system, $\Omega^k(X, U)$.

We can now say how these constructions are related to differential equations. We begin by giving some illustrative examples. In view of the definition of the jet bundle, we will think of a differential equation as a relationship between the coordinates of an appropriate jet bundle. Such a relationship defines a subset of the jet bundle. Indeed, we shall specialise this somewhat and consider differential equations which define embedded submanifolds.

Example 2. First order ordinary differential equation, $y' = F(x, y)$. Consider the embedded submanifold of $J^1(\mathbb{R}, \mathbb{R})$ defined by

$$\mathcal{R} = \{(x, u, u_1) \in J^1(\mathbb{R}, \mathbb{R}) \mid u_1 - F(x, u) = 0\}.$$

The manifold $\mathcal{R}$ is a surface in J^1 which is also a graph over the xu-plane with local coordinates x, u. A curve $(x, y(x))$ in the xu-plane defines a solution of the differential equation $y' = F(x, y)$ if it "lifts to $\mathcal{R}$" under the 1-graph $x \mapsto (x, y(x), y'(x))$. That is, $y(x)$ is a solution of the differential equation if and only if the image of the 1-graph of $y(x)$ is a curve whose height above each point $(x, y(x))$ in the xu-plane is $F(x, y)$. This, after all, is precisely what we mean when we say that a function $y(x)$ "satisfies" the differential equation. We can say this a little more precisely by introducing the natural inclusion

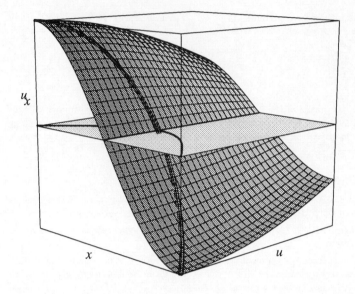

Figure 1: Submanifold $\mathcal{R} \subset J^1(\mathbb{R}, \mathbb{R})$ with a solution of a first order ordinary differential equation and its lift under its 1-graph.

map $i : \mathcal{R} \to J^1(\mathbb{R}, \mathbb{R})$ and letting any curve $S \subset \mathcal{R}$ be the image of a map $h : X \to \mathcal{R}$. Then S defines a solution of the differential equation if and only if $i \circ h : X \to J^1(\mathbb{R}, \mathbb{R})$ is the 1-graph of a differentiable function. This will be the case if and only if $(i \circ h)^*\Omega^1(\mathbb{R}, \mathbb{R}) = 0$. Thus, the image of a map $h : X \to \mathcal{R}$ will define a solution of the differential equation if and only if it is an integral submanifold of the differential system $i^*\Omega^1$, the contact system restricted to the differential equation manifold $\mathcal{R}$. Thus, if we wish to study the solutions of a first order ordinary differential equation we can equivalently study the integral submanifolds of a certain differential system, namely the restricted contact system $i^*\Omega^1 = \{du - F(x, u)dx\}$.

Example 3. First order partial differential equation, $u_y = F(x, y, u, u_x)$. Once again, introduce the embedded submanifold

$$\mathcal{R} = \{(x_1, x_2, u, u_1, u_2) \in J^1(\mathbb{R}^2, \mathbb{R}) \mid u_2 - F(x_1, x_2, u, u_1) = 0\},$$

of $J^1(\mathbb{R}^2, \mathbb{R})$ and inclusion map i. As in the above example, a submanifold S in $\mathcal{R}$ will define a solution of the equation if and only if S is an integral submanifold of the restricted contact system $i^*\Omega^1(\mathbb{R}^2, \mathbb{R}) = \{du - u_1 dx_1 - F dx_2\}$. Here the pictorial analogy is harder to pursue since S is a two-dimensional submanifold of the four-dimensional manifold $\mathcal{R}$. However, the geometric situation is exactly the same. A surface in xyu-space defines a solution (in fact, it's the 0-graph of a solution) if and only if it "lifts to $\mathcal{R}$" under its 1-graph. In the case of a first order ordinary differential equation, this lift assigns the slope of the tangent

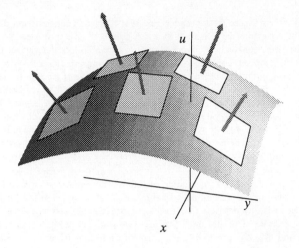

Figure 2: Graph of a solution of a first order partial differential equation showing the orientation of some tangent planes.

line at each point (x, u) of the curve to a length along the u_1-axis, forming a curve in a three-dimensional space. In the case of a first order partial differential equation, the lift assigns the value of u_x, u_y at each point (x, y, u) of a solution surface to the u_1u_2-plane, thereby specifying the orientation of each tangent plane. Thus, the 0-graph of a function is the 0-graph of a solution if and only if the orientation of each of its tangent planes is constrained by $\mathcal{R}$.

Example 4. Second order partial differential equation, $u_{yy} = F(x, y, u, u_x, u_y, u_{xx}, u_{xy})$. The embedded submanifold is

$$\mathcal{R} = \{(x_1, x_2, u, u_1, u_2, u_{11}, u_{12}, u_{22}) \in J^2(\mathbb{R}^2, \mathbb{R}) \mid$$
$$u_{22} - F(x_1, x_2, u, u_1, u_2, u_{11}, u_{12}) = 0\},$$

being a subset of $J^2(\mathbb{R}^2, \mathbb{R})$. As before if i denotes the inclusion, then the solutions of the second order partial differential equation may be studied by studying the integral submanifolds of the restricted contact system

$$i^*\Omega^2(\mathbb{R}^2, \mathbb{R}) = \{du - u_1 dx_1 - u_2 dx_2, du_1 - u_{11} dx_1 - u_{12} dx_2, du_2 - u_{21} dx_1 - F dx_2\}.$$

Here the pictorial analogy is even harder to pursue than in the previous case since the solutions are two-dimensional submanifolds of a seven-dimensional manifold, $\mathcal{R}$. However, the geometric situation is similar. The 0-graph of a function $X \to U$ is the 0-graph of a solution of the second order partial differential equation if and only if its 2-graph lifts to $\mathcal{R}$. This gives a constraint on the orientation of the tangent planes (the 1-jets) as well as the nature of the "local quadric surfaces" determined by the second order derivatives at each point (the 2-jets).

It will now be clear to the reader how this geometric construction describes differential equations and their solutions. A subset of an appropriate jet bundle determines the "constraints" imposed on a function and its derivatives in order that the function be a solution. In the above examples, this subset is always taken to be a submanifold; in general this need not be the case. In addition, these submanifolds have defined upon them, in a canonical way, distinguished differential systems whose integral submanifolds are in correspondence with the solutions of the original systems of differential equations. These differential systems (the restricted contact systems) have useful properties that one may exploit to gain insight into the nature of the differential equations which one would not be able to gain simply by studying the differential equations themselves.

We are now able to complete the analogy, promised earlier, between Riemannian geometry and a "geometry of differential equations". A *differential equations manifold* is an embedded submanifold of an appropriate jet bundle, $\mathcal{R}$, together with its restricted contact structure, $\mathcal{C}$, obtained by restricting the canonical contact structure on J^k to $\mathcal{R}$. Two differential equations manifolds $(\mathcal{R}_1, \mathcal{C}_1)$ and $(\mathcal{R}_2, \mathcal{C}_2)$ are *locally equivalent* if and only if there is a diffeomorphism $\phi : \mathcal{R}_1 \to \mathcal{R}_2$ which identifies their contact structures: $\phi^* \mathcal{C}_2 = \mathcal{C}_1$.[1] This is nothing more than a geometric way of saying that two differential equations can be transformed, one into the other, by a change of dependent and independent variables. The association of the contact structure to the differential equation is the essential step in the "geometrisation". One can now study the pair $(\mathcal{R}, \mathcal{C})$ as a geometric object and seek its invariants in the same way as one studies Riemannian structures to uncover isometric invariants. The resulting class of transformations are called *contact transformations*. Two differential equations manifolds which are locally contact equivalent essentially describe the same differential equation. Thus, as in Riemannian geometry, one can set down the broad ultimate goal of the geometry of differential equations: *describe the local and global structure of differential equations manifolds and their integral submanifolds (solutions)*. Thus, much of the geometry of differential equations "reduces" to the problem of studying exterior differential systems.

Though the geometry of differential equations has a long history going back to the work of Monge, the subject has had its most significant impetus from the work of late 19th and early 20th century mathematicians such as G. Darboux (1842–1917), S. Lie (1842–1899), E. Goursat (1858–1936), W. Killing (1847–1923), A. Tresse (1868–1958), E. Cartan (1869–1951) and E. Vessiot (1865–1952), among others. The influence of Cartan to the modern formulation of the subject is the equal of Lie's. It was Cartan who introduced the theory of exterior differential systems to the study of various problems in differential equations and differential geometry. A review of Cartan's work is therefore of great importance in the history of the modern theory of the geometry of differential equations. In this regard, the book [AR] and paper [ChC], giving broad overviews of Cartan's life and work are helpful. In addition, because of its very significant implications for the geometric aspects of differential equations, we also mention the book by

---

[1]This is not the only type of exterior differential system that one may associate to a differential equation, however, it is a very important one, (see [BGH], [BC3G]).

R. Sharpe, [Sh], which provides, for the first time, a modern account of Cartan's generalisation of differential geometry and F. Klein's "Erlangen Program".

In this book, a number of aspects of the above ideas are discussed. Niky Kamran gives a concise and elementary introduction to the basic theory of exterior differential systems, including the basic local existence and uniqueness theorem for arbitrary analytic systems of differential equations, the Cartan–Kähler theorem. The reader will find, in addition, a guide to the recent as well as classical literature on the subject including Cartan's classic exposition of exterior differential systems which is still well worth reading.

This chapter then prepares the reader for the applications presented in latter chapters. Earlier, we mentioned Lie's discovery of Lie pseudogroups. We remark that, as a student, Cartan was deeply influenced by Lie and he took up the problem of classifying Lie algebras in his doctoral thesis, building on the foundational work of Killing. Subsequently, Lie groups formed a large part of Cartan's preoccupations over the next 60 years of his professional life, advancing the subject profoundly. In their chapter, Ian Lisle and Greg Reid give an elementary account with numerous examples of the basic elements of Cartan's theory of infinite Lie pseudogroups. They then use this theory and apply it to differential equations which admit such groups as symmetries.

As shown explicitly by Reid, Lisle, Boulton and Wittkopf in an earlier work, [RLBW], if a differential equation admits a finite Lie group of symmetries, then the structure of the group can be discovered without recourse to any integrations, that is, without solving differential equations. It is known, however, that many differential equations admit an infinite Lie pseudogroup of symmetries. Owing to the lack of an abstract structure in the infinite case, it is difficult to extend this result to the infinite case. However, by exploiting the Cartan structure of the infinite Lie pseudogroups, Lisle and Reid show in this chapter how to do the same, at least in the transitive case. Several nontrivial examples are given including the steady boundary layer equations and the Kadomtsev–Petviashvili equation.

One of Cartan's most important discoveries is that a great many problems in geometry and differential equations may be fruitfully studied via the "method of moving frames". This method seems to have a complicated history but as Cartan himself says, he took his inspiration from Darboux's use of the method. Indeed, Cartan wrote an expository paper, [Ca], which outlines in an elementary way his thoughts on the subject and a little of his motivation and inspiration. A moving frame on a manifold is a frame field chosen in such a way so as to be "well suited" to the problem one wishes to study. The prototypical example is the Serret–Frenet frame for a curve in $\mathbb{R}^3$. By choosing a framing of (a portion of) $\mathbb{R}^3$ "adapted" to the curve, the derivatives of the frame along the curve (expressed in terms of the frame field itself – the moving frame) quickly give rise to the basic invariants of the curve, the curvature and torsion. Cartan discovered the fundamental fact that this idea has very extensive generalisation so that even though a given problem may not at first appear to be amenable to this method, it can often be so expressed. Cartan expresses a given problem as a *coframe* on a manifold. Differentiation is replaced in this context by the exterior derivative. The simple observation that the exterior derivative is a natural geometric operation (commutes with pullback) gives rise to a powerful general paradigm in geometry, known as the Cartan equivalence method which solves the so-called "local equivalence problem for G-structures". Thus, the second application of exterior differential systems contained in this book is David Hartley's exposition of this method. Hartley gives a careful and clear introduction to the basic geometric ideas associated with the method showing, in particular, why it has such wide applicability. Attention is paid to how the method can be applied, explicitly describing the role played by exterior differential systems. It turns out that Cartan equivalence is a decisive tool in analysing the local structure of differential equations manifolds, which, as we mentioned earlier, is one of the central problems in the geometry of differential equations. Hartley chooses two fundamental examples to illustrate the theoretical considerations: Lagrangian mechanics and Riemannian geometry. These two examples are studied in detail throughout the chapter.

Lagrangians within the calculus of variations is the main theme of the chapter contributed by Geoff Prince. To explain this in slightly more detail, let us recall that the calculus of variations deals with the following type of question. Suppose we have a function I defined on some family of curves on a manifold M. We want to know whether or not there is one or more curves C of this family which optimises the function I. Usually the family of curves is subjected to some extra condition. The prototypical problem within the calculus of variations is the "shortest distance problem". Fix two points in a Riemannian manifold. What is the shortest distance between them? Attempting to answer this question leads to the theory of geodesics.

Abstracting from this example leads to more general situations such as: find the curves $C = (x, y(x))$, where y is vector-valued, which optimises the functional

$$I(C) = \int_a^b L(x, y, y') dx$$

for some function L, the Lagrangian. As is well-known, the critical points $y(x)$ of I satisfy the Euler–Lagrange equation

$$\frac{\partial L}{\partial y} - \frac{d}{dx} \frac{\partial L}{\partial y'} = 0.$$

A classic problem within the calculus of variations, going back to the 19th century, is the *inverse problem:* given a system of differential equations, $\mathcal{D}$, when is $\mathcal{D}$ the Euler–Lagrange equation for some Lagrangian? In his contribution to this volume, Prince outlines recent progress in this problem in the case when $\mathcal{D}$ is a system of second order ordinary differential equations

$$\ddot{\mathbf{x}} = F(t, \mathbf{x}, \dot{\mathbf{x}})$$

where the dot represents t-differentiation and $F, \mathbf{x}$ take values in $\mathbb{R}^n$. The solution of the problem in the analytic case when $n = 2$ is a famous result due to J. Douglas [Do]. Douglas's main tool was the Riquier–Janet existence and uniqueness theorem (see Stormark, [Stor]; Reid, [Re]).[2] This is a coordinate dependent version of the Cartan–Kähler theorem discussed by Kamran and by Hartley in this volume. The use of the Riquier–Janet theorem is also discussed in the contribution of Lisle and Reid to this volume.

---

[2] For an interesting recent application of the Riquier–Janet theorem as well as of Cartan's method of moving frames to a problem in Riemannian and pseudo-Riemannian geometry the reader is referred to Paus, [Pa].

Even in this simplest nontrivial case, the computational complexity of the problem is prohibitive. Prince shows in his chapter that it is very illuminating to approach the problem more geometrically, using the theory of differential systems and attendant results from the calculus on manifolds. As he indicates, this yields a number of important new results as well as making the problem for $n = 3$ a more reasonable proposal for future development. In addition to describing the inverse problem and sketching aspects of its solution, the author also describes some well-known important results from the geometric theory of the calculus of variations such as the relationship between symmetries and first integrals. The chapter concludes with a description of the extended Liouville–Arnol'd theorem and a proof of the local part of the theorem is given. Finally, the full theorem is applied to some illustrative examples.

Over the last few years an increasing amount of attention has been paid to the relationship between completely integrable systems and the twistor theory construction due to R. Penrose. As an example, we mention the work of Mason and Sparling, [MS]. Twistor theory was originally introduced to make use of techniques from complex algebraic geometry to study problems in general relativity. As Michael Murray points out in his contribution to this volume, one can think of twistor theory as the use of complex analytic methods to solve problems in real differential geometry. In his chapter, Murray describes the so called *mini-twistor space*, defined as the set of all oriented lines in $\mathbb{R}^3$. Its geometry is explored and shown to be a two-dimensional complex manifold. This is then used to show that classical results, that can be found, for instance, in Whittaker and Watson [WW] for solving linear constant coefficient differential equations, can be given a geometrical interpretation via mini-twistor space. This result is extended in the body of the chapter and a twistor theoretic approach is developed for solving any constant coefficient, linear, homogeneous differential equation in terms of an integral transform, called a *Penrose transform*. The chapter develops the necessary theory of complex line bundles in order to carry out the construction. Numerous references are offered to provide background in algebraic geometry and complex manifolds. The chapter concludes with a discussion of how twistor theory can be used to study the solutions of nonlinear differential equations, in this case an equation related to the self-duality equations describing instantons in $\mathbb{R}^4$, the Bolgomolny equations. These are the Euler–Lagrange equations for the Yang–Mills–Higgs functional. Here, too, the discussion is supplemented with references allowing the reader to pursue the topic further.

References

[GGKM] C.S. Gardner, J.M. Greene, M.D. Kruskal, R.M. Miura, Method for solving the Korteweg–de Vries equation, *Phys. Rev. Lett.* **19**, (1967), 1095–1097; Korteweg–de Vries equation and generalisations VI. Methods for exact solution, *Comm. Pure Appl. Math.* **27**, (1974), 97–133

[AC] M.J. Ablowitz, P.A. Clarkson, *Solitons, Nonlinear Evolution Equations and Inverse Scattering*, Cambridge University Press, 1991

[AN] A.C. Newell, *Solitons in Mathematics and Physics*, SIAM, 1985

[DEGM] R.K. Dodd, J.C. Eilbeck, J.D. Gibbon, H.C. Morris, *Solitons and Non-*

linear Wave Equations, Academic Press, 1982

[AC] M.J. Ablowitz, P.A. Clarkson, *Solitons, Nonlinear Evolution Equations and Inverse Scattering*, Cambridge University Press, 1991

[Pal] R.S. Palais, Symmetries of solitons, *Bull. Amer. Math. Soc.* **34**(4), (1997), 339–403

[Da] G. Darboux, *Leçons sur la Théorie des Surfaces*, Gauthier-Villars, Vol 1, 1887, Vol 2, 1889, Vol 3, 1894, Vol 4, 1896

[Sp] M. Spivak, *A Comprehensive Introduction to Differential Geometry*, Vol 3, 2nd Edition, Publish or Perish Press, 1979

[KT95] N. Kamran, K. Tenenblat, Laplace transformation of Cartan submanifolds, *Mathemática Contemporánea* **9**, (1995), 117–138

[Ch] S.S. Chern, Laplace transforms of a class of higher dimensional varieties in a projective space of n-dimensions, *Proc. Nat. Acad. Sci. U.S.A.* **30**, (1944), 95–97

[KT96] N. Kamran, K. Tenenblat, Laplace transformations in higher dimensions, *Duke Mathematical Journal* **84**(1), (1996), 237–266

[Ts] S.P. Tsarev, The geometry of Hamiltonian systems of hydrodynamic type: the generalized hodograph method, *Math. USSR Izv.* **37**, (1991), 397–419

[Ten] K. Tenenblat, *Transformations of Manifolds and Applications to Differential Equations*, Chapman & Hall, 1998

[Va] P.J. Vassiliou, On some geometry associated with a generalised Toda lattice, *Bull. Austral. Math. Soc.* **49**, (1994), 439–462

[Wh] E.T. Whittaker, *Analytical Dynamics of Particles and Rigid Bodies*, 4th Edition, Cambridge University Press, 1937

[WE] H.D. Wahlquist, F.B. Estabrook, Bäcklund transformation for solutions of the KdV equation, *Phys. Rev. Lett.* **31**, (1973), 1386–1390

[BK] G.W. Bluman, S. Kumei, *Symmetries and Differential Equations*, Applied Mathematical Sciences 81, Springer Verlag, 1989

[O] P.J. Olver, *Applications of Lie Groups to Differential Equations*, 2nd Edition, Graduate Texts in Mathematics, 107, Springer Verlag, 1993

[BGH] R.L. Bryant, P.A. Griffiths, L. Hsu, Towards a geometry of differential equations, in *Conference Proceedings and Lecture Notes in Geometry and Topology, Volume VI*, International Press, 1995

[KV] I.S. Krasilsh'chik, A.M. Vinogradov , *Symmetries and Conservation Laws for Differential Equations of Mathematical Physics*, Translations of Mathematical Monographs, Volume 182, American Mathematical Society, 1999

[BC3G] R.L. Bryant, S.S. Chern, R.B. Gardner, H.L. Goldschmidt, P.A. Griffiths, *Exterior Differential Systems*, Mathematical Sciences Research Institute Publications, 18, Springer-Verlag, 1991

[AR] M.A. Akvis, B.A. Rosenfeld, *Elie Cartan, (1869-1951)*, Translations of Mathematical Monographs, Volume 123, American Mathematical Society, 1993

[ChC] S.S. Chern, C. Chevalley, Élie Cartan and his mathematical work, *Bull. Amer. Math. Soc.* **58**, (1952), 217–250

[Sh] R.W. Sharpe, *Differential Geometry: Cartan's Generalisation of Klein's Erlangen Program*, Graduate Texts in Mathematics, 166, Springer Verlag, 1997

[RLBW] G. Reid, I. Lisle, A. Boulton, A. Witkopf, Algorithmic determination of commutation relations for Lie symmetry algebras of PDEs, in *Proc. ISSAC '92*, ACM Press, 1992

[Ca] É. Cartan, *La Methode du Repère Mobile, la Théorie des Groupes Continus et les Espace Généralisés*, Exposés de Géométrie, V, Hermann, 1935

[Do] J. Douglas, Solution of the inverse problem of the calculus of variations, *Trans. Amer. Math. Soc.* **50**, (1941), 71–128

[Stor] O. Stormark, *Formal and Local Solvability of Partial Differential Equations*, TRITA-MAT-1989-11, Royal Swedish Institute of Technology, Stockholm, Sweden, 1989

[Re] G.J. Reid, Algorithms for reducing a system of PDEs to standard form, determining the dimension of its solution space and calculating its Taylor series solution, *Euro. J. Appl. Math.*, **2**, (1991), 293–318

[Pa] W.H. Paus, *Sum of Squares Manifolds: The Expressibility of the Laplace–Beltrami Operator on Pseudo-Riemannian Manifolds as a Sum of Squares of Vector Fields*, Ph.D Thesis, University of New South Wales, 1996; *Trans. Amer. Math. Soc.* **350**(10), 1998, 3943-3966

[MS] L.J. Mason, G.A.J. Sparling, Twistor correspondences for soliton hierarchies, *J. Geom. Phys.* **8**, (1992), no. 1-4, 243-271-000

[WW] E.T. Whittaker, G.N. Watson, *A Course of Modern Analysis*, 4th Edition, Cambridge University Press, 1927

Bäcklund and his Works:
Applications in Soliton Theory

COLIN ROGERS, WOLFGANG K. SCHIEF AND MARK E. JOHNSTON

Keywords: Bäcklund transformations, completely integrable systems, inverse
scattering

The name of the Swedish mathematical physicist and geometer Albert Victor
Bäcklund (1845–1922) is nowadays commonly associated with two distinct types
of surface transformations. Thus, in addition to the Bäcklund transformations
which have proven to possess remarkable physical applications in Soliton Theory there are also higher-order tangent transformations which have come to be
termed Lie–Bäcklund transformations. The latter have important connections,
in particular, with Noether's seminal work on the theory of conservation laws.
These two fields have a common source in attempts to extend Lie's theory of
contact transformations. However, it is interesting to record that Bäcklund also
made a significant but little-known incursion into the theory of characteristics
which originated in the work of Monge and Ampère. Indeed, Bäcklund, with his
extension to many independent variables, was regarded by both Goursat and
Hadamard as the founder of the modern theory of characteristics.

In this review, we celebrate the 150th anniversary of Bäcklund's birth. The
classical foundations of what are now termed in the literature Lie–Bäcklund
transformations have been described in Anderson and Ibragimov [1]. Here,
attention is accordingly restricted to an account of the geometric origins of
Bäcklund transformations and applications in modern Soliton Theory. This
augments material to be found in the monograph on Bäcklund transformations
by Rogers and Shadwick [2].

The result for which Bäcklund is, perhaps, best known relates to a parameter-
dependent invariant transformation of the classical sine-Gordon equation. The
latter is here set in the geometric context of a more general integrable system due to Bianchi which describes a class of hyperbolic surfaces and which
is gauge-equivalent to an equation cognate to the important Ernst equation
of General Relativity. The sine-Gordon equation arises as the reduction of
the Bianchi system corresponding to surfaces of constant negative curvature,
namely, pseudo-spherical surfaces. The auto-Bäcklund transformation for the
sine-Gordon equation is associated with a simple geometric construction of such
surfaces. However, the most important aspect of the Bäcklund transformation is
that there is a concomitant nonlinear superposition principle whereby, in particular, multi-soliton solutions may be generated by purely algebraic procedures.
Indeed, it turns out that the possession of an auto-Bäcklund transformation
known as a Permutability Theorem is generic to all nonlinear integrable solitonic equations. Here, the Bäcklund transformation and nonlinear superposition
principle are used, by way of illustration, to construct pseudo-spherical surfaces
corresponding to breather solutions of the sine-Gordon equation.

In a natural development of Bäcklund's work, we next consider the motion of pseudo-spherical surfaces. This introduces, in a geometric context, the concept of compatible integrable systems. In particular, it is shown that the mKdV equation and a classical Weingarten system descriptive of a class of triply orthogonal surfaces are both compatible with the sine-Gordon equation. Auto-Bäcklund transformations are presented for these compatible systems. The important role of Bäcklund transformations in Soliton Theory, in general, is then summarised. Soliton surfaces are constructed whose Gauss equations provide linear representations for, in turn, the Bianchi system, the Tzitzeica equation and the nonlinear Schrödinger equation. Such surfaces admit Bäcklund transformations analogous to that for the pseudo-spherical surfaces linked to the classical sine-Gordon equation. The important roles of the Permutability Theorem as an integrable discretisation and of the auto-Bäcklund transformation as an integrable differential-difference equation are outlined in the next section.

To conclude, we turn to recent remarkable results which have their roots in quite a separate development involving the application of Bäcklund transformations in Continuum Mechanics. Thus, Loewner [3] in 1950 introduced a generalisation of the classical notion of Bäcklund transformations in a study of the reduction of the hodograph equations of gas dynamics to appropriate canonical forms in subsonic, transonic and supersonic régimes. Subsequently, in 1952 Loewner extended his discussion to a class of infinitesimal Bäcklund transformations [4]. It was not until 1991 that, in work of Konopelchenko and Rogers [5], the two separate developments in the application of Bäcklund transformations, namely in Soliton Theory and Continuum Mechanics were brought together. Thus, it was shown that the class of infinitesimal Bäcklund transformations, as originally introduced in a gas dynamics context, suitably re-interpreted and generalised produce a novel linear triad representation for a 2+1-dimensional integrable system of considerable generality. Indeed, most recently, compatibility between hierarchies of these LKR systems and the multi-component mKP and KP hierarchies of Sato Theory has been established [6].

1 Historical Background

Albert Victor Bäcklund was born on January 11th 1845 in a small village in Malmöhus län in southern Sweden. He received his tertiary education at the University of Lund. In 1864, he was offered a position at the Astronomical Observatory where he became a student of Professor Axel Möller. In 1868, Bäcklund received his PhD for a thesis concerning a method for measuring latitude from astronomical observations. Shortly after, Bäcklund commenced his work on geometry and became aware of the work of the Norwegian mathematician Sophus Lie.[1]

In 1874, Bäcklund was awarded a travel grant from the government to pursue his studies on the Continent for six months. He spent most of his time in Leipzig and Erlangen where he met Klein and Lindemann. Ideas he gained in this period inspired his later work in geometry on what have come to be known as Bäcklund transformations.

In 1878, Bäcklund was awarded the title of Associate Professor of Mechanics and Mathematical Physics at the University of Lund. In 1888, he was

---

[1]Bäcklund alluded to the work of Lie for the first time in 1872 in his *Jugendarbeit*.

elected a Fellow of the Swedish Academy of Science but it was not until 1897 that Bäcklund finally obtained a Professorship at Lund. During the period 1907–1909, Bäcklund was the rector of the University of Lund. Following his retirement in 1910, Bäcklund resumed his studies in Differential Geometry and participated in the debate on the recently introduced Theory of General Relativity. Bäcklund died on February 23rd 1922. The publications of Bäcklund are listed in the Appendix.

The historical origin of Bäcklund transformations resides in the attempts to extend the pioneering work of Lie on contact transformations. Lie had raised the question of the existence of transformations for which tangency of higher order is an invariant condition. Bäcklund independently studied this problem. An account of higher-order tangent transformations and their history is given in Anderson and Ibragimov [1]. In a series of papers published in Mathematische Annalen between the years 1875–1882, Bäcklund not only made a major contribution to the theory of tangent transformations but, importantly, was led to introduce a second class of transformations of surfaces which, together with their modern extensions, have become known as Bäcklund transformations. In particular, the celebrated auto-Bäcklund transformation associated with the generation of pseudo-spherical surfaces was set down. It is with this result that we commence our present introductory account of the role of Bäcklund transformations in Soliton Theory. A detailed treatment of the subject is given in Rogers and Shadwick [2].

2 The Classical Bäcklund Transformation. A Nonlinear Superposition Principle

The classical Bäcklund transformation and its modern extensions have, remarkably, proven to be of fundamental importance in Soliton Theory. Thus, if a nonlinear equation is solitonic, it generically admits an auto-Bäcklund transformation and associated nonlinear superposition principle whereby, in particular, multi-soliton solutions may be generated. In the solitonic context, the action of the Bäcklund transformation at the nonlinear level is associated with the action at a linear level of a Darboux-type transformation [7]. Here, to place Bäcklund transformations in their original geometric setting, we describe, in detail, the classical result for the sine-Gordon equation. In terms of applications, it was the work of Seeger *et al.* [8]–[10] that demonstrated that this Bäcklund transformation has important application in the theory of crystal dislocations. Indeed, in [10], within the context of Frenkel's and Kontorova's dislocation theory, the superposition of so-called 'eigen-motions' was obtained by means of the classical Bäcklund transformation. The interaction of what today is called a breather with a kink-type dislocation was both described analytically and displayed graphically. The typical solitonic features to be later discovered in 1965 for the Korteweg–de Vries equation, namely preservation of velocity and shape following interaction as well as the concomitant phase shift were all recorded for the sine-Gordon equation in this remarkable paper of 1953.[2]

---

[2] "Man sieht ..., daß beim Durchdringen von Wellengruppe und Versetzung weder die Energie noch die Geschwindigkeit beider geändert wird. Es tritt lediglich eine Verschiebung des Versetzungsmittelpunktes ... und des Schwerpunktes der Wellengruppe ... auf." [10, p. 189]

Subsequently, Lamb [11] showed that the associated nonlinear superposition principle has application in the theory of ultra-short pulse propagation. In particular, decomposition phenomena observed experimentally in Rb vapour [12] were thereby reproduced theoretically. In addition, the classical Bäcklund transformation has been shown to have application in the theory of long Josephson junctions [13].

2.1   Geometric Preliminaries

Here, it proves convenient to set the sine-Gordon equation in the more general context of a system due to Bianchi which describes hyperbolic surfaces. If $r = r(u, v)$ is the position vector of a surface Σ in $\mathbb{R}^3$ then [14]

$$\begin{aligned} \text{I}: \quad d\boldsymbol{r} \cdot d\boldsymbol{r} &= E\,du^2 + 2F\,du\,dv + G\,dv^2, \\ \text{II}: \quad -d\boldsymbol{r} \cdot d\boldsymbol{N} &= e\,du^2 + 2f\,du\,dv + g\,dv^2, \end{aligned} \tag{1}$$

are termed the *1st* and *2nd fundamental forms* of Σ where $\boldsymbol{N} = \boldsymbol{r}_u \times \boldsymbol{r}_v / |\boldsymbol{r}_u \times \boldsymbol{r}_v|$ is the unit normal to Σ and

$$E = g_{11} = \boldsymbol{r}_u \cdot \boldsymbol{r}_u > 0, \quad F = g_{12} = \boldsymbol{r}_u \cdot \boldsymbol{r}_v, \quad G = g_{22} = \boldsymbol{r}_v \cdot \boldsymbol{r}_v > 0.$$

The definition of II and the identities $\boldsymbol{r}_u \cdot \boldsymbol{N} = \boldsymbol{r}_v \cdot \boldsymbol{N} = 0$ imply that

$$e = \boldsymbol{r}_{uu} \cdot \boldsymbol{N}, \quad f = \boldsymbol{r}_{uv} \cdot \boldsymbol{N}, \quad g = \boldsymbol{r}_{vv} \cdot \boldsymbol{N}.$$

The celebrated Gauss equations associated with Σ are

$$\begin{aligned} \boldsymbol{r}_{uu} &= \Gamma_{11}^1 \boldsymbol{r}_u + \Gamma_{11}^2 \boldsymbol{r}_v + e\boldsymbol{N}, \\ \boldsymbol{r}_{uv} &= \Gamma_{12}^1 \boldsymbol{r}_u + \Gamma_{12}^2 \boldsymbol{r}_v + f\boldsymbol{N}, \\ \boldsymbol{r}_{vv} &= \Gamma_{22}^1 \boldsymbol{r}_u + \Gamma_{22}^2 \boldsymbol{r}_v + g\boldsymbol{N}, \end{aligned} \tag{2}$$

where $\Gamma_{\alpha\beta}^\gamma$ are the usual Christoffel symbols given by

$$\Gamma_{\alpha\beta}^\gamma = \tfrac{1}{2} g^{\gamma\mu}(g_{\alpha\mu,\beta} + g_{\beta\mu,\alpha} - g_{\alpha\beta,\mu}).$$

The compatibility conditions

$$(\boldsymbol{r}_{uu})_v = (\boldsymbol{r}_{uv})_u, \quad (\boldsymbol{r}_{vv})_u = (\boldsymbol{r}_{uv})_v \tag{3}$$

applied to the *linear* Gauss system (2) produce the generally *nonlinear* Mainardi–Codazzi equations

$$\begin{aligned} \frac{\partial}{\partial v}\left(\frac{e}{\mathcal{H}}\right) - \frac{\partial}{\partial u}\left(\frac{f}{\mathcal{H}}\right) + \Gamma_{22}^2 \frac{e}{\mathcal{H}} - 2\Gamma_{12}^2 \frac{f}{\mathcal{H}} + \Gamma_{11}^2 \frac{g}{\mathcal{H}} &= 0, \\ \frac{\partial}{\partial u}\left(\frac{g}{\mathcal{H}}\right) - \frac{\partial}{\partial v}\left(\frac{f}{\mathcal{H}}\right) + \Gamma_{22}^1 \frac{e}{\mathcal{H}} - 2\Gamma_{12}^1 \frac{f}{\mathcal{H}} + \Gamma_{11}^1 \frac{g}{\mathcal{H}} &= 0, \end{aligned} \tag{4}$$

and the *Theorema Egregrium* of Gauss which delivers the expression

$$\mathcal{K} = \frac{1}{\mathcal{H}}\left[\frac{\partial}{\partial v}\left(\frac{\mathcal{H}\Gamma_{11}^2}{E}\right) - \frac{\partial}{\partial u}\left(\frac{\mathcal{H}\Gamma_{12}^2}{E}\right)\right] \tag{5}$$

for the *total curvature* $\mathcal{K} = (eg - f^2)/\mathcal{H}^2$ of Σ. In the above, it has been convenient to introduce $\mathcal{H}^2 = EG - F^2$.

At this point, the reader familiar with Soliton Theory will note that this derivation of a nonlinear system, that is (4)–(5) via compatibility conditions applied to a linear system, namely the Gauss equations (2) is strongly reminiscent of the generation of soliton equations via compatibility conditions applied to a linear representation [15, 16]. However, there is at this stage no spectral parameter present in the Gauss equations as there is in the linear representation of 1+1-dimensional soliton equations. Accordingly, it is natural to enquire if constraints can be imposed on the geometry so that a spectral parameter can be infiltrated naturally into the linear Gauss system and an integrable nonlinear system (4)–(5) obtained. It is in this connection that we next introduce a classical system due to Bianchi.

2.2   The Bianchi System

If the asymptotic lines on the surface Σ are taken as parametric curves then $e = g = 0$ and the Mainardi–Codazzi equations (4) reduce to [14]

$$\frac{\partial}{\partial u}\left(\frac{f}{\mathcal{H}}\right) + 2\Gamma_{12}^2 \frac{f}{\mathcal{H}} = 0, \quad \frac{\partial}{\partial v}\left(\frac{f}{\mathcal{H}}\right) + 2\Gamma_{12}^1 \frac{f}{\mathcal{H}} = 0, \tag{6}$$

where

$$\Gamma_{12}^1 = (GE_v - FG_u)/(2\mathcal{H}^2), \quad \Gamma_{12}^2 = (EG_u - FE_v)/(2\mathcal{H}^2)$$

while

$$\mathcal{K} = -f^2/\mathcal{H}^2 = -1/\rho^2 < 0. \tag{7}$$

Thus, the total curvature is always negative and Σ is termed a *hyperbolic* surface.[3] The angle ω between the parametric lines is such that

$$\cos\omega = F/\sqrt{EG}, \quad \sin\omega = \mathcal{H}/\sqrt{EG} \tag{8}$$

and if a, b are introduced such that

$$E = \rho^2 a^2, \quad G = \rho^2 b^2 \tag{9}$$

then the fundamental forms I and II for Σ become

$$\mathrm{I} = \rho^2[a^2 du^2 + 2ab\cos\omega\, du\, dv + b^2 dv^2],$$
$$\mathrm{II} = 2\rho ab\sin\omega\, du\, dv. \tag{10}$$

The Mainardi–Codazzi equations (4) together with the Theorema Egregium produce a nonlinear system descriptive of the above hyperbolic surfaces, namely

$$\omega_{uv} + \frac{1}{2}\left(\frac{\rho_u}{\rho}\frac{b}{a}\sin\omega\right)_u + \frac{1}{2}\left(\frac{\rho_v}{\rho}\frac{a}{b}\sin\omega\right)_v - ab\sin\omega = 0,$$

$$a_v + \frac{1}{2}\frac{\rho_v}{\rho}a - \frac{1}{2}\frac{\rho_u}{\rho}b\cos\omega = 0, \tag{11}$$

$$b_u + \frac{1}{2}\frac{\rho_u}{\rho}b - \frac{1}{2}\frac{\rho_v}{\rho}a\cos\omega = 0.$$

---

[3]Only hyperbolic surfaces can be parametrized in terms of real asymptotic coordinates.

Levi and Sym [17] in 1990 considered the system (11) and were led to the additional constraint $\rho_{uv} = 0$ for integrability so that

$$\rho = f(u) + g(v). \tag{12}$$

Thus, this assumption allows the insertion of a spectral 'parameter'

$$\zeta = \zeta(u, v, k) = \frac{1}{2}\sqrt{\frac{k - g(v)}{k + f(u)}}, \qquad k = \text{const} \tag{13}$$

into a 2×2 matrix version of the Gauss equations (2). This was achieved via Lie group methods. It turns out, remarkably, that the nonlinear system (11) with the adjoined condition (12) had received extensive attention in the classical literature of the nineteenth century and, indeed, is originally due to Bianchi. It may be shown to admit an auto-Bäcklund transformation whereby multi-soliton solutions can be generated.

In what follows, we concentrate on the special case when ρ is constant and the surfaces Σ are pseudo-spherical. The auto-Bäcklund transformation for such surfaces will be derived 'ab initio' by geometric means.

2.3 Pseudo-Spherical Surfaces. Generation via an Auto-Bäcklund Transformation

If $\mathcal{K} = -1/\rho^2 < 0$ is a constant then Σ is termed a pseudo-spherical surface. In that case, the Bianchi system (11) reduces to

$$\omega_{uv} = ab \sin \omega \tag{14}$$

where $a = a(u), b = b(v)$. If Σ is now parametrized by arclength along asymptotic lines (corresponding to the transformation $du \rightarrow du' = \sqrt{E(u)}\,du$, $dv \rightarrow dv' = \sqrt{G(v)}\,dv$) then the fundamental forms become, on dropping the primes,

$$\text{I} = du^2 + 2\cos\omega\,du\,dv + dv^2,$$

$$\text{II} = \frac{2}{\rho}\sin\omega\,du\,dv \tag{15}$$

and (14) reduces to the *sine-Gordon* equation

$$\omega_{uv} = \frac{1}{\rho^2}\sin\omega. \tag{16}$$

To construct pseudo-spherical surfaces $\Sigma : \boldsymbol{r} = \boldsymbol{r}(u, v)$, it is required to generate solutions of the nonlinear equation (16) and then determine the associated position vector $\boldsymbol{r} = \boldsymbol{r}(u, v)$ via the Gauss equations

$$\boldsymbol{r}_{uu} = \omega_u \cot\omega\,\boldsymbol{r}_u - \omega_u \text{cosec}\,\omega\,\boldsymbol{r}_v,$$

$$\boldsymbol{r}_{uv} = \frac{1}{\rho}\sin\omega\boldsymbol{N}, \tag{17}$$

$$\boldsymbol{r}_{vv} = -\omega_v \text{cosec}\,\omega\,\boldsymbol{r}_u + \omega_v \cot\omega\,\boldsymbol{r}_v$$

augmented by the Weingarten system

$$N_u = \frac{1}{\rho} \cot \omega \, r_u - \frac{1}{\rho} \mathrm{cosec}\, \omega \, r_v,$$

$$N_v = -\frac{1}{\rho} \mathrm{cosec}\, \omega \, r_u + \frac{1}{\rho} \cot \omega \, r_v. \tag{18}$$

2.3.1 The Classical Bäcklund Transformation

The auto-Bäcklund transformation for (16) will be seen to provide:

- A simple geometric construction for pseudo-spherical surfaces Σ' : $r' = r'(u,v)$ given a seed pseudo-spherical surface Σ : $r = r(u,v)$

- An invariant transformation for the sine-Gordon equation which contains a key Bäcklund (spectral) parameter

- An associated nonlinear superposition principle known as a *Permutability Theorem* whereby, in particular, multi-soliton solutions may be generated by purely algebraic procedures well-suited to symbolic algebra packages such as MAPLE.

The Geometric Construction. If a point P is taken on an initial pseudo-spherical surface Σ, then a line segment PP' of constant length and tangential to Σ at P may be so constructed that the locus of such points P' as P traces out Σ is another pseudo-spherical surface Σ' with the same total curvature as Σ. This procedure may then be iterated to generate a sequence of pseudo-spherical surfaces.

The above result is readily established. To this end, let Σ be a pseudo-spherical surface with total curvature $\mathcal{K} = -1/\rho^2$ and with generic position vector $r = r(u,v)$ where u,v correspond to the parametrization by arclength along asymptotic lines. In this parametrization, r_u, r_v and N are all unit vectors but the vectors r_u and r_v in the tangent plane are not orthogonal. It proves convenient, in what follows, to introduce an orthonormal triad $\{A, B, C\}$ where

$$A = r_u, \quad B = -r_u \times N = \mathrm{cosec}\, \omega \, r_v - \cot \omega \, r_u, \quad C = N. \tag{19}$$

The Gauss–Weingarten equations (17)–(18) now show that

$$
\begin{pmatrix} A \\ B \\ C \end{pmatrix}_u =
\begin{pmatrix} 0 & -\omega_u & 0 \\ \omega_u & 0 & 1/\rho \\ 0 & -1/\rho & 0 \end{pmatrix}
\begin{pmatrix} A \\ B \\ C \end{pmatrix},
$$

$$
\begin{pmatrix} A \\ B \\ C \end{pmatrix}_v =
\begin{pmatrix} 0 & 0 & (1/\rho)\sin\omega \\ 0 & 0 & -(1/\rho)\cos\omega \\ -(1/\rho)\sin\omega & (1/\rho)\cos\omega & 0 \end{pmatrix}
\begin{pmatrix} A \\ B \\ C \end{pmatrix}. \tag{20}
$$

This linear system is compatible if and only if ω satisfies the sine-Gordon equation (16).

We seek a new pseudo-spherical surface Σ' with position vector r' in the form

$$r' = r + L \cos\varphi \, A + L \sin\varphi \, B \tag{21}$$

where $L = |r' - r|$ is constant. Here, φ is a function of u and v to be so chosen that Σ' is pseudo-spherical and parametrized in a similar manner to Σ. A necessary condition for this to be the case is that Σ' have a first fundamental form of the type $(15)_I$. In particular, this requires that

$$r'_u \cdot r'_u = 1, \quad r'_v \cdot r'_v = 1 \tag{22}$$

where, on use of (20) and (21),

$$r'_u = [1 - L(\varphi_u - \omega_u)\sin\varphi]\,A + L(\varphi_u - \omega_u)\cos\varphi\,B + \frac{L}{\rho}\sin\varphi\,C,$$
$$r'_v = (\cos\omega - L\varphi_v \sin\varphi)\,A + (\sin\omega + L\varphi_v \cos\varphi)\,B + \frac{L}{\rho}\sin(\omega - \varphi)\,C. \tag{23}$$

The conditions (22) now yield, in turn,

$$\varphi_u = \omega_u + \frac{\beta}{\rho}\sin\varphi,$$
$$\varphi_v = \frac{1}{\beta\rho}\sin(\varphi - \omega), \tag{24}$$

where

$$\beta = \frac{\rho}{L}\left(1 \pm \sqrt{1 - \frac{L^2}{\rho^2}}\right). \tag{25}$$

Accordingly, the system (24), which is compatible modulo the sine-Gordon equation (16), constitutes necessary conditions on φ for Σ' to be a pseudo-spherical surface parametrized by arclength along asymptotic lines.[4] It turns out that the system (24) is sufficient in this regard. Thus, on use of (24), the relations (23) show that $r'_u \cdot r'_v = \cos(2\varphi - \omega)$ whence the first fundamental form of Σ' becomes

$$I' = du^2 + 2\cos(2\varphi - \omega)\,du\,dv + dv^2. \tag{26}$$

Moreover, the unit normal N' to Σ' is given by

$$N' = \frac{r'_u \times r'_v}{|r'_u \times r'_v|} = -\frac{L}{\rho}\sin\varphi\,A + \frac{L}{\rho}\cos\varphi\,B + \left(1 - \frac{L\beta}{\rho}\right)C \tag{27}$$

and, on use of (21), it is seen that $(r' - r) \cdot N' = 0$. Accordingly, the vector $r' - r$ joining corresponding points on Σ and Σ' is tangential to Σ' at the point of correspondence. It is recalled that it is tangential to Σ by construction.

It is now readily shown that

$$r'_u \cdot N'_u = 0, \quad r'_u \cdot N'_v = r'_v \cdot N'_u = -\frac{1}{\rho}\sin(2\varphi - \omega), \quad r'_v \cdot N'_v = 0$$

leading to

$$II' = \frac{2}{\rho}\sin(2\varphi - \omega)\,du\,dv. \tag{28}$$

---

[4]In fact, the system (24) may be linearized, leading to a β-dependent linear representation for the sine-Gordon equation (16).

Accordingly, Σ' : $r' = r'(u,v)$ with r' given by (21) and the 'swivel' angle φ constrained by the system (24) does indeed represent a pseudo-spherical surface parametrized in the same manner as Σ. The angle ω' between the asymptotic lines on Σ' given by

$$\omega' = 2\varphi - \omega \tag{29}$$

plays the same role in relation to Σ' as ω does for Σ so that

$$\omega'_{uv} = \frac{1}{\rho^2}\sin\omega'. \tag{30}$$

Use of the relation (29) to eliminate φ in the system (24) now yields

$$\mathcal{B}_\beta : \quad \begin{aligned} \left(\frac{\omega'-\omega}{2}\right)_u &= \frac{\beta}{\rho}\sin\left(\frac{\omega'+\omega}{2}\right), \\ \left(\frac{\omega'+\omega}{2}\right)_v &= \frac{1}{\beta\rho}\sin\left(\frac{\omega'-\omega}{2}\right). \end{aligned} \tag{31}$$

This is the standard form of the classical Bäcklund transformation linking the sine-Gordon equations (16) and (30). In that $\mathcal{B}_\beta$ represents a correspondence between solutions of the same equation, it is commonly termed an *auto*-Bäcklund transformation.

It is noted that, under $\mathcal{B}_\beta$,

$$N' \cdot N = 1 - \frac{L\beta}{\rho} = \text{const}, \tag{32}$$

that is the tangent planes at corresponding points on Σ and Σ' meet at a constant angle ζ where $\beta = \tan(\zeta/2)$. In Bianchi's original geometric construction of which the Bäcklund procedure is an extension, it was required that the tangent planes be orthogonal with $\beta = 1$ and $L = \rho$. Bäcklund's relaxation of the orthogonality requirement allows the key parameter β to be inserted into the Bianchi transformation. In fact, the Bäcklund transformation (31) may be viewed as a composition of Bianchi's transformation $\mathcal{B}_{\beta=1}$ with the simple Lie group invariance $u \to \beta u$, $v \to v/\beta$ of the sine-Gordon equation (16).

In terms of the construction of pseudo-spherical surfaces, the Bäcklund transformation corresponds to the following result:

Let r be the coordinate vector of the pseudo-spherical surface Σ corresponding to a solution ω of the sine-Gordon equation (16). Let ω' denote the Bäcklund transform of ω via $\mathcal{B}_\beta$. Then, the coordinate vector r' of the pseudo-spherical surface Σ' corresponding to ω' is given by

$$r' = r + \frac{\rho\sin\zeta}{\sin\omega}\left[\sin\left(\frac{\omega-\omega'}{2}\right)r_u + \sin\left(\frac{\omega+\omega'}{2}\right)r_v\right]. \tag{33}$$

Bianchi's Permutability Theorem. Multi-Soliton Solutions. We next turn to the application of the auto-Bäcklund transformation $\mathcal{B}_\beta$ to construct multi-soliton solutions of the sine-Gordon equation.

Let us start with the seed vacuum solution $\omega = 0$ of (16). The Bäcklund transformation (31) shows that a second nontrivial solution ω' of (30) may be constructed by integration of the pair of first order equations

$$\omega'_u = \frac{2\beta}{\rho}\sin\left(\frac{\omega'}{2}\right), \quad \omega'_v = \frac{2}{\beta\rho}\sin\left(\frac{\omega'}{2}\right)$$

leading to the new *single soliton* solution

$$\omega' = 4\arctan\left[\exp\left(\frac{\beta}{\rho}u + \frac{v}{\beta\rho} + \alpha\right)\right] \tag{34}$$

where α is an arbitrary constant of integration. It is noted that here, it is the quantities ω'_u, ω'_v which have the characteristic hump shape associated with a soliton.

It turns out, remarkably, that analytic expressions for multi-soliton solutions which encapsulate their nonlinear interaction may now be obtained by an entirely *algebraic procedure*. This is a consequence of an elegant result derived from the auto-Bäcklund transformation $\mathcal{B}_\beta$ which embodies a *nonlinear superposition principle* and which is known as *Bianchi's Permutability Theorem*.

Thus, suppose ω is a seed solution of the sine-Gordon equation (16) and that ω_1 and ω_2 are the Bäcklund transforms of ω via $\mathcal{B}_{\beta_1}$ and $\mathcal{B}_{\beta_2}$, that is $\omega_1 = \mathcal{B}_{\beta_1}\omega$, $\omega_2 = \mathcal{B}_{\beta_2}\omega$. Let $\omega_{12} = \mathcal{B}_{\beta_2}\omega_1$ and $\omega_{21} = \mathcal{B}_{\beta_1}\omega_2$. The situation may be represented schematically by a *Bianchi diagram* as given below:

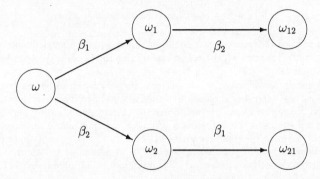

Figure 1: A Bianchi diagram.

It is natural to enquire if there are any circumstances under which the commutative condition $\omega_{12} = \omega_{21}$ applies. To investigate this matter, we set down the u-parts of the Bäcklund transformations associated with the above Bianchi diagram. Thus,

$$\omega_{1u} = \omega_u + \frac{2\beta_1}{\rho}\sin\left(\frac{\omega_1 + \omega}{2}\right), \quad \omega_{12u} = \omega_{1u} + \frac{2\beta_2}{\rho}\sin\left(\frac{\omega_{12} + \omega_1}{2}\right),$$
$$\omega_{2u} = \omega_u + \frac{2\beta_2}{\rho}\sin\left(\frac{\omega_2 + \omega}{2}\right), \quad \omega_{21u} = \omega_{2u} + \frac{2\beta_1}{\rho}\sin\left(\frac{\omega_{21} + \omega_2}{2}\right). \tag{35}$$

If we now set $\omega_{12} = \omega_{21} = \Omega$ then the operations $(35)_1 + (35)_2 - (35)_3 - (35)_4$ produce the necessary requirement

$$\Omega = \omega + 4\arctan\left[\frac{\beta_2 + \beta_1}{\beta_2 - \beta_1}\tan\left(\frac{\omega_2 - \omega_1}{4}\right)\right]. \tag{36}$$

If the expression (36) for Ω is substituted back into $(35)_{2,4}$ in place of ω_{12} and ω_{21} then these equations may be seen to be satisfied modulo $(35)_{1,3}$. Moreover, the corresponding equations in the v-part of the Bäcklund transformation are also satisfied by the expression (36). These considerations allow closure of the Bianchi diagram as indicated below:

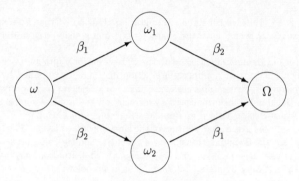

Figure 2: A commutative Bianchi diagram.

The relation (36) represents a nonlinear superposition principle known as a Permutability Theorem which acts on the solution set $\{\omega, \omega_1, \omega_2\}$ to produce a new solution Ω.

The commutative property now allows a *Bianchi lattice* to be constructed corresponding to iterated application of the Permutability Theorem. N-soliton solutions of the sine-Gordon equation may be thereby generated by purely algebraic procedures. These represent a nonlinear superposition of single soliton solutions (34) with Bäcklund parameters $\beta_1, \ldots, \beta_N$. Thus, at each application of the Bäcklund transformation, a new Bäcklund parameter β_i is introduced and an ith order solution generated. The procedure is indicated below in Fig 3 by means of a Bianchi lattice.

2.3.2 Physical Applications

Seeger *et al.* [10] exploited the Permutability Theorem for the sine-Gordon equation to investigate interaction properties of kink and breather-type solutions in connection with a crystal dislocation model. Later, the procedure was adopted by Lamb [11] and subsequently by Barnard [18] in the analysis of ultra-short optical pulse propagation in a resonant medium. Therein, analytic expressions for '$2N\pi$' light pulses were obtained via the above nonlinear superposition principle. These $2N\pi$ pulses exhibit the distinctive property that they ultimately decompose into N stable 2π pulses. Experimental evidence for this is provided, in particular, by the work of Gibbs and Slusher [12] which describes the decomposition of a 6π pulse into three 2π pulses in a Rb vapour. An account of such decomposition in ultra-short pulse propagation is presented in [11].

2.3.3 Solitonic Pseudo-Spherical Surfaces. Breathers

Single-Soliton Surfaces In the case of the single soliton (34), it may be shown that the position vector $\boldsymbol{r}$ of the associated pseudo-spherical surface is

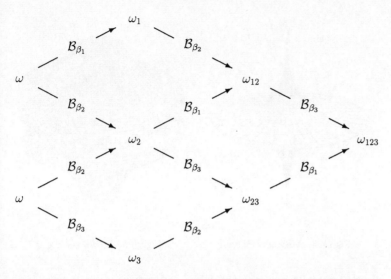

Figure 3: A Bianchi lattice.

given in lines of curvature coordinates $x = u + v$, $y = u - v$ by

$$r = \begin{pmatrix} \rho \sin \zeta \operatorname{sech} \chi \cos(y/\rho) \\ \rho \sin \zeta \operatorname{sech} \chi \sin(y/\rho) \\ x - \rho \sin \zeta \tanh \chi \end{pmatrix} \tag{37}$$

where $\chi = (x - y \cos \zeta)/(\rho \sin \zeta)$ and we have set $\alpha = 0$ without loss of generality. In the stationary soliton case $\zeta = \pi/2$, this gives a *pseudo-sphere* while in the non-stationary case $\zeta \neq \pi/2$, a *Dini surface* is obtained. These are illustrated in Fig 4 and Fig 5 respectively.

The sine-Gordon equation admits an important subclass of entrapped periodic two-soliton solutions known as *breathers*.

The two-soliton solution Ω given by (36) is

$$\Omega = 4 \arctan \left[\frac{\beta_2 + \beta_1}{\beta_2 - \beta_1} \sinh \left(\frac{\chi_1 - \chi_2}{2} \right) \Big/ \cosh \left(\frac{\chi_1 + \chi_2}{2} \right) \right] \tag{38}$$

with seed solution $\omega = 0$ and constituent one-soliton solutions $\omega_i = 4 \arctan e^{\chi_i}$ where $\chi_i = [(1 + \beta_i^2)x - (1 - \beta_i^2)y]/(2\beta_i \rho)$.

To obtain a periodic solution, complex conjugate parameters $\beta_1 = c + id$ and $\beta_2 = c - id$ are introduced to deliver the breather

$$\omega = 4 \arctan \left(\frac{c}{d} \frac{\sin \xi}{\cosh \eta} \right) \tag{39}$$

where ξ and η are linear functions in x and y, namely

$$\begin{aligned} \xi &= \frac{d}{2\rho(c^2 + d^2)}[(1 - c^2 - d^2)x - (1 + c^2 + d^2)y], \\ \eta &= \frac{c}{2\rho(c^2 + d^2)}[(1 + c^2 + d^2)x - (1 - c^2 - d^2)y]. \end{aligned} \tag{40}$$

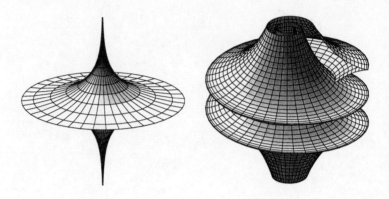

Figure 4: A pseudo-sphere.   Figure 5: A Dini surface.

Hence, the breather solution (39) is real and periodic in the variable ξ. In particular, if we require that $|\beta_1| = 1$ then a solution that is periodic in y is obtained, viz

$$\omega = -2\arctan\left[\frac{c}{d}\frac{\sin(dy/\rho)}{\cosh(cx/\rho)}\right]. \tag{41}$$

This is known as a *stationary* breather since, if x is regarded as a spatial variable and y as a time variable then the breather is not translated as time evolves.

The Bäcklund construction (33) may now be used to show that the position vector r of the pseudo-spherical surface associated with the stationary breather solution (41) is given by [19]

$$r_{\text{breather}} = \begin{pmatrix} 0 \\ 0 \\ x \end{pmatrix} + \frac{2d}{c}\frac{\sin(dy)\cosh(cx)}{d^2\cosh^2(cx) + c^2\sin^2(dy)}\begin{pmatrix} \sin y \\ -\cos y \\ 0 \end{pmatrix}$$

$$+ \frac{2d^2}{c}\frac{\cosh(cx)}{d^2\cosh^2(cx) + c^2\sin^2(dy)}$$

$$\times \frac{\cosh^2(cx) - \sin^2(dy)}{\cos^2(dy)\cosh^2(cx) + \sin^2(dy)\sinh^2(cx)}\begin{pmatrix} \cos y\cos(dy) \\ \sin y\cos(dy) \\ -\sinh(cx) \end{pmatrix}$$

where $c = \sqrt{1 - d^2}$ and $\rho = 1$. Thus, to every rational number d between zero and unity there corresponds a pseudo-spherical breather surface which is periodic in the y-parameter. If we write $d = p/q$ where p and q are co-prime integers with $p < q$ then the period of the breather solution is $2\pi q/p$ while that of the surface is $2\pi q$ when $p+q$ is odd and πq when $p+q$ is even. Stationary breather pseudo-spherical surfaces corresponding to various values of the parameter d are displayed in Fig 6.

The Bäcklund construction (33) may now be iterated to produce pseudo-spherical surfaces associated with N-soliton solutions as generated by the Permutability Theorem.

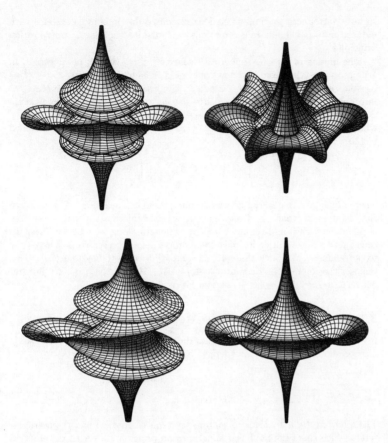

Figure 6: Stationary breather surfaces: $\frac{p}{q} = \frac{1}{4}, \frac{3}{4}, \frac{1}{5}, \frac{1}{2}$.

It turns out that there are other nonlinear solitonic systems of physical importance which may be linked, in a natural geometric manner, to the classical sine-Gordon equation and which, indeed, share its nonlinear superposition principle. Such systems *compatible* with the sine-Gordon equation are derived by suitably constrained *motions* of pseudo-spherical surfaces.

3   The Motion of Pseudo-Spherical Surfaces

Here, we consider certain motions of pseudo-spherical surfaces. One type of motion produces a continuum version of an anharmonic lattice model which, in a specialization, reduces to the well-known *modified Korteweg–de Vries* (mKdV) equation. Another type of motion leads to a classical system due to Weingarten corresponding to a subclass of the Lamé system descriptive of triply orthogonal surfaces. In general terms, consideration of the motion of soliton surfaces introduces the important idea of compatible integrable systems with induced auto-Bäcklund transformations and inherited nonlinear superposition principles.

The motion of a pseudo-spherical surface $\Sigma : \boldsymbol{r} = \boldsymbol{r}(u, v, t)$ parametrized in asymptotic coordinates u, v is now considered. The total curvature $\mathcal{K} = \mathcal{K}(t)$ is constant and negative at each instant t. The orthonormal basis $\{\boldsymbol{A}, \boldsymbol{B}, \boldsymbol{C}\}$ as introduced in (19) is employed whence the linear system (20) applies wherein now, however, $\rho = \rho(t)$ and $\omega = \omega(u, v, t)$. The most general time evolution which will maintain the orthonormality of the triad $\{\boldsymbol{A}, \boldsymbol{B}, \boldsymbol{C}\}$ is now adjoined to (20), namely

$$\begin{pmatrix} \boldsymbol{A} \\ \boldsymbol{B} \\ \boldsymbol{C} \end{pmatrix}_t = \begin{pmatrix} 0 & \alpha & \beta \\ -\alpha & 0 & \gamma \\ -\beta & -\gamma & 0 \end{pmatrix} \begin{pmatrix} \boldsymbol{A} \\ \boldsymbol{B} \\ \boldsymbol{C} \end{pmatrix}, \tag{42}$$

where α, β, γ are, in general, dependent on u, v and t. The system (20), in which time t enters only parametrically, encapsulates the information that the surface is pseudo-spherical and parametrized by arclength along asymptotic lines. To construct a time evolution in which these properties are preserved, it is necessary to determine α, β, γ such that (42) is compatible with the system (20). This imposes the compatibility conditions $\boldsymbol{A}_{ut} = \boldsymbol{A}_{tu}$, $\boldsymbol{A}_{vt} = \boldsymbol{A}_{tv}, \ldots$ which, in turn, deliver a *linear* inhomogeneous system for α, β and γ.

3.1 Particular Compatible Motions

If $\boldsymbol{v} = \boldsymbol{r}_t = \lambda \boldsymbol{A} + \mu \boldsymbol{B} + \nu \boldsymbol{C}$ represents the velocity of $\Sigma : \boldsymbol{r} = \boldsymbol{r}(u, v, t)$ then it may be shown that the triads $\{\alpha, \beta, \gamma\}$ and $\{\lambda, \mu, \nu\}$ are related via

$$\alpha = -\lambda \omega_u + \mu_u - \nu/\rho, \quad \beta = \nu_u + \mu/\rho,$$
$$\gamma = -(\nu_u + 2\mu/\rho) \cot \omega + \lambda/\rho + \nu_v \operatorname{cosec} \omega. \tag{43}$$

The following $\{\lambda, \mu, \nu\}$ leads to solitonic systems of physical and geometric interest which share the nonlinear superposition principle (36) with the classical sine-Gordon equation:

3.1.1 A Continuum Version of an Anharmonic Lattice Model. The mKdV Equation

One possible motion $\{\lambda, \mu, \nu\}$ is given by [20]

$$\{\lambda, \mu, \nu\} = \{\rho\Big(\frac{\omega_u^2}{4} + \delta \cos \omega - \zeta\Big), \rho\Big(\delta \sin \omega - \frac{\omega_{uu}}{2}\Big), \omega_u\} \tag{44}$$

where

$$\omega_t = \frac{\rho}{2}\omega_{uuu} + \frac{\rho}{4}\omega_u^3 + \Big(\frac{3}{2\rho} - \rho\zeta\Big)\omega_u + \delta\rho\omega_v,$$
$$\omega_{uv} = \frac{1}{\rho^2}\sin \omega, \tag{45}$$

and $\delta = \delta(t)$, $\zeta = \zeta(t)$ are arbitrary, while $\rho_t = 0$ so that, in this case, the total curvature $\mathcal{K} = -1/\rho^2$ remains constant on Σ in the motion.

Elimination of ω_v in (45) yields

$$\omega_{ut} = \frac{\rho}{2}\omega_{uuuu} + \frac{3}{4}\omega_u^2\omega_{uu} + \left(\frac{3}{2\rho} - \rho\zeta\right)\omega_{uu} + \frac{\delta}{\rho}\sin\omega. \tag{46}$$

This nonlinear soliton equation with $\delta \neq 0$ and $\zeta = 3/(2\rho^2)$ was derived by Konno *et al.* [21, 22] as the continuum limit of a model of wave propagation in an anharmonic lattice. The specialisation $\delta = 0$ and $\zeta = 3/(2\rho^2)$ produces the mKdV equation[5]

$$\Omega_t + 6\Omega^2\Omega_u + \Omega_{uuu} = 0, \tag{47}$$

in $\Omega = \omega_u$ with $u \to 4u$ and $t \to -t$. The mKdV equation arises, in particular, in the analysis of nonlinear Alfvén waves in a collisionless plasma [23]. Its connection with acoustic wave propagation in anharmonic lattices has been described by Zabusky [24].

3.1.2 A Classical Weingarten System

Here, we consider a purely normal motion in which the total curvature $\mathcal{K} = -1/\rho^2$ of Σ is now allowed to evolve in time and

$$\{\lambda, \mu, \nu\} = \{0, 0, \rho\theta_t\}. \tag{48}$$

In lines of curvature coordinates x, y, the compatibility conditions then yield

$$\begin{aligned} &\theta_{xyt} - \theta_x\theta_{yt}\cot\theta + \theta_y\theta_{xt}\tan\theta = 0, \\ &\left(\frac{\theta_{xt}}{\cos\theta}\right)_x - \frac{1}{\rho}\left(\frac{1}{\rho}\sin\theta\right)_t - \frac{1}{\sin\theta}\theta_y\theta_{yt} = 0, \\ &\left(\frac{\theta_{yt}}{\sin\theta}\right)_y + \frac{1}{\rho}\left(\frac{1}{\rho}\cos\theta\right)_t + \frac{1}{\cos\theta}\theta_x\theta_{xt} = 0, \\ &\theta_{xx} - \theta_{yy} = \frac{1}{\rho^2}\sin\theta\cos\theta. \end{aligned} \tag{49}$$

The nonlinear Weingarten system (49) appears in Eisenhart [25] in connection with a special subclass of triply orthogonal surfaces known as triple systems of Bianchi.[6] Such systems were studied extensively by Bianchi and Darboux [26, 27]. Here, Σ undergoes a purely normal propagation, whence it follows that the surfaces swept out by the parametric lines on Σ as t evolves will be orthogonal to Σ at all times. However, lines of curvature on a surface are mutually orthogonal, so that the family of surfaces swept out by the x-parametric lines will be orthogonal to that swept out by the y-parametric lines. Thus, the system which consists of these two families, augmented by the family whose members are the pseudo-spherical surfaces formed at each time t constitute a triply orthogonal system of surfaces. This explains the appearance of the Weingarten system (49), since a triple system of Bianchi is simply a triply orthogonal

---

[5]The mKdV equation (47) is linked to the celebrated KdV equation

$$\Omega_t + 6\Omega\Omega_u + \Omega_{uuu} = 0$$

via a Bäcklund transformation of Miura type [2].

[6]It should be noted that, in the Weingarten system (49), each of (49)$_2$ and (49)$_3$ is a consequence of (49)$_4$ and the other. Consequently, one of (49)$_2$ or (49)$_3$ may be discarded.

system of surfaces in which members of one family are pseudo-spherical. Triple systems associated with multi-soliton solutions of the Weingarten system (49) may be readily constructed. These are generated by a nonlinear superposition principle of the type (36) associated with an auto-Bäcklund transformation of the above Weingarten system.

3.2   The Bäcklund Transformations

Given the position vector $r = r(u, v, t)$ of a generic point on a moving pseudo-spherical surface Σ, a Bäcklund transformation may be applied at each time t to generate a new pseudo-spherical surface. If this is done in such a way that the constraints on the motion associated, in turn, with the anharmonic lattice model and Weingarten system are preserved then auto-Bäcklund transformations for these systems are induced. These are embodied in the following:

Theorem 3.1. *The system*

$$\omega_t = \frac{\rho}{2}\omega_{uuu} + \frac{\rho}{4}\omega_u^3 + \delta\rho\omega_v$$

$$\omega_{uv} = \frac{1}{\rho^2}\sin\omega \tag{50}$$

is invariant under the Bäcklund transformation

$$\left(\frac{\omega' - \omega}{2}\right)_u = \frac{\beta}{\rho}\sin\left(\frac{\omega' + \omega}{2}\right),$$

$$\left(\frac{\omega' + \omega}{2}\right)_v = \frac{1}{\beta\rho}\sin\left(\frac{\omega' - \omega}{2}\right),$$

$$(\omega' - \omega)_t = \frac{\beta^2}{\rho}\omega_u + \beta\omega_{uu}\cos\left(\frac{\omega' + \omega}{2}\right) + \left(\frac{\beta}{2}\omega_u^2 + \frac{\beta^3}{\rho^2}\right)\sin\left(\frac{\omega' + \omega}{2}\right) \tag{51}$$

$$+ \delta\left[-2\rho\omega_v + \frac{2}{\rho}\sin\left(\frac{\omega' - \omega}{2}\right)\right]$$

where $\rho_t = 0, \beta_t = 0$.

Theorem 3.2. *The Weingarten system (49) in asymptotic coordinates* u, v *and with* $\omega = 2\theta$, *namely*

$$\omega_{uut} - \omega_u\omega_{ut}\cot\omega + \omega_u\omega_{vt}\mathrm{cosec}\,\omega - \frac{1}{\rho^2}\omega_t = 0,$$

$$\omega_{vvt} - \omega_v\omega_{vt}\cot\omega + \omega_v\omega_{ut}\mathrm{cosec}\,\omega - \frac{1}{\rho^2}\omega_t = 0, \tag{52}$$

$$\omega_{uv} = \frac{1}{\rho^2}\sin\omega$$

is invariant under the Bäcklund transformation

$$\left(\frac{\omega' - \omega}{2}\right)_u = \frac{\beta}{\rho}\sin\left(\frac{\omega' + \omega}{2}\right),$$

$$\left(\frac{\omega' + \omega}{2}\right)_v = \frac{1}{\beta\rho}\sin\left(\frac{\omega' - \omega}{2}\right), \tag{53}$$

$$\omega'_t = \frac{1 + \beta^2}{1 - \beta^2}\omega_t + \frac{2\rho\beta}{1 - \beta^2}\mathrm{cosec}\,\omega\left[\omega_{vt}\sin\left(\frac{\omega' + \omega}{2}\right) - \omega_{ut}\sin\left(\frac{\omega' - \omega}{2}\right)\right]$$

where $\mathcal{K} = -1/\rho^2(t)$ and $\beta = \beta(t)$ *is connected to* $\rho(t)$ *by the relation (25) with L being an arbitrary non-zero constant.*

It is observed that the *spatial* parts of the Bäcklund transformations (51) and (53) coincide with the Bäcklund transformation $\mathcal{B}_\beta$ given by (31) for the classical sine-Gordon equation. This has the important implication that the Permutability Theorem (36) associated with the original Bäcklund transformation $\mathcal{B}_\beta$ also applies to both the lattice system (50) and the Weingarten system (52). Indeed, this nonlinear superposition principle is generic to all integrable systems compatible with the classical sine-Gordon equation in the sense that they derive from compatible motions of pseudo-spherical surfaces.

The Permutability Theorem has been exploited by Konno and Sanuki [22] to generate kink and soliton solutions of the continuum anharmonic lattice model (50). An account of that work is given in Rogers and Shadwick [2].

It is noted that the specialisation $\delta = 0$ in Theorem 3.1 provides an auto-Bäcklund transformation given by $(51)_{1,3}$ for the potential mKdV equation $(50)_1$ and thereby for the mKdV equation (47). Multi-soliton solutions of the latter have been constructed via the Permutability Theorem by Hirota and Satsuma [28].

The position vector r of the moving Dini surface corresponding to the single-soliton solution of the system (50) with $\delta = 0$ is given by [20]

$$
r = r(u,v,t) = \begin{pmatrix} \dfrac{2\rho\beta}{1+\beta^2}\operatorname{sech}\chi\cos\left(\dfrac{u-v}{\rho} - \dfrac{t}{2\rho^2}\right) \\[2mm] \dfrac{2\rho\beta}{1+\beta^2}\operatorname{sech}\chi\sin\left(\dfrac{u-v}{\rho} - \dfrac{t}{2\rho^2}\right) \\[2mm] u+v - \dfrac{3t}{\rho} - \dfrac{2\rho\beta}{1+\beta^2}\tanh\chi \end{pmatrix} \tag{54}
$$

where $\chi = (\beta u + v/\beta)/\rho + \beta^3 t/(2\rho^2) + \varepsilon$ and ε is an arbitrary constant.

If we set $v = $ const in (54) then the position vector for a single-soliton surface of the mKdV equation is obtained. A particular such mKdV surface is depicted in Fig 7. In Fig 8, compatible single soliton mKdV and sine-Gordon surfaces are shown together.

In the case of the classical Weingarten system (52) we have the intriguing result that triply orthogonal systems of surfaces may be constructed associated with multi-soliton solutions. These may be readily generated by repeated application of the Permutability Theorem (36) wherein the Bäcklund parameters are assigned values $\beta_i(t) = \rho(t)[1 \pm \sqrt{1 - L_i^2/\rho^2(t)}]/L_i$ for specified Gaussian curvature $\mathcal{K} = -1/\rho^2(t)$. The basic components in this nonlinear superposition principle are single-soliton solutions of the Weingarten system. The position vector r of the triply orthogonal system associated with the single-soliton

$$\omega = 4\arctan\exp\chi$$

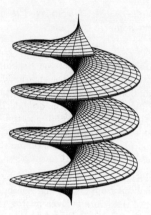

Figure 7: A one-soliton mKdV surface.

is given in terms of lines of curvature coordinates x, y by

$$r = r(x, y, t) = \begin{pmatrix} L\operatorname{sech}\chi \cos\left(\dfrac{y}{\rho(t)} + q(t)\right) \\ L\operatorname{sech}\chi \sin\left(\dfrac{y}{\rho(t)} + q(t)\right) \\ x - L\tanh\chi \end{pmatrix}$$

where $\chi = [x \pm y\sqrt{1 - L^2/\rho^2(t)}]/L + f(t)$, while $\rho(t), f(t)$ are arbitrary and

$$q(t) = -\frac{1}{L} \int \sqrt{\rho^2(t) - L^2}\, \frac{df}{dt}\, dt.$$

The coordinate surfaces $t = \text{const}$ in this triply orthogonal system are pseudo-spherical surfaces of Dini-type (Fig 5).

4   Bäcklund Transformations in 1+1-Dimensional Soliton Theory. General Connections

It is natural to enquire as to whether the admittance of an auto-Bäcklund transformation and concomitant nonlinear superposition principle is restricted to the classical sine-Gordon equation and its compatible solitonic systems. Thus, since the pioneering work of Kruskal and Zabusky in 1965 which unexpectedly rediscovered the KdV equation[7] in the context of the celebrated Fermi–Pasta–Ulam problem, an extraordinarily diverse range of nonlinear physical models have been revealed to be amenable to the IST and to admit multi-soliton solutions. It is

---

[7]The KdV equation seems to have been originally set down by Boussinesq in his extensive memoire of 1877 [29, p. 360].

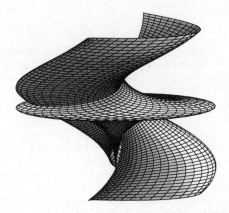

Figure 8: A one-soliton mKdV surface compatible with a pseudo-sphere.

a remarkable fact that these so-called *integrable* systems characteristically possess an auto-Bäcklund transformation and associated nonlinear superposition principle whereby, in particular, multi-soliton solutions may be generated algorithmically by purely algebraic means. Moreover, in geometric terms, just as the sine-Gordon equation is linked to pseudo-spherical surfaces so, in general, other 1+1-dimensional integrable equations may be associated via the Gauss equations with *soliton surfaces* [30]. These, in turn, may be generated by iterated Bäcklund transformations which act at the linear level. Particular single-soliton surfaces associated with the integrable Bianchi system (11)–(12) are displayed in Fig 9.

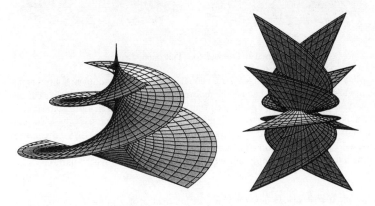

Figure 9: Bianchi surfaces.

The classical sine-Gordon equation, interestingly, arises in an alternative geometric context, namely as a specialization of the celebrated two-dimensional

Toda lattice system

$$(\ln h_n)_{uv} = -h_{n-1} + 2h_n - h_{n+1}, \qquad n \in \mathbb{Z} \qquad (55)$$

derived, in the last century by Darboux, in the theory of conjugate nets and rediscovered in its one-dimensional version by Toda in 1967 [31]. An auto-Bäcklund transformation for the Toda lattice equation was subsequently obtained by Wadati and Toda [32].

The *Tzitzeica* equation

$$\omega_{uv} = e^\omega - e^{-2\omega} \qquad (56)$$

is another soliton equation of physical interest which may be obtained as a reduction of the Toda lattice system (55). It was originally derived in 1910 as the compatibility condition for the linear Gauss system (2) in asymptotic coordinates u, v for surfaces Σ with the *Tzitzeica Property*[8] [33]. It was rediscovered in a solitonic context by Dodd and Bullough [34] in 1977. It was subsequently linked to an integrable anisentropic gas dynamics system in [35]. In Fig 10, single-soliton Tzitzeica surfaces are displayed as recently generated by an auto-Bäcklund transformation [36]. Fig 11 shows a Tzitzeica surface associated with a breather solution of (56).

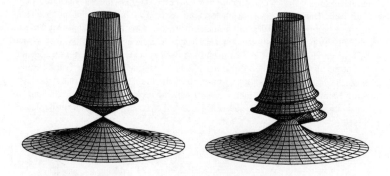

Figure 10: One-soliton Tzitzeica surfaces.

The nonlinear Schrödinger (NLS) equation is another canonical soliton equation which has a simple geometric genesis and yet has widespread physical applications ranging from the description of self-focussing in optical beams in nonlinear media [37]–[39] to the analysis of deep water gravity waves [40] and capillarity models [41]. It was the work of Hasimoto [42] in 1972 on the self-induced motion of a thin non-stretching vortex travelling through an inviscid fluid that provided a simple but elegant derivation of the NLS equation

$$iq_t + q_{ss} + 2|q|^2 q = 0. \qquad (57)$$

Therein, if $\boldsymbol{r} = \boldsymbol{r}(s, t)$ denotes the position vector of a point P on the vortex filament then the velocity relation

$$\boldsymbol{r}_t = \kappa \boldsymbol{b} \qquad (58)$$

---

[8]Namely the total curvature $\mathcal{K} \sim d^4$ where d is the distance from the origin to the tangent plane at the current point P on Σ.

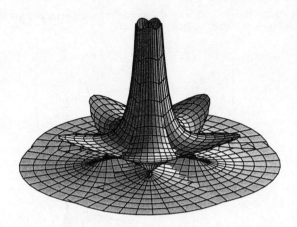

Figure 11: A breather Tzitzeica surface.

was derived where s denotes arclength along the vortex and $\kappa = \kappa(s,t)$ is its curvature. The compatibility conditions between the Serret–Frenet equations

$$\begin{pmatrix} t \\ n \\ b \end{pmatrix}_s = \begin{pmatrix} 0 & \kappa & 0 \\ -\kappa & 0 & \tau \\ 0 & -\tau & 0 \end{pmatrix} \begin{pmatrix} t \\ n \\ b \end{pmatrix}, \tag{59}$$

and the time evolution

$$\begin{pmatrix} t \\ n \\ b \end{pmatrix}_t = \begin{pmatrix} 0 & \alpha & \beta \\ -\alpha & 0 & \gamma \\ -\beta & -\gamma & 0 \end{pmatrix} \begin{pmatrix} t \\ n \\ b \end{pmatrix} \tag{60}$$

of the orthonormal triad $\{t, n, b\}$ admit the solution $\{\alpha, \beta, \gamma\} = \{-\kappa\tau, \kappa_s, -\tau^2 + \kappa_{ss}/\kappa\}$ where

$$q = \frac{\kappa}{2} \exp\left(i \int \tau \, ds \right), \tag{61}$$

satisfies the NLS equation (57). The latter admits the auto-Bäcklund transformation [2]

$$(q + q')_s = (q - q')\sqrt{4\beta^2 - |q + q'|^2},$$

$$(q + q')_t = i(q - q')_s \sqrt{4\beta^2 - |q + q'|^2} + \frac{i}{2}(q + q')(|q + q'|^2 + |q - q'|^2) \tag{62}$$

with associated Permutability Theorem derived via its spatial part. Multi-soliton solutions of (57) may be generated thereby as a nonlinear superposition of single-soliton solutions

$$q = \nu \operatorname{sech}\left[\nu(s - 2\tau_0 t)\right] \exp[i(\tau_0 s + (\nu^2 - \tau_0^2)t)] \tag{63}$$

with associated position vector r on the soliton surface determined via integration of the system (59)–(60) to obtain

$$
r(s,t) = \begin{pmatrix}
s - \dfrac{2\nu}{\nu^2 + \tau_0^2}\tanh(\nu\xi) \\[2ex]
-\dfrac{2\nu}{\nu^2 + \tau_0^2}\operatorname{sech}(\nu\xi)\cos[\tau_0 s + (\nu^2 - \tau_0^2)t] \\[2ex]
-\dfrac{2\nu}{\nu^2 + \tau_0^2}\operatorname{sech}(\nu\xi)\sin[\tau_0 s + (\nu^2 - \tau_0^2)t]
\end{pmatrix}
\tag{64}
$$

where $\xi = s - 2\tau_0 t$ and τ_0, ν are arbitrary parameters. Particular stationary and moving single-soliton surfaces of the NLS equation are depicted in Fig 12.

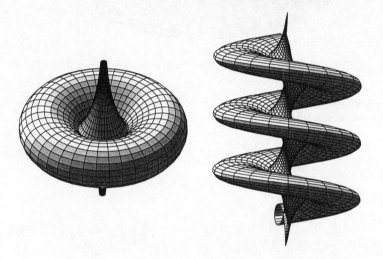

Figure 12: Stationary and moving one-soliton surfaces.

In Fig 13, 'smoke-ring' surfaces corresponding to spatially periodic solutions of the NLS equation are displayed. Smoke-ring solutions have been discussed by Cieśliński *et al.* [43], Rogers and Schief [44] and Calini in this volume. Surfaces associated with breather solutions of the NLS equation which are periodic in time are shown in Fig 14.

The NLS equation is linked to a number of other well-known nonlinear physical models which admit multi-soliton solutions. Thus, in particular, if we identify $t = r_s$ in the above with the classical spin vector S then the *Heisenberg spin* equation

$$
S_t = S \times S_{ss}, \qquad S^2 = 1
\tag{65}
$$

results [45, 46]. On the other hand, it may be shown that the *unpumped Maxwell-Bloch* system [47]

$$
E_x = P, \quad P_t = EN,
$$
$$
N_t = -\tfrac{1}{2}(E^*P + EP^*), \quad N^2 + PP^* = 1
\tag{66}
$$

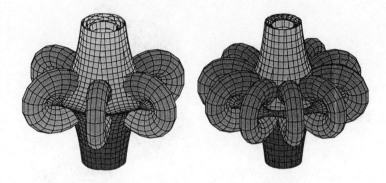

Figure 13: NLS 'smoke-ring' surfaces.

which generates the sharpline *self-induced transparency* (SIT) system

$$\chi_{xt} = \sin\chi + \nu_x\nu_t \tan\chi,$$
$$\nu_{xt} = -\nu_x\chi_t \cot\chi - \nu_t\chi_x(\sin\chi\cos\chi)^{-1} \tag{67}$$

is compatible with the NLS equation in the same way that the classical sine-Gordon equation and the mKdV equation are compatible. The SIT system is, in turn, linked to the Pohlmeyer–Lund–Regge system descriptive of the dynamics of relativistic vortices [48, 49] as well as to an integrable stimulated Raman scattering (SRS) system [50].

The sine-Gordon, NLS and KdV equations constitute canonical members of the celebrated AKNS system of 1+1-dimensional soliton equations [51]. The members of the AKNS system generically admit auto-Bäcklund transformations and associated nonlinear superposition principles. Indeed, these remarkable properties extend to all soliton systems be they nonlinear integro-differential such as the Benjamin–Ono equation which models internal water wave propagation in deep water [52, 53] or nonlinear differential-difference equations such as the Toda lattice system which, in one dimension, models longitudinal vibration in a chain of masses interconnected by nonlinear springs [31]. Auto-Bäcklund transformations may also be constructed for ordinary differential equations of Painlevé-type. These have recently been used to reveal novel hidden symmetries of, in particular, Painlevé IV [54].

5 Discrete Systems via Permutability Theorems

There is currently a strong resurgence of interest in integrable differential-difference and fully discrete equations. It has been shown that, typically, these (differential) difference equations translate to integrable continuous counterparts by taking appropriate continuum limits. The limiting processes involved are usually quite subtle, with continuous variables entering via a multiple scales process. A recent review of this work can be found in [55].

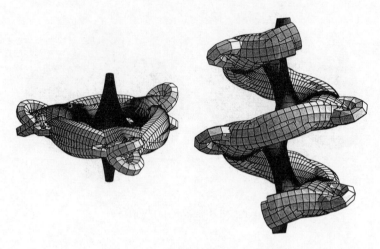

Figure 14: NLS breather surfaces.

There exists, however, a distinct method of constructing integrable discrete systems which reduce to continuous ones by means of a natural limiting procedure in which the discrete independent variables may be regarded as truly discretized versions of the continuous variables. This technique has its origin in work by Levi and Benguria [56] and has been refined and generalized by several authors [57]–[60]. The main idea is to reinterpret Bäcklund transformations and associated nonlinear superposition principles as, in turn, integrable differential-difference and pure difference versions of their continuous counterparts. Once again, the best illustration is by way of the classical sine-Gordon equation and the corresponding Bianchi Permutability Theorem.

Thus, let us recall the classical Permutability Theorem (36). It is noted that the sine-Gordon equation is invariant under $\omega \rightarrow -\omega$ so that the signs in front of the solutions $\omega_{12}, \omega_2, \omega_1$ and ω in the Permutability Theorem may be chosen arbitrarily. For convenience, we consider the form

$$\tan\left(\frac{\omega_{12} + \omega}{4}\right) = \frac{\beta_2 + \beta_1}{\beta_2 - \beta_1}\tan\left(\frac{\omega_2 + \omega_1}{4}\right). \tag{68}$$

Now, it is assumed that ω represents a point on a discrete lattice labelled by n_1 and n_2, that is

$$\omega = \omega(n_1, n_2). \tag{69}$$

Then, regard the transformed objects ω_i as its neighbours on this lattice, namely

$$\omega_1 = \omega(n_1 + 1, n_2), \quad \omega_2 = \omega(n_1, n_2 + 1). \tag{70}$$

Analogously, we make the identification

$$\omega_{12} = \omega(n_1 + 1, n_2 + 1). \tag{71}$$

In this way, the Permutability Theorem (68) becomes a *discrete* equation with discrete independent variables n_1 and n_2. In fact, (68) constitutes a discrete sine-Gordon equation as may be seen as follows:

If one introduces the continuous variable $\alpha = n_1 \epsilon_1$ for small ϵ_1 then

$$\omega_1 = \omega + \epsilon_1 \omega_\alpha + O(\epsilon_1^2) \tag{72}$$

according to Taylor's Theorem. Hence, on setting $\beta_1 = \epsilon_1 \gamma_1$, the lowest order in ϵ_1 of (68) vanishes identically so that the limit $\epsilon_1 \to 0$ produces a nontrivial result, viz

$$(\omega_2 - \omega)_\alpha = 4\frac{\gamma_1}{\beta_2} \sin\left(\frac{\omega_2 + \omega}{2}\right). \tag{73}$$

This is nothing but the Bäcklund equation $(31)_1$ and represents a differential-difference sine-Gordon equation. A fully continuous equation is obtained via setting $\beta = n_2 \epsilon_2$, $\beta_2 = \gamma_2/\epsilon_2$ and performing the limit $\epsilon_2 \to 0$. Equation (73) then becomes the sine-Gordon equation

$$\omega_{\alpha\beta} = 4\frac{\gamma_1}{\gamma_2} \sin\omega. \tag{74}$$

Thus, we have established the important result that the Permutability Theorem (68) and the Bäcklund equation $(31)_1$ represent discrete and differential-difference versions respectively of the classical sine-Gordon equation.

It is important to point out that the (differential) difference equations obtained in the manner sketched above inherit the integrability property from the underlying differential equations. For example, if one regards the transformation (21) of the position vector r as shift on a lattice of position vectors then the corresponding expressions for (A_1, B_1, C_1) and (A_2, B_2, C_2) deliver a *discrete* Gauss–Weingarten system. Indeed, (algebraic) compatibility produces the discrete sine-Gordon equation (68). Moreover, it may be shown that the Bäcklund transformation for the continuous sine-Gordon equation induces a Bäcklund transformation for the discrete one. In this respect, integrability is preserved by construction.

Remarkably, discrete models for pseudo-spherical surfaces were developed independently by Sauer [61] and Wunderlich [62] as early as 1950 using elementary difference geometric tools. However, it is only recently that the connection between the discrete sine-Gordon equation (68) and such discrete *K-surfaces* has been established [63]. Since then both surfaces of constant mean curvature [64] and isothermic surfaces [65] have been discretized in a purely geometric manner to preserve integrability. Moreover, it has been shown that the discrete Tzitzeica equation as set down in [66] admits a simple geometric interpretation in terms of *discrete affine spheres* [67]. Additionally, convex affine spheres governed by the elliptic version of the Tzitzeica equation (56) have been successfully discretized in [68].

Bäcklund transformations may also be exploited to discretize other important integrable systems such as the Darboux and Lamé systems descriptive of conjugate nets and triply orthogonal systems of surfaces respectively [69]–[72], the Ernst equation of General Relativity and the 2+1-dimensional sine-Gordon system (88). The latter is amenable to the classical Moutard transformation which produces the simple discrete version

$$\tau\tau_{123} = \tau_1\tau_{23} - \tau_2\tau_{13} + \tau_3\tau_{12}. \tag{75}$$

This result has been recently established in [73].

In conclusion, we note a connection with the concept of characteristics which plays an important role in both the analytical and numerical treatment of hyperbolic differential equations [74, 75]. Thus, the sine–Gordon and Tzitzeica equations may be formulated in terms of Monge–Ampère equations which arise naturally in a geometrical and physical context. It turns out that the classical Bäcklund transformation and the Moutard transformation provide integrable discretizations of the characteristic equations associated with these Monge–Ampère equations [76, 77].

6   Bäcklund Transformations in Continuum Mechanics

There were, in fact, important independent developments in the application of Bäcklund transformation theory in Continuum Mechanics which predated those in Soliton Theory. Thus, in 1950, Loewner, who was best known, perhaps, for his contributions on the Bieberbach conjecture and the theory of continuous groups, introduced a generalization of the notion of the classical Bäcklund transformation to systems involving pairs of dependent variables [3]. Transformations of the form[9]

$$\mathcal{B}_i\left(x^a, \phi, \frac{\partial \phi}{\partial x^a}, \psi, \frac{\partial \psi}{\partial x^a}; x'^a, \phi', \frac{\partial \phi'}{\partial x'^a}, \psi', \frac{\partial \psi'}{\partial x'^a}\right) = 0,$$

$$i = 1, \dots, 6; \quad a = 1, 2 \tag{76}$$

were introduced which relate surface elements $\{x^a, \phi, \partial\phi/\partial x^a\}$, $\{x^a, \psi, \partial\psi/\partial x^a\}$ and $\{x'^a, \phi', \partial\phi'/\partial x'^a\}$, $\{x'^a, \psi', \partial\psi'/\partial x'^a\}$ associated with pairs of surfaces $\phi = \phi(x^a)$, $\psi = \psi(x^a)$ and $\phi' = \phi'(x'^a)$, $\psi' = \psi'(x'^a)$.

Loewner's investigation was concerned with a search for multi-parameter pressure-density laws for which the hodograph equations of plane gas dynamics may be reduced by a subclass of Bäcklund transformations of the type (76) to tractable canonical forms in subsonic, transonic and supersonic flow régimes. Thus, it was shown therein that the matrix Bäcklund transformations

$$\Lambda'_{x^1} = A\Lambda_{x^1} + H'\tilde{C}\Lambda + H'\tilde{D}\Lambda' + H'\tilde{E},$$

$$\Lambda'_{x^2} = A\Lambda_{x^2} + \tilde{C}\Lambda + \tilde{D}\Lambda' + \tilde{E}, \tag{77}$$

$$x'^a = x^a$$

link the 2×2 matrix systems

$$\Lambda_{x^1} = A^{-1}H'A\Lambda_{x^2} \tag{78}$$

and

$$\Lambda'_{x'^1} = H'\Lambda'_{x'^2} \tag{79}$$

---

[9]It was asserted in [3] that the choice of six relations in (76) is natural in that if the x^a are regarded as independent variables then the number of equations coincides with the number of unknowns.

with $\mathbf{\Lambda} = (\phi, \psi)^T$, $\mathbf{\Lambda}' = (\phi', \psi')^T$, subject to the constraints

$$\tilde{C}_{x^1} - (H'\tilde{C})_{x^2} + [\tilde{D}, H']\tilde{C} = 0,$$
$$\tilde{D}_{x^1} - (H'\tilde{D})_{x^2} + [\tilde{D}, H']\tilde{D} = 0,$$
$$\tilde{E}_{x^1} - (H'\tilde{E})_{x^2} + [\tilde{D}, H']\tilde{E} = 0,$$
$$A_{x^1} - H'\tilde{C} - H'\tilde{D}A - (A_{x^2}\tilde{C} - \tilde{D}A)H' = 0$$

(80)

where $A, \tilde{C}, \tilde{D}$ with $|A_{x^2} - \tilde{C} - \tilde{D}A| \neq 0$ and $\tilde{E}$ are 2×2 matrices dependent, in general, on the x^α. In the gas dynamics context considered by Loewner [3], (78) is taken as the well-known hodograph system

$$\begin{pmatrix} \phi \\ \psi \end{pmatrix}_q = \begin{pmatrix} 0 & -(1 - M^2)/(\rho q) \\ \rho/q & 0 \end{pmatrix} \begin{pmatrix} \phi \\ \psi \end{pmatrix}_\theta,$$

(81)

while (79) is an appropriate target canonical form, viz:

$$M < 1 : \begin{pmatrix} \phi' \\ \psi' \end{pmatrix}_s = \begin{pmatrix} 0 & -1 \\ 1 & 0 \end{pmatrix} \begin{pmatrix} \phi' \\ \psi' \end{pmatrix}_\theta, \qquad \text{The Cauchy–Riemann System}$$

$$M \sim 1 : \begin{pmatrix} \phi' \\ \psi' \end{pmatrix}_s = \begin{pmatrix} 0 & -s \\ 1 & 0 \end{pmatrix} \begin{pmatrix} \phi' \\ \psi' \end{pmatrix}_\theta, \qquad \text{The Tricomi System}$$

$$M > 1 : \begin{pmatrix} \phi' \\ \psi' \end{pmatrix}_s = \begin{pmatrix} 0 & 1 \\ 1 & 0 \end{pmatrix} \begin{pmatrix} \phi' \\ \psi' \end{pmatrix}_\theta, \qquad \text{The Classical 1+1-Dimensional Wave System}$$

In the above, ρ is the gas density, $\boldsymbol{q} = (q\cos\theta, q\sin\theta)$ the gas velocity and M the local Mach number; ϕ, ψ and ϕ', ψ' denote potential and stream functions in the real and approximated gasdynamic systems respectively; here $s = s(q)$.

In fact, the linear Bäcklund transformations (77) have turned out to have widespread physical applications. Thus, while in gas dynamics they may be used, in particular, to analyse pulse propagation in shock tubes, in nonlinear elastodynamics they have application to the study of the transmission of large amplitude disturbances through bounded media. Applications to boundary value problems involving electromagnetic wave propagation in nonlinear dielectric media and to crack, punch and torsion problems in elastostatics together with the work in gas dynamics and electrodynamics are described, in detail, in [2]. In each case, the underlying idea is to use Bäcklund transformations of the type (77) subject to the constraints (80) to develop tractable multi-parameter representations of physical models which may be used to approximate real material behaviour over appropriate ranges of the physical variables.

It is noted that, whereas the Bäcklund transformations used in Soliton Theory, Miura transformations apart, are concerned with invariance, the Bäcklund transformations (77) as introduced by Loewner are involved with reduction to canonical form. Indeed, Loewner's original work on Bäcklund transformations and its subsequent application in Continuum Mechanics have proceeded entirely independently of Soliton Theory. However, remarkably, it has recently been shown in [5] that an infinitesimal version of the Bäcklund transformations (77) have a dramatic and unexpected connection with Soliton Theory.[10]

---

[10] It has also been recently established that the class of finite Bäcklund transformations (77)

This link and its implications for current research in Soliton Theory in higher dimensions are reviewed in the following section which ends our brief review of how Bäcklund transformations impact upon Soliton Theory.

7 Infinitesimal Bäcklund Transformations. The LKR Integrable System

Loewner [4] in 1952, again in the gas dynamics context of reduction of the hodograph system (81) introduced the notion of an *infinitesimal Bäcklund transformation*. Therein, the increment $\Lambda' - \Lambda = \Delta\Lambda$ was entered instead of the Λ' in (77) and all matrices were assumed to contain a continuous parameter ε. A class of infinitesimal Bäcklund transformations

$$\Lambda_{y\varepsilon} = U\Lambda_\varepsilon + V\Lambda_y + W\Lambda,$$
$$\Lambda_{x\varepsilon} = \tilde{U}\Lambda_\varepsilon + \tilde{V}\Lambda_x + \tilde{W}\Lambda \tag{82}$$

resulted which act on a hodograph system

$$\Lambda_x = S\Lambda_y. \tag{83}$$

Loewner stated in [4] that applications of these transformations in gas dynamics would be forthcoming. However, an examination of his collected works as edited by his sometime doctoral student, Lipman Bers in 1988 reveals that Loewner, regrettably, never again published on this subject. Indeed, Loewner's paper, although known and cited, lay dormant for some forty years. It was not until 1991 that, suitably reinterpreted and extended, the class of infinitesimal Bäcklund transformations introduced by Loewner were shown to have profound connections with Soliton Theory. Thus, Konopelchenko and Rogers in [5, 79] proposed an $N \times N$ matrix generalization and reinterpretation of the Loewner system, viz

$$(\partial_x - S\partial_y - P)\psi = 0,$$
$$(\partial_y\partial_z - U\partial_z - V\partial_y - W)\psi = 0, \tag{84}$$
$$(\partial_x\partial_z - \tilde{U}\partial_z - \tilde{V}\partial_x - \tilde{W})\psi = 0.$$

Therein, the continuous parameter ε in Loewner's original infinitesimal Bäcklund transformation is now to be regarded as a third independent variable z. Accordingly, the system (84) is viewed from the modern IST standpoint as a linear triad representation in the three independent variables x, y, z. Therein, $S, P, U, V, W, \tilde{U}, \tilde{V}$ and $\tilde{W}$ are $N \times N$ matrices dependent, in general, on x, y, z.

The compatibility conditions for the linear system (84) may be shown to

---

may be represented as compound gauge and Darboux-type transformations [78]. Iterated versions of the Loewner transformations were thereby constructed by established methods of Soliton Theory.

generate the following nonlinear system [79]:

$$S_z = [V, S],$$

$$P_z = VP + \tilde{W} - SW,$$

$$UVS + V_x - V_y S + WS - \tilde{W} - \tilde{U}V = 0,$$

$$UVP + U\tilde{W} + W_x + WP - \tilde{U}W - V_y P - \tilde{W}_y = 0, \tag{85}$$

$$U_x - \tilde{U}_y + [U, \tilde{U}] = 0,$$

$$\tilde{U} = SU + P, \quad \tilde{V} = V.$$

Since its introduction in 1991, the so-called LKR system (85) has been shown to lead to a wide class of solitonic equations, some new, either by way of inclusion or compatibility. Thus, included in the system as special reductions are the Ragnisco–Bruschi system [80]

$$S_z + [S, \partial_x^{-1}[S, W]] = 0, \quad W_x = 0 \tag{86}$$

which incorporates the Toda lattice system (55) as a special reduction and hence, 'a fortiori' the classical sine-Gordon equation (16) and the Tzitzeica equation (56). Moreover, the specialisation

$$S = \begin{pmatrix} -\cos\chi & -e^{-i\nu}\sin\chi \\ -e^{i\nu}\sin\chi & \cos\chi \end{pmatrix}, \quad W = \begin{pmatrix} a & 0 \\ 0 & b \end{pmatrix} \tag{87}$$

with $a - b = -1/2$ produces the SIT system (67) with $t \to z$. Both the Pohlmeyer–Lund–Regge and SRS systems are accordingly incorporated in the LKR system. The NLS equation (57) and Heisenberg spin model (65) are captured by compatibility. The classical Darboux system descriptive of conjugate coordinate systems [81] together with its integrable matrix generalization, the Zakharov–Manakov system are likewise embedded in the LKR system [82]. Integrable constrained versions of the Darboux system which are contained in the LKR system include, remarkably, the classical Lamé system descriptive of triply orthogonal surfaces. The Weingarten triply orthogonal surface system (49), like the mKdV equation (47) and the nonlinear system (45), is captured by the LKR system via compatibility with the sine-Gordon equation. The principal chiral field model may also be shown to arise as a reduction of the LKR system. Novel 2+1-dimensional integrable systems of interest discovered via the LKR system include integrable extensions of the Bianchi system (11)–(12) and its elliptic $su(2, 1)$ analogue, namely the Ernst equation of General Relativity [83].

An important particular reduction of the 2+1-dimensional Bianchi system is that given by the pair of equations

$$\left(\frac{\Theta_{xt}}{\sin\Theta}\right)_x - \left(\frac{\Theta_{yt}}{\sin\Theta}\right)_y - \frac{\Theta_y \Phi_{xt} - \Theta_x \Phi_{yt}}{\sin^2\Theta} = 0,$$

$$\left(\frac{\Phi_{xt}}{\sin\Theta}\right)_x - \left(\frac{\Phi_{yt}}{\sin\Theta}\right)_y - \frac{\Theta_y \Theta_{xt} - \Theta_x \Theta_{yt}}{\sin^2\Theta} = 0, \tag{88}$$

originally discovered by Konopelchenko and Rogers [5]. This represents the integrable 2+1-dimensional extension of the classical sine-Gordon equation analogous to the Davey–Stewartson and Nizhnik–Veselov–Novikov (NVN) extensions of the NLS and KdV equations respectively. The auto-Bäcklund transformation for (88) was constructed in [84]. This represents a natural extension of the classical Bäcklund transformation $\mathcal{B}_\beta$ for the sine-Gordon equation. Certain plane wave soliton and breather solutions of (88) have been generated by Nimmo [85], while spatially localized solutions have been constructed by Konopelchenko and Dubrovsky [86] and Schief [87]. The geometry of the sine-Gordon system has been discussed in [88]. The classical and non-classical Lie symmetries of the Konopelchenko–Rogers system have been extensively investigated by Clarkson et al. [89]. Therein, it was shown via a non-classical reduction that the system (88) admits the interesting class of solutions

$$\Theta = \vartheta(x,t) + \tilde{\vartheta}(y,t), \quad \Phi = \phi(x,t) + \tilde{\phi}(y,t) \tag{89}$$

where

$$\vartheta_{xt} = T(t)\sin\vartheta, \quad \phi_{xt} = -\tilde{T}(t)\sin\phi,$$
$$\tilde{\vartheta}_{yt} = \tilde{T}(t)\sin\tilde{\vartheta}, \quad \tilde{\phi}_{yt} = -T(t)\sin\tilde{\phi}. \tag{90}$$

Thus, we have the remarkable result that M-soliton and N-soliton solutions of the classical sine-Gordon equation as generated by their respective nonlinear superposition principles of the type (36) may be *linearly superposed* to produce novel solutions of the 2+1-dimensional sine-Gordon system (88).

Remarkably, in a particular case, this superposition principle admits a simple geometric interpretation. Thus, if we let the Gaussian curvature of the pseudospherical surfaces in a triple system of Bianchi tend to zero, that is, if we consider triply orthogonal systems of surfaces in which members of one family are developable then such systems are governed by a degenerate Weingarten system of the form $(49)_{\rho\to\infty}$, namely[11]

$$\theta_{x'y't} - \theta_{x'}\theta_{y't}\cot\theta + \theta_{y'}\theta_{x't}\tan\theta = 0,$$

$$\left(\frac{\theta_{x't}}{\cos\theta}\right)_{x'} - \frac{1}{\sin\theta}\theta_{y'}\theta_{y't} = 0,$$

$$\left(\frac{\theta_{y't}}{\sin\theta}\right)_{y'} + \frac{1}{\cos\theta}\theta_{x'}\theta_{x't} = 0, \tag{91}$$

$$\theta_{x'x'} - \theta_{y'y'} = 0.$$

As usual, x' and y' are the lines of curvature coordinates. If one now makes the identifications $x = x' + y'$, $y = x' - y'$ then $(91)_1$ constitutes an integrable reduction of (88) where $\Theta = \Phi = 2\theta$. In this case, $(91)_4$ yields the linear superposition (89) while $(91)_{2,3}$ reduce to the sine-Gordon equations (90) with $\tilde{T} = -T$.

8 General Perspectives

The role of auto-Bäcklund transformations in the construction of multi-soliton solutions integrable systems is, by now, well established. Particularly striking in this regard was the use of an auto-Bäcklund transformation by Boiti et

---

[11]The single 2+1-dimensional sine-Gordon equation $(91)_1$ turns out to have been derived by Darboux in connection with the Lamé system.

al. [90] in 1988 to discover the coherent structure solutions of the integrable 2+1-dimensional NLS system due to Davey and Stewartson [91]. Such lump solutions, now known as *dromions* have become an integral part of higher-dimensional Soliton Theory.

The search for a universal system which, in some sense, generates all known integrable systems is a subject of current research. The celebrated Self-Dual Yang–Mills (SDYM) system has received much attention in this connection [92]–[94]. In view of recent developments, the question arises as to whether the LKR system, with its roots in the theory of infinitesimal Bäcklund transformations, has a role in this regard. Indeed, it has recently been shown by Oevel and Schief [6] that a hierarchy of LKR systems are readily constructed which are compatible with the very general multi-component Kadomtsev–Petviashvili (KP) and modified Kadomtsev–Petviashvili (mKP) hierarchies of Sato theory.

The LKR system possesses the strong advantage that it admits a natural parametrization with canonical subclasses which may be subjected to separate investigation. Moreover, by virtue of its roots in the theory of infinitesimal Bäcklund transformations, the LKR system allows a geometric interpretation in terms of the deformation of surfaces.

In summary, it may be truly said that the works of Bäcklund, only some 15 in total, were 'pauca sed matura'. Like the contributions of his great European contemporaries, Lie, Darboux and Bianchi, they survive as a testament to the key role geometry can play in our understanding of nonlinear models of physical phenomena. Thus, Bäcklund's name on this 150th anniversary of his birth lives on in Soliton Theory.

Acknowledgements

The authors are indebted to Per Gunnar Martinsson for a translation of the biographical details on Bäcklund abstracted from the Yearbook of the Swedish Royal Academy of Science, 1924. Enlightening discussions with Professor Nail Ibragimov are also gratefully acknowledged.

Appendix

A.V. BÄCKLUND (1845–1922)

PUBLICATIONS

[B1] Nragro satser on plana algebraiska kurvors normaler, *Lunds Univ. rArsskr* **V** (1869).

[B2] Ett bidrag till kul-komplernas theori, *Lunds Univ. rArsskr* **IX** (1873).

[B3] Einiges über Kurven- und Flächen-Transformationen, *Lunds Univ. rArsskr* **X** 1–12 (1874).

[B4] Über Flachentransformationen, *Math. Ann.* **IX** 298–320 (1875).

[B5] Über partielle Differentialgleichungen höherer Ordnung, die intermediäre erste Integrale besitzen, *Math. Ann.* **XI** 199–241 (1877).

[B6] Über Systeme partieller Differentialgleichungen erster Ordnung, *Math. Ann.* **XI** 412–433 (1877).

[B7] Über partielle Differentialgleichungen höherer Ordnung, die intermediäre erste Integrale besitzen. Zweite Abhandlung, *Math. Ann.* **XIII** 69–108 (1878).

[*B*8] Zur Theorie der Charakteristiken der partiellen Differentialgleichungen zweiter Ordnung, *Math. Ann.* **XIII** 411–428 (1878).

[*B*9] Zur Theorie der partiellen Differentialgleichungen zweiter Ordnung, *Math. Ann.* **XV** 39–88 (1879).

[*B*10] Zur Theorie der partiellen Differentialgleichungen erster Ordnung, *Math. Ann.* **XVII** 285–328 (1880).

[*B*11] Zur Theorie der Flächentransformationen, *Math. Ann.* **XIX** 387–422 (1882).

[*B*12] On ytor med konstant negativ krökning, *Lunds Univ. rArsskr* **XIX** (1883).

[*B*13] Einiges über Kugelkomplexe, *Annali di Matematica Ser III* **XX** 65–107 (1913).

[*B*14] Ein Satz von Weingarten über auf einander abwickelbare Flächen, *Lunds Univ. rArsskr* **XIV** (1918).

[*B*15] Zur Transformationstheorie partieller Differentialgleichungen zweiter Ordnung, *Lunds Univ. rArsskr* **XVI** (1920).

References

[1] R.L. Anderson and N.H. Ibragimov, *Lie–Bäcklund Transformations in Applications,* SIAM, Philadelphia (1979).

[2] C. Rogers and W.F. Shadwick, *Bäcklund Transformations and Their Applications,* Academic Press, New York (1982).

[3] C. Loewner, A transformation theory of partial differential equations of gasdynamics, *Nat. Advis. Comm. Aeronaut. Tech. Notes* **2065**, 1–56 (1950).

[4] C. Loewner, Generation of solutions of systems of partial differential equations by composition of infinitesimal Bäcklund transformations, *J. Anal. Math.* **2**, 219–242 (1952).

[5] B.G. Konopelchenko and C. Rogers, On 2+1-dimensional nonlinear systems of Loewner-type, *Phys. Lett.* **A152**, 391–397 (1991).

[6] W. Oevel and W.K. Schief, Squared eigenfunctions of the (modified) KP hierarchy and scattering problems of Loewner type, *Rev. Math. Phys.* **6** 1301–1338 (1994).

[7] V.B. Matveev and M.A. Salle, *Darboux Transformations and Solitons,* Springer Series in Nonlinear Dynamics, Springer Verlag (1991).

[8] A. Kochendörfer and A. Seeger, Theorie der Versetzungen in eindimensionalen Atomreihen. I. Periodisch angeordnete Versetzungen, *Zeit. Physik* **127**, 533–550 (1950).

[9] A. Seeger and A. Kochendörfer, Theorie der Versetzungen in eindimensionalen Atomreihen. II Beliebig angeordnete und beschleunigte Versetzungen, *Zeit. Physik* **130**, 321–336 (1951).

[10] A. Seeger, H. Donth and A. Kochendörfer, Theorie der Versetzungen in eindimensionalen Atomreihen. III Versetzungen, Eigenbewegungen und ihre Wechselwirkung, *Zeit. Physik* **134**, 173–193 (1953).

[11] G.L. Lamb Jr, Analytical descriptions of ultrashort optical pulse propagation in a resonant medium, *Rev. Mod. Phys.* **43**, 99–124 (1971).

[12] H.M. Gibbs and R.E. Slusher, Peak amplification and pulse breakup of a coherent optical pulse in a simple atomic absorber, *Phys. Rev. Lett.* **24**, 638–641 (1970).

[13] A.C. Scott, Propagation of magnetic flux on a long Josephson junction, *Nuovo Cimento* **B69**, 241–261 (1970).

[14] D.J. Struick, *Lectures on Classical Differential Geometry*, 2nd ed., Addison-Wesley Publishing Company, Inc. Reading, Mass, USA (1961).

[15] M.J. Ablowitz and P.A. Clarkson, *Solitons, Nonlinear Evolution Equations and Inverse Scattering*, Cambridge University Press, Cambridge (1991).

[16] B.G. Konopelchenko, *Introduction to Multi-Dimensional Integrable Equations*, Plenum Press, New York (1992).

[17] D. Levi and A. Sym, Integrable systems describing surfaces of nonconstant curvature, *Phys. Lett.* **A149**, 381–387 (1990).

[18] T.W. Barnard, $2N\pi$ ultrashort light pulses, *Phys. Rev.* **A7**, 373–376 (1973).

[19] M.E. Johnston, *Geometry and the Sine Gordon Equation*, M.Sc. Thesis, University of New South Wales (1994).

[20] M.E. Johnston, C. Rogers, W.K. Schief and W.M. Seiler, On moving pseudospherical surfaces: A generalised Weingarten system and its formal analysis, *Lie Groups and Their Applications*, **1**, 124–136 (1994).

[21] K. Konno, W. Kameyama and H. Sanuki, Effect of weak dislocation potential on nonlinear wave propagation in anharmonic crystal, *J. Phys. Soc. Japan* **37**, 171–176 (1974).

[22] K. Konno and H. Sanuki, Bäcklund transformation for equation of motion for nonlinear lattice under weak dislocation potential, *J. Phys. Soc. Japan* **39**, 22–24 (1975).

[23] T. Kakutani and H. Ono, Weak nonlinear hydromagnetic waves in a cold collisionless plasma, *J. Phys. Soc. Japan* **26**, 1305–1318 (1969).

[24] N.J. Zabusky, A synergetic approach to problems of nonlinear dispensive wave propagation and interaction in *Nonlinear Partial Differential Equations*, W.F. Ames ed., Academic Press, New York (1967).

[25] L.P. Eisenhart, *A Treatise on the Differential Geometry of Curves and Surfaces*, Dover Publications, Inc. New York (1960).

[26] L. Bianchi, *Lezioni di Geometria Differentiale,* Spoerri, Pisa (1922).

[27] G. Darboux, *Leçons sur la Théorie Générale des Surfaces et les Applications Géométrique du Calcul Infinitésimal,* T1-4 Paris (1887–1896).

[28] R. Hirota and J. Satsuma, A simple structure of superposition formula of the Bäcklund transformation, *J. Phys. Soc. Japan* **45**, 1741–1750 (1978).

[29] M.J. Boussinesq, Essai sur la théorie des eaux courantes, *Mémoires présentés par divers savants à l'Académie des Sciences Inst. France (séries 2)* **23**, 1–680 (1877).

[30] A. Sym, Soliton surfaces and their applications (Soliton geometry from spectral problems), *Lecture Notes in Physics* **239**, Springer Verlag (1985).

[31] M. Toda, Vibration of a chain with nonlinear interaction, *J. Phys. Soc. Japan* **22**, 431–436 (1967).

[32] M. Wadati and M. Toda, Bäcklund transformation for the exponential lattice, *J. Phys. Soc. Japan* **39**, 1196–1203 (1975).

[33] G. Tzitzeica, Sur une nouvelle classe de surfaces, *C.R. Acad. Sci. Paris* **150**, 1227–1229 (1910).

[34] R.K. Dodd and R.K. Bullough, Polynomial conserved densities for the sine-Gordon equations, *Proc. R. Soc. London A* **352**, 481–503 (1977).

[35] B. Gaffet, A class of 1-D gas flows solvable by the inverse scattering transform, *Physica* **26D**, 123–139 (1987).

[36] W.K. Schief and C. Rogers, The Affinsphären equation. Moutard and Bäcklund transformations, *Inverse Problems* **10**, 711–731 (1994).

[37] P.L. Kelley, Self focussing of optic beams, *Phys. Rev. Lett.* **15**, 1005–1008 (1965).

[38] V.I. Talanov, Self focussing of wave beams in nonlinear media, *JETP Engl. Transl.* **2**, 138–141 (1965).

[39] V.I. Bespalov and V.I. Talanov, Filamentary structure of light beams in nonlinear liquids, *JETP Engl. Transl.* **3**, 307–310 (1966).

[40] V.E. Zakharov, Stability of periodic waves of finite amplitute on the surface of a deep fluid, *J. Appl. Mech. Tech. Phys.* **9**, 86–94 (1968).

[41] L.K. Antonovskii, C. Rogers and W.K. Schief, A note on a capillarity model and the nonlinear Schrödinger equation, *J. Phys. A: Math. Gen.* **30**, L555-L557 (1997).

[42] H. Hasimoto, A soliton on a vortex filament, *J. Fluid. Mech.* **51**, 477–485 (1972).

[43] J. Cieśliński, P.K.H. Gragert and A. Sym, Exact solutions to localized-induction-approximation equation modelling smoke ring motion, *Phys. Rev. Lett.* **57**, 1507–1510 (1986).

[44] C. Rogers and W.K. Schief, Intrinsic geometry of the NLS equation and its auto-Bäcklund transformation, to be published in *Stud. Appl. Math.* (1998).

[45] M. Lakshmanan, Continuum spin system as an exactly solvable dynamical system, *Phys. Lett.* **61A**, 53–54 (1977).

[46] L.A. Takhtajan, Integration of the continuous Heisenberg spin chain through the inverse scattering method, *Phys. Lett.* **64A**, 235–237 (1977).

[47] J.C. Eilbeck, J.D. Gibbon, P.J. Caudrey and R.K. Bullough, Solitons in nonlinear optics I. A more accurate description of the 2π-pulse in self induced transparency, *J. Phys.* **A6**, 1337–1347 (1973).

[48] K. Pohlmeyer, Integrable Hamiltonian systems and interactions through quadratic constraints, *Comm. Math. Phys.* **46**, 207–221 (1976).

[49] F. Lund and T. Regge, Unified approach to strings and vortices with soliton solutions, *Phys. Rev.* **D14**, 1524–1535 (1976).

[50] H. Steudel, Solitons in Stimulated Raman Scattering and Resonant Two-Photon Propagation, *Physica* **6D**, 155–178 (1983).

[51] M.J. Ablowitz, D.J. Kaup, A.C. Newell and H. Segur, The inverse scattering transform – Fourier analysis for nonlinear problems, *Stud. Appl. Math.* **53**, 249–315 (1974).

[52] T.B. Benjamin, Internal waves of permanent form in fluids of great depth, *J. Fluid. Mech.* **29**, 559–592 (1967).

[53] H. Ono, Algebraic solitary waves in stratified fluids, *J. Phys. Soc. Japan* **39**, 1082–1091 (1975).

[54] A.P. Bassom, P.A. Clarkson and A.C. Hicks, Bäcklund transformations and solution hierarchies for the fourth Painléve equation, *Stud. Appl. Math.* **95**, 1–71 (1995).

[55] F.W. Nijhoff and H.W. Capel, The discrete Korteweg–de Vries equation, *Acta. Applic. Math.* **39**, 133–158 (1995).

[56] D. Levi and R. Benguria, Bäcklund transformations and nonlinear differential difference-equations, *Proc. Natl. Acad. Sci. USA,* **77**, 5025–5027 (1980).

[57] D. Levi, Nonlinear differential difference-equations as Bäcklund transformations, *J. Phys. A* **14**, 1082–1098 (1981).

[58] B.G. Konopelchenko, Elementary Bäcklund transformations, nonlinear superposition principle and solutions of the integrable equations, *Phys. Lett.* **A87**, 443–448 (1982).

[59] G.R.W. Quispel, F.W. Nijhoff, H.W. Capel and J. van der Linden, Linear integral equations and nonlinear difference-difference equations, *Physica* **A125**, 344–380 (1984).

[60] F.W. Nijhoff, H.W. Capel, G.L. Wiersma and G.R.W. Quispel, Bäcklund transformations and three-dimensional lattice equations, *Phys. Lett.* **A105**, 267–272 (1984).

[61] R. Sauer, Parallelogrammgitter als Modelle Pseudosphärischer Flächen, *Math. Z.* **52**, 611–622 (1950).

[62] W. Wunderlich, Zur Differenzengeometrie der Flächen Konstanter Negativer Krümmung, *Österreich. Akad. Wiss. Math.-Nat. Kl. S.-B. II* **160**, 39–77 (1951).

[63] A. Bobenko and U. Pinkall, Discrete surfaces with constant negative Gaussian curvature and the Hirota equation, *J. Diff. Geom.* **43** 527–611 (1996).

[64] A. Bobenko and U. Pinkall, Discretization of surfaces and integrable systems, in A. Bobenko and R. Seiler eds., *Discrete Integrable Geometry and Physics*, Oxford University Press, Oxford (1999).

[65] A. Bobenko and U. Pinkall, Discrete isothermic surfaces, *J. R. Angew. Math.* **475**, 187–208 (1996).

[66] W.K. Schief, Self-dual Einstein spaces and a discrete Tzitzeica equation. A permutability theorem link, to appear in *Proc. Symmetries and Integrability of Difference Equations II*, P. Clarkson and F. Nijhoff eds., Cambridge University Press, Cambridge (1997).

[67] A. Bobenko and W.K. Schief, Discrete indefinite affine spheres, in A. Bobenko and R. Seiler eds., *Discrete Integrable Geometry and Physics*, Oxford University Press, Oxford (1999).

[68] A. Bobenko and W.K. Schief, Affine Spheres: Discretization via Duality Relations, *J. Exp. Math.*, to appear.

[69] L.V. Bogdanov and B.G. Konopelchenko, Lattice and q-difference Darboux–Zakharov–Manakov Systems via $\bar{\partial}$-dressing method, *J. Phys. A: Math. Gen.* **28**, L173-L178 (1995).

[70] A. Doliwa, Geometric discretisation of the Toda system, *Phys. Lett. A* **234**, 187–192 (1997).

[71] A. Bobenko, Discrete holomorphic maps, to appear in *Proc. Symmetries and Integrability of Difference Equations II*, P. Clarkson and F. Nijhoff eds., Cambridge University Press, Cambridge (1997).

[72] B.G. Konopelchenko and W.K. Schief, Three-dimensional integrable lattices in Euclidean spaces: Conjugacy and orthogonality, to be published in *Proc. R. Soc. London A* (1998).

[73] J.J.C. Nimmo and W.K. Schief, Superposition principles associated with the Moutard transformation. An integrable discretization of a $(2 + 1)$-dimensional sine-Gordon system, *Proc. R. Soc. London A* **453**, 255–297 (1997).

[74] R. Courant and D. Hilbert, *Methods of Mathematical Physics II*, Interscience Publishers, New York (1962).

[75] G.D. Smith, *Numerical Solution of Partial Differential Equations*, Oxford University Press, Oxford (1965).

[76] C. Rogers and W.K. Schief, The classical Bäcklund transformation and integrable discretisation of characteristic equations, *Phys. Lett. A* **232**, 217–223 (1997).

[77] W.K. Schief, An integrable discretization of the characteristic equations associated with a gasdynamics system, *Phys. Lett. A* **238**, 278–282 (1998).

[78] W.K. Schief and C. Rogers, Loewner transformations. Adjoint and binary Darboux connections, *Stud. Appl. Math.* **100**, 391–422 (1998).

[79] B.G. Konopelchenko and C. Rogers, On generalised Loewner systems: Novel integrable equations in $2+1$ dimensions, *J. Math. Phys.* **34**, 214–242 (1993).

[80] M. Bruschi and O. Ragnisco, Nonlinear evolution equations associated with the chiral field spectral problem, *Nuovo Cimento* **88B**, 119–139 (1985).

[81] G. Darboux, *Leçons sur les Systèmes Orthogonaux et les Coordonnées Curvilignes*, Paris (1910).

[82] W.K. Schief, C. Rogers and S.P. Tsarev, On a $2 + 1$-dimensional Darboux system: Integrable and geometric connections. *Chaos Solitons Fractals* **5**, 2357–2366 (1995).

[83] W.K. Schief, On a $2+1$-dimensional Ernst equation, *Proc. R. Soc. London A* **446**, 381–398 (1994).

[84] B.G. Konopelchenko, W.K. Schief and C. Rogers, The $2 + 1$-dimensional sine-Gordon system: Its auto-Bäcklund transformation, *Phys. Lett.* **172A** 39–48 (1992).

[85] J.J.C. Nimmo, A class of solutions of the Konopelchenko–Rogers equation, *Phys. Lett.* **A168**, 113–119 (1992).

[86] B.G. Konopelchenko and V.G. Dubrovsky, A $2 + 1$-dimensional integrable generalization of the sine-Gordon equation I. $\bar{\partial}$-∂ dressing and the initial-value problem. *Stud. Appl. Math.* **90**, 189–223 (1993).

[87] W.K. Schief, On localised solitonic solutions of a $2 + 1$-dimensional sine-Gordon system, *J. Phys. A: Math. Gen.* **25**, L1351-L1354 (1992).

[88] W.K. Schief, On the geometry of a $2+1$-dimensional sine-Gordon system, to appear in *Proc. R. Soc. London A* (1997).

[89] P.A. Clarkson, E.L. Mansfield and A.E. Milne, Symmetries and exact solutions of a $2+1$-dimensional sine-Gordon system, *Phil. Trans. R. Soc. London A* **354**, 1807–1835 (1996).

[90] M. Boiti, J. Leon, L. Martina and F. Pempinelli, Scattering of localised solitons in the plane, *Phys. Lett.* **A132**, 432–439 (1988).

[91] A. Davey and K. Stewartson, On three-dimensional packets of surface waves, *Proc. Roy. Soc. London A* **338**, 101–110 (1974).

[92] L.J. Mason and G.A.J. Sparling, Nonlinear Schrödinger and Korteweg–de Vries reductions of self-dual Yang–Mills, *Phys. Lett.* **137A**, 29–33 (1989).

[93] M.J. Ablowitz, S. Chakravarty and L.A. Takhtajan, A self-dual Yang–Mills hierarchy and its reductions to integrable systems in $1 + 1$ and $2 + 1$ dimensions, *Commun. Math. Phys.* **158**, 289–314 (1993).

[94] T.A. Ivanova and A.D. Popov, Some new integrable equations from self-dual Yang–Mills equations, *Phys. Lett.* **A205**, 158–166 (1995).

Recent Developments in Integrable Curve Dynamics

ANNALISA CALINI

Keywords: Vortex filaments, integrable systems, solitons

---

1 Background

1.1 The Physical Model

Whirlpools and smoke rings are common examples of vortex filaments: approximately one-dimensional regions where the velocity distribution of a fluid has a rotational component. We give below an idealised description of the self-induced dynamics of a closed line vortex in a Eulerian fluid, based on Batchelor's approach [3].

Let $\mathbf{u}$ be the velocity distribution of an incompressible (div $\mathbf{u} = 0$) fluid filling an unbounded region in space. We suppose that the vorticity $\mathbf{w} = \operatorname{curl} \mathbf{u}$ (measuring the rotational component of the velocity field) is zero at points not on the line vortex. Because div $\mathbf{u} = 0$, and assuming the infinite domain to be simply connected, we can introduce a vector potential $\mathbf{A}$ and write $\mathbf{u} = \operatorname{curl} \mathbf{A}$. The choice of $\mathbf{A}$ is not unique, in fact $\operatorname{curl} \mathbf{A} = \operatorname{curl}(\mathbf{A} + \operatorname{grad} \phi)$ for any arbitrary scalar function ϕ. We use this freedom to satisfy div $\mathbf{A} = 0$ and to derive the following vector Poisson equation,

$$\mathbf{w} = \operatorname{curl}(\operatorname{curl} \mathbf{A}) = -\Delta \mathbf{A}, \tag{1}$$

where Δ is the Laplacian operator. In an infinite domain, its solution is

$$\mathbf{A} = \frac{1}{4\pi} \int d^3 \mathbf{x}' \frac{\mathbf{w}(\mathbf{x}')}{\|\mathbf{x} - \mathbf{x}'\|},$$

and the corresponding velocity distribution is

$$\mathbf{u} = -\frac{1}{4\pi} \int d^3 \mathbf{x}' \frac{(\mathbf{x} - \mathbf{x}') \times \mathbf{w}(\mathbf{x}')}{\|\mathbf{x} - \mathbf{x}'\|^3}. \tag{2}$$

We now make an essential simplification: we model the line vortex with a tube of infinitesimal cross-sectional area dA in which the vorticity $\mathbf{w}$ (everywhere tangent to the line vortex) has constant magnitude w. We then have

$$\mathbf{u} = -\frac{1}{4\pi} \int dA \frac{\delta l(\mathbf{x} - \mathbf{x}') \times \mathbf{w}(\mathbf{x})}{\|\mathbf{x} - \mathbf{x}'\|^3} = -\frac{1}{4\pi} \int w \, dA \oint \frac{(\mathbf{x} - \mathbf{x}') \times \delta l}{\|\mathbf{x} - \mathbf{x}'\|^3}.$$

Introducing the circulation $\Gamma = \oint \mathbf{u} \cdot \delta l = \int w \, dA$, we obtain the expression

$$\mathbf{u} = -\frac{\Gamma}{4\pi} \oint \frac{(\mathbf{x} - \mathbf{x}') \times \delta l}{\|\mathbf{x} - \mathbf{x}'\|^3}, \tag{3}$$

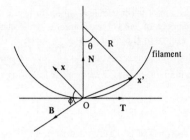

Figure 1: Idealised portion of a vortex filament.

which describes a velocity distribution that becomes singular along the line vortex. Evidently, the fluid around the line vortex circulates at an increasing speed as it approaches the core of the vortex, causing local cross-sections of the vortex tube to rotate about its center without translating. Therefore, we need to compute the asymptotic contribution of the non-circulatory part of the velocity at points infinitesimally close to the vortex filament in order to describe its dynamics.

We approximate a portion of the vortex filament with a circular arc as shown in Figure 1, where the Frenet–Serret frame of the curve $(\mathbf{T}, \mathbf{N}, \mathbf{B})$ defines a local coordinate system. We choose $\mathbf{x} = x_2\mathbf{N} + x_3\mathbf{B}$ in the plane normal to the line vortex, $\mathbf{x}' = x_1'\mathbf{T} + x_2'\mathbf{N}$ to be the position vector of a point along the curve, and we let $\sigma = \sqrt{x_2^2 + x_3^2}$ be the distance of $\mathbf{x}$ from the filament. Near the origin O (for small values of the angle θ), we write

$$\mathbf{x}' \simeq l\mathbf{T} + \frac{1}{2}kl^2\mathbf{N}, \qquad \delta\mathbf{l} = d\mathbf{x}' \simeq (\mathbf{T} + kl\mathbf{N})dl, \qquad (4)$$

where $l = R\theta$ is the arclength and $k = R^{-1}$ is the curvature of the filament. In this approximation, the integrand in equation (3) becomes

$$\frac{(\mathbf{x} - \mathbf{x}') \times \delta\mathbf{l}(\mathbf{x}')}{\|\mathbf{x} - \mathbf{x}'\|^3} \simeq \frac{-klx_3'\mathbf{T} + x_3\mathbf{N} - (x_2 + \frac{1}{2}kl^2)\mathbf{B}}{[l^2 + \sigma^2 - x_2kl^2 + \frac{1}{4}k^2l^4]^{\frac{3}{2}}}dl.$$

This expression is valid at a point $\mathbf{x} = \sigma\cos\phi\mathbf{N} + \sigma\sin\phi\mathbf{B}$ close enough to the vortex filament, and for l varying in a given interval $[-L, L]$; introducing the new variable $m = l/\sigma$, we rewrite equation (3) as

$$\mathbf{u} = \frac{\Gamma}{4\pi}\int_{-\frac{L}{\sigma}}^{\frac{L}{\sigma}} \frac{\sigma^{-1}[\cos\phi\mathbf{B} - \sin\phi\mathbf{N}] + \frac{1}{2}m^2k\mathbf{B}}{\left[1 + m^2 - k\sigma m^2\cos\phi + \frac{1}{4}\sigma^2k^2m^4\right]^{\frac{3}{2}}}dl. \qquad (5)$$

(We observe that there is no tangential component, the corresponding term in the integrand being an odd function of m.) As $\sigma \to 0$, the denominator of the integrand converges to $(1 + m^2)^{\frac{3}{2}}$ and we obtain the following asymptotic expansion

$$\mathbf{u} \simeq \frac{\Gamma}{2\pi\sigma}(\sin\phi\mathbf{B} - \cos\phi\mathbf{N}) + k\frac{\Gamma}{4\pi}\log\frac{L}{\sigma}\mathbf{B}. \qquad (6)$$

The $\mathcal{O}(\sigma^{-1})$-term represents the rotational component of the motion which does not produce a displacement of the vortex filament, while the logarithmic term causes the filament to move in the fluid at a large speed in the direction of the binormal. By rescaling the time variable $t \to \dfrac{\Gamma}{4\pi} \log \dfrac{L}{\sigma} t$, and introducing the position vector to the filament $\gamma(x,t)$, we write the following evolution equation

$$\frac{\partial \gamma}{\partial t} = k\mathbf{B}. \tag{7}$$

We observe that the vector field is purely local (a point along the filament only "feels" the effect of nearby points) and non-zero only when the filament is curved (there is no self-induced motion of a straight line vortex).

1.2   The Hamiltonian Formulation

Following a treatment contained in [5], we describe the vortex filament equation (7) derived above as a Hamiltonian system on a suitable infinite-dimensional symplectic manifold.

We first need to specify the appropriate phase space. We will restrict consideration to closed curves in space; let then $\mathcal{L}\mathbb{R}^3 = \{\gamma : S^1 \to \mathbb{R}^3\}$ be the space of smooth maps from the circle into $\mathbb{R}^3$ (the *loop space of* $\mathbb{R}^3$). The infinite group $Diff^+(S^1)$ of orientation preserving diffeomorphisms of the circle acts on $\mathcal{L}\mathbb{R}^3$ simply by changing the parametrisation of any given element. For our purposes, we wish to identify knots which "look the same" as global objects, however, we want to distinguish identical pairs that have opposite orientations. To this end, we introduce the quotient space $\mathcal{Y} = \mathcal{L}\mathbb{R}^3 / Diff^+(S^1)$ as the space of unparametrised oriented loops. $\mathcal{Y}$ is a very singular space, for example it contains curves with an infinite number of self-crossings, or with self-tangencies of infinite order, and at such points the tangent space to $\mathcal{Y}$ cannot be defined. Let $\widehat{\mathcal{X}}$ be the space of loops with finitely many self-crossings and contacts of finite order, then

Theorem 1.1. *[5] The space of oriented singular knots*

$$\widehat{\mathcal{Y}} = \widehat{\mathcal{X}} / Diff^+(S^1)$$

is a smooth infinite-dimensional manifold modelled on the Fréchet space $\mathcal{C}^\infty(S^1, \mathbb{R}^2)$.

The tangent space of $\widehat{\mathcal{Y}}$ can be characterised in the following way: a tangent vector at a point γ in $\widehat{\mathcal{Y}}$ is a smooth vector field along the curve γ that is everywhere normal to γ (we quotient out the tangential component which purely reparametrises the curve).

We now define a 2-form on $T\widehat{\mathcal{Y}}$ which endows $\widehat{\mathcal{Y}}$ with a symplectic structure, and allows us to construct Hamiltonian flows on it. The following 2-form was introduced by J. Marsden and A. Weinstein in their study of vortex dynamics in Eulerian fluids [42]

$$\omega_\gamma(\mathbf{u}, \mathbf{v}) = \int_0^{2\pi} \left(\frac{d\gamma}{dx} \times \mathbf{u} \cdot \mathbf{v} \right) dx, \tag{8}$$

where **u**, **v** are arbitrary tangent vectors at γ (i.e. two normal vector fields along γ) and x is the arclength parameter (we can choose an arbitrary parametrisation of γ). The integrand in equation (8) is simply the oriented volume of the parallelepiped constructed on the triple $(\gamma, \mathbf{u}, \mathbf{v})$.

It is immediate to verify that ω is a closed 2-form; in fact, we can carry the exterior differentiation inside the integral and observe that the exterior derivative of a volume form on a three-dimensional manifold is zero. Nondegeneracy also follows easily.

In the case of a finite-dimensional symplectic manifold M (necessarily of even dimension) a symplectic form ω determines a natural isomorphism

$$\begin{aligned} TM &\longrightarrow T^*M \\ X &\longrightarrow \omega(X, \cdot) \end{aligned}$$

between tangent vectors and 1-forms (see for example [2]). In particular, ω associates to any smooth function H on M (a Hamiltonian) a unique vector field X_H (and thus a Hamiltonian flow on M) defined by

$$\omega(\cdot, X_H) = dH(\cdot). \tag{9}$$

In an infinite-dimensional setting the correspondence $TM \to T^*M$ may fail to be onto, i.e. not for every Hamiltonian functional can we construct an associated Hamiltonian vector field. This technical point is addressed in [5] for the space $\hat{\mathcal{Y}}$ and we refer the interested reader to that book for details. Here we simply remark that, among those functionals which admit Hamiltonian vector fields constructed via the correspondence (9) (Brylinski calls such functionals supersmooth) is the *total length* functional

$$\mathcal{L}(\gamma) = \int_0^{2\pi} \left\| \frac{d\gamma}{dx} \right\| dx, \tag{10}$$

defined on arbitrary smooth curves (not necessarily arclength parametrised). The Fréchet derivative of $\mathcal{L}(\gamma)$, restricted to the representative of γ that is arclength parametrised is

$$d\mathcal{L}(\gamma) = -\frac{d^2\gamma}{dx^2}.$$

We check using (9) that the Hamiltonian vector field for the total length functional $\mathcal{L}$ is given by

$$X_{\mathcal{L}} = \frac{d\gamma}{dx} \times \frac{d^2\gamma}{dx^2}.$$

In fact, using the Marsden–Weinstein symplectic form we compute, for arbitrary $\boldsymbol{\xi}$

$$\omega(\boldsymbol{\xi}, X_{\mathcal{L}}) = \int_0^{2\pi} \frac{d\gamma}{dx} \times \boldsymbol{\xi} \cdot \left(\frac{d\gamma}{dx} \times \frac{d^2\gamma}{dx^2} \right) dx = \int_0^{2\pi} \left(-\frac{d^2\gamma}{dx^2} \cdot \boldsymbol{\xi} \right) dx = d\mathcal{L}[\gamma](\boldsymbol{\xi}).$$

Given the Hamiltonian vector field $X_{\mathcal{L}}$, we write the associated Hamiltonian evolution equation

$$\frac{\partial\gamma}{\partial t} = \frac{\partial\gamma}{\partial x} \times \frac{\partial^2\gamma}{\partial x^2}, \tag{11}$$

which can be reduced to the vortex filament equation (7) using the Frenet equations for γ in the case of nowhere vanishing curvature.

Remark. An immediate consequence of the Hamiltonian form (11) is the invariance of the total length during the curve evolution; a stronger result can be proven with a short calculation: local arclength is also preserved during the motion, i.e. $\partial_t \|\gamma_x\| = 0$, therefore the vortex filament moves in time without stretching (an implicit assumption in the study of the physical model). It follows that x and t are independent variables: we will see that the ability to "commute mixed partials" is at the basis of the most interesting properties of the vortex filament equation.

1.3   Completely Integrable Infinite-Dimensional Systems

This special class of nonlinear partial differential equations gained its importance in the mid-sixties with the coining of the word "soliton" by Kruskal and Zabuski (although "solitonic" behaviour was first observed in 1953 in a seldom acknowledged article by Seeger, Donth and Kochendörfer [51]; we thank an anonymous referee for this historical remark) and with the development of the inverse scattering method for solving the Korteweg–de Vries equation (KdV) by Gardner, Green, Kruskal and Miura.

The several features shared by these nonlinear equations make them the infinite-dimensional analogues of integrable Hamiltonian systems in finite dimensions.

In this section, we give a brief exposition of the most important properties of integrable PDE's. A good overview of the subject can be found in the book by A. Newell [44] and a very clear exposition is contained in the book by G.L. Lamb [33]; we also refer to M.J. Ablowitz and H. Segur's monograph [1].

Integrable PDE's possess a class of special solutions widely known as solitons. These are solitary waves in the form of pulses whose behaviour has many particle-like features. During their evolution, solitons propagate without change of shape and with no energy loss. When two or more solitons with different propagation speed collide, after a highly nonlinear interaction the pulses emerge with the same initial form and no energy is lost in radiation processes in the course of the interaction.

All the known soliton equations can be considered as infinite-dimensional counterparts of completely integrable finite-dimensional Hamiltonian systems in the following sense. They possess an infinite sequence $\{I_k\}_{k=1}^{\infty}$ of constants of motion, whose gradients are linearly independent and whose associated Hamiltonian flows commute (i.e. the I_k's are pairwise in involution with respect to a given Poisson bracket on the phase space). For example, the focussing cubic nonlinear Schrödinger equation

$$i\psi_t + \psi_{xx} + 2|\psi|^2\psi = 0, \tag{12}$$

which will play an important role in the course of this article, can be rewritten as a Hamiltonian system

$$\psi_t = \{H, \psi\},$$

with Hamiltonian

$$H[\psi] = \int_{\mathcal{D}} (|\psi_x|^2 - |\psi|^4)\, dx$$

with respect to the following Poisson bracket

$$\{F, G\} = i \int_{\mathcal{D}} \left(\frac{\partial F}{\partial \psi} \frac{\partial G}{\partial \bar{\psi}} - \frac{\partial F}{\partial \bar{\psi}} \frac{\partial G}{\partial \psi} \right) dx.$$

The first few invariants are

$$I_1 = \frac{1}{2i} \int_{\mathcal{D}} (\psi_x \bar{\psi} - \bar{\psi}_x \psi) \, dx, \quad I_2 = \int_{\mathcal{D}} |\psi|^2 \, dx, \quad I_3 = \int_{\mathcal{D}} \left(|\psi_x|^2 - |\psi|^4 \right) \, dx, \dots$$

They satisfy $\{I_k, I_j\} = 0$, $\forall k, j$ and their gradients are linearly independent.

As a consequence of the existence of a hierarchy $\{I_k\}_{k=1}^{\infty}$ of invariants, the phase trajectories are restricted to lie on the intersection of the level sets $I_k = c_k$, $k = 1 \dots \infty$, which forms a submanifold of infinite codimension. In the case of a finite-dimensional phase space, the preimage of a regular value $\vec{c}$ of the map $\vec{I} = (I_1, \dots, I_n)$ is diffeomorphic to a product of circles and lines (see [2]) and the dynamical system can be described in terms of the linear evolution of a collection of action-angle variables. For soliton PDE's, the inverse scattering method explicitly constructs a nonlinear change of variables that linearises the flow. The KdV equation was the first soliton equation for which the inverse scattering method was developed: in this case, the analogues of action-angle variables are the scattering data of a related linear Schrödinger operator. The initial value problem for the KdV equation can thus be solved exactly by mapping the initial condition to its scattering data, evolving the scattering data according to a linear evolution up to time t, and by reconstructing the solution at time t of the original PDE by means of the inverse transform.

We mentioned that at the heart of the inverse scattering method for the KdV equation is an underlying linear operator L. Peter Lax [38] proved that the spectrum of the Schrödinger operator L does not vary in time, and recast the nonlinear PDE into a general framework. If the spectrum of L is independent of time, then its evolution can be written as

$$L(t) = U(t)L(0)U^{-1}(t) \tag{13}$$

for some time-dependent unitary operator $U(t)$. It follows directly from equation (13) that, if some function ϕ solves the eigenvalue problem $L\phi = \lambda\phi$, then ϕ is also a solution of the linear system $\phi_t = B\phi$, where $B = U_t U^{-1}$. Differentiating equation (13) with respect to time gives Lax form of the KdV equation

$$L_t = [B, L],$$

where the operators L and B are called a *Lax pair* for the soliton equation. Lax also showed the existence of an infinite sequence of operators B that give rise to spectrum preserving evolutions of L: the corresponding hierarchy of Lax equations coincide with the infinite sequence of commuting Hamiltonian KdV flows mentioned above.

In fact, all completely integrable PDE's arise from a Lax pair, i.e. as solvability conditions of an auxiliary pair of linear systems. The Lax pair for the focussing cubic NLS equation has the following form

$$\begin{aligned} \mathbf{F}_x &= U\mathbf{F} \\ \mathbf{F}_t &= V\mathbf{F}, \end{aligned} \tag{14}$$

where $\mathbf{F}$ is an auxiliary complex vector-valued function and

$$U = i\lambda\sigma_3 + \begin{pmatrix} 0 & i\bar{\psi} \\ i\psi & 0 \end{pmatrix}, V = (2i\lambda^2 - i|\psi|^2)\sigma_3 + \begin{pmatrix} 0 & 2i\lambda\bar{\psi} + \bar{\psi}_x \\ 2i\lambda\psi - \psi_x & 0 \end{pmatrix},$$

with $\sigma_3 = \begin{pmatrix} 1 & 0 \\ 0 & -1 \end{pmatrix}$ (the parameter λ is called the *spectral parameter*).

We conclude this section with a geometric interpretation of the Lax pair (14) given in [20] that is appropriate to this context. The solvability condition for the overdetermined system (14) is obtained by differentiating the first equation with respect to t, the second equation with respect to x and equating the mixed partial derivatives, giving the equation

$$\frac{\partial U}{\partial t} - \frac{\partial V}{\partial x} + [U, V] = 0. \tag{15}$$

We can check that equation (15) is identically true for every λ provided the NLS equation (12) is satisfied. We can give to this equation and to the associated linear system (14) a natural geometrical interpretation. The system (14) can be regarded as equations of parallel transport in the trivial vector bundle $\mathbb{R}^2 \times \mathbb{C}^2$, where the vector function $\mathbf{F}$ takes values in the fibre $\mathbb{C}^2$ and the matrices U and V are interpreted as local connection coefficients. Then, equations (14) express the fact that the covariant derivative of $\mathbf{F}$ is zero and equation (15) is equivalent to the (U, V)-connection being a flat connection on $\mathbb{R}^2 \times \mathbb{C}^2$. For this reason, Lax equations of the form (15) are also called the *zero curvature formulation* of the soliton equation.

1.4   The Hasimoto Transformation

In 1972, R. Hasimoto [26] constructed the complex function $\psi = k \exp{(i \int^x \tau ds)}$ of the curvature k and the torsion τ of a space curve, and showed that if the curve evolves according to the vortex filament equation (7), then ψ solves the focussing cubic nonlinear Schrödinger equation

$$i\psi_t + \psi_{xx} + \frac{1}{2}|\psi|^2\psi = 0, \tag{16}$$

which can be reduced to equation (12) of section 1.3 by rescaling ψ to 2ψ. The most direct way to prove this result is to rewrite the filament equation in terms of a new orthonormal frame $(\mathbf{T}, \mathbf{U}, \mathbf{V})$, where

$$\mathbf{U} = \cos(\int^x \tau \, ds)\mathbf{N} - \sin(\int^x \tau \, ds)\mathbf{B},$$

$$\mathbf{V} = \sin(\int^x \tau \, ds)\mathbf{N} + \cos(\int^x \tau \, ds)\mathbf{B}.$$

The role of $(\mathbf{T}, \mathbf{U}, \mathbf{V})$ will be the main theme of section 4, for our purposes here we just need the Darboux equations for the new frame

$$\mathbf{T}_x = k_1\mathbf{U} + k_2\mathbf{V}; \qquad \mathbf{U}_x = -k_1\mathbf{T}, \qquad \mathbf{V}_x = -k_2\mathbf{T}, \tag{17}$$

with $k_1 = k\cos(\int^x \tau ds)$, $k_2 = k\sin(\int^x \tau ds)$. The filament equation (7) can be rewritten in the following form

$$\gamma_t = -k_2\mathbf{U} + k_1\mathbf{V}. \tag{18}$$

Differentiating (18) twice with respect to the arclength parameter x, we obtain

$$\gamma_{txx} = -k_{2xx}\mathbf{U} + k_{1xx}\mathbf{V} + (k_1 k_{2x} - k_2 k_{1x})\mathbf{T}. \tag{19}$$

On the other hand, $\gamma_{xx} = k_1\mathbf{U} + k_2\mathbf{V}$ from which we derive

$$\gamma_{xxt} = k_{1t}\mathbf{U} + k_{2t}\mathbf{V} + k_1\mathbf{U}_t + k_2\mathbf{V}_t. \tag{20}$$

In order to compare equations (19) and (20), we observe that $\mathbf{T}_t = -k_{2x}\mathbf{U} + k_{1x}\mathbf{V}$ (this follows directly from differentiating (18) with respect to x), and that $\mathbf{U}_t \cdot \mathbf{V} = -\mathbf{U} \cdot \mathbf{V}_t$. Using the Darboux equations for the new frame, we compute

$$\begin{aligned}(\mathbf{U}_t \cdot \mathbf{V})_x =& \mathbf{U}_{xt} \cdot \mathbf{V} + \mathbf{U}_t \cdot \mathbf{V}_x = \\ & - k_1 k_{1x} - k_2 \mathbf{U}_t \cdot \mathbf{T} = -k_1 k_{1x} + k_2 \mathbf{U} \cdot \mathbf{T}_t = -k_1 k_{1x} - k_2 k_{2x}.\end{aligned}$$

It follows that $\mathbf{U}_t \cdot \mathbf{V} = -(k_1^2 + k_2^2)/2 + A(t)$, where $A(t)$ is some arbitrary x-independent function. By equating the right-hand sides of (19) and (20), we derive the following pair of equations

$$k_{1t} + \left[\frac{1}{2}(k_1^2 + k_2^2) - A(t)\right]k_2 = -k_{2xx}$$

$$k_{2t} - \left[\frac{1}{2}(k_1^2 + k_2^2) - A(t)\right]k_1 = k_{1xx}$$

which can be written as

$$i\psi_t + \psi_{xx} + \left[\frac{1}{2}|\psi|^2\psi - A(t)\right]\psi = 0, \tag{21}$$

in terms of the complex function $\psi = k_1 + ik_2$, and can be reduced to the focussing nonlinear Schrödinger equation (16) by changing ψ to $\exp(-i\int^t A(s)ds)\psi$. Therefore, the vortex filament evolution is described by a completely integrable soliton equation. A consequence of this fact is the existence of an infinite hierarchy of global geometric functionals that remain constant throughout the evolution and that are obtained by re-expressing the constants of motion of the NLS equation in terms of the curvature and torsion of the corresponding curve. The first few global invariants of the filament equation are listed below (the first two are not obtained from NLS invariants, as we will see in the next section):

$$I_{-1} = \int_0^{2\pi} \|\gamma_s\|\, ds, \quad I_0 = \int_0^{2\pi} \tau\, ds, \quad I_1 = \int_0^{2\pi} \frac{1}{2}k^2 ds,$$

$$I_2 = \int_0^{2\pi} k^2\tau\, ds, \quad I_3 = \int_0^{2\pi} \left[(k_x)^2 + k^2\tau^2 - k^4\right] ds \dots .$$

J. Langer and R. Perline [34] unveiled the geometrical significance of the Hasimoto transformation by showing that the NLS equation and the vortex filament equation can be regarded as the same Hamiltonian system written with respect to different Poisson structures. We will explain this interesting result using an intermediate dynamical system introduced in the next section.

1.5 The Continuous Heisenberg Model

If we differentiate the vortex filament equation (11) once with respect to the variable x and permute t and x derivatives (the fact that arclength is locally preserved allows us to do that), we obtain the following equation for the unit tangent vector $\mathbf{T}$,

$$\mathbf{T}_t = \mathbf{T} \times \mathbf{T}_{xx}. \tag{22}$$

This equation was derived by M. Lakshmanan [32] and it is known as the Continuous Heisenberg Model (HM). It describes the integrable evolution of the continuum approximation of a discrete spin chain with nearest neighbour interaction (the analogue of the localised potential for the vortex filament dynamics).

The Lax pair for the Heisenberg Model is more conveniently written using matrices rather than spin vectors. If we represent the unit vector $\mathbf{T}^T = (t_1, t_2, t_3)$ with the Hermitian trace zero matrix

$$S = \begin{pmatrix} t_3 & t_1 - it_2 \\ t_1 + it_2 & -t_3 \end{pmatrix} \qquad S^2 = I,$$

the equation for the continuous spin chain can then be expressed in the following commutator form

$$S_t = \frac{1}{2i}[S, S_{xx}], \qquad S(x + 2\pi, t) = S(x, t). \tag{23}$$

It can be easily checked that (23) is the compatibility condition of the following pair of linear systems

$$\begin{aligned} \mathbf{F}_x &= i\lambda S\mathbf{F} \\ \mathbf{F}_t &= (2i\lambda^2 S - \lambda S_x S)\mathbf{F}, \end{aligned} \tag{24}$$

where λ is the complex spectral parameter. The main aspects of the complete integrability of HM, the Hamiltonian formulation, its inverse scattering transform and the hierarchy of conserved functionals are discussed in [20].

Of particular interest to us is its relation with the vortex filament equation and the nonlinear Schrödinger equation. In a short communication, J. Langer and R. Perline [35] point out that the HM is to be regarded as the intermediate model between these two evolution equations. In [20] the HM and the NLS equation are shown to be related by a gauge transformation between the respective Lax pairs; since we will provide several geometrical interpretations of this gauge equivalence, we briefly describe it here as a purely algebraic fact and leave its verification to the reader.

Proposition 1.2 (The gauge transformation). *Let V be the unitary matrix which solves the NLS Lax pair (14) at $\lambda = 0$ and which satisfies the additional relation $VSV^{-1} = \sigma_3$. Then, if $\mathbf{F}$ is an eigenvector for the Lax pair (24) of the HM, the vector $V\mathbf{F}$ solves the linear problem for the NLS equation.*

In the course of this article, we will also discuss various properties of the infinite geometry of the HM as they manifest themselves in the context of curve dynamics. We conclude by mentioning two of the advantages of using the HM for modelling closed vortex filaments (besides being itself a model of curve evolution on the 2-sphere). The space of periodic curves is characterised by the following invariant subspace of the HM phase space

$$\mathcal{P}_0 = \left\{ \mathbf{T} \ : \ \int_0^{2\pi} \mathbf{T}(x,t)dx = 0, \quad \mathbf{T}(2\pi,t) = \mathbf{T}(0,t) \right\}.$$

On the other hand, establishing which subspace of the NLS phase space corresponds to closed curves is a difficult task: deriving a simple set of conditions on the curvature k and the torsion τ of a space curve for the curve to be closed is still an open problem.

Moreover, the curve γ is easily reconstructed as $\gamma(x,t) = \int^x \mathbf{T}(s,t)ds + \mathbf{C}$ (where $\mathbf{C}$ can be chosen to be time-independent), while the reconstruction of the curve from its curvature and torsion (and thus from a solution of the NLS equation) involves solving an inverse problem (the Frenet equations of the curve).

2 Special Solutions: Methods of Algebraic Geometry

Soliton equations, when considered on periodic domains, admit large classes of special solutions which are the analogues of solitary waves for rapidly decreasing initial data on an infinite domain. These solutions are named N-phase solutions since they depend on x and t through a finite number of linear phases. Like the solitary waves, they interact nonlinearly in a particle-like manner and superimpose nonlinearly to produce more general solutions of the boundary value problem.

N-phase solutions of soliton equations such as KdV, NLS and general AKNS systems were first constructed by S.P. Novikov and I.M. Krichever [31]. Their method is a rediscovery of classical Riemann surface techniques developed by J.L. Burchnall and T.W. Chaundy [6, 7] for classifying commutative algebras of scalar differential operators.

A different characterisation of N-phase solutions was given by P.D. Lax (in particular for the KdV equation) [39]: N-phase solutions are critical points for a linear combination of integrals of motion. This characterisation, although non-constructive except for a very low number of phases, provides an interesting point of view: N-phase solutions lie on low-dimensional sets which topologically are N-tori in function space and which foliate the phase space in a way similar to a finite-dimensional integrable system.

In this section we construct the explicit formulas for N-phase solutions of the HM and the associated curves using methods of algebraic geometry. These curves are interesting from a geometrical point of view: they are critical points of the global geometric invariants $\int_0^{2\pi} \tau \, ds, \int_0^{2\pi} k^2 ds, \int_0^{2\pi} k^2 \tau \, ds \dots$ (among them we find the classical elastic curves). We find the construction of the N-phase solutions of the HM itself interesting, since it gives a clear interpretation of the gauge transformation between the HM and the NLS equation.

2.1 The Reconstruction of the Curve

Instead of taking the anti-derivative of the tangent vector, we derive a formula for the position vector of the curve in terms of the eigenfunction of the HM linear problem. In the context of N-phase solutions, this provides one with a straightforward way to obtain the curve. A analogous formula (involving the NLS eigenfunctions) was obtained by A. Sym [54] via a different argument. We start with the following result, whose proof follows from standard ODE theory.

Proposition 2.1. *The solution of the linear problem*

$$\frac{\partial \mathbf{F}}{\partial x} = i\lambda S\mathbf{F} \tag{25}$$

is analytic in $\lambda \in \mathbb{C}$ *for an analytic initial condition* $\mathbf{F}(0; \lambda)$.

Let Φ be the fundamental solution matrix of equation (25). Since it is analytic in the eigenvalue parameter, we can differentiate both sides of the linear system with respect to λ and evaluate them at $\lambda = 0$, obtaining the following formula for S:

$$S = -i \left. \frac{\partial}{\partial x} \frac{\partial \Phi}{\partial \lambda} \right|_{\lambda=0} \Phi^{-1}|_{\lambda=0}. \tag{26}$$

This expression is a perfect derivative since the columns of the fundamental solution matrix at $\lambda = 0$ are independent of x. Integrating (26) with respect to x we obtain the Hermitian matrix

$$\Gamma =: \int^x S(s)ds = -i \left. \frac{\partial \Phi}{\partial \lambda} \right|_{\lambda=0} \Phi^{-1}|_{\lambda=0}, \tag{27}$$

which represents the position vector of the curve in terms of the eigenfunction matrix Φ of the HM linear problem.

2.2 The Baker–Akhiezer Function

Multi-phase solutions are associated to a set of data on a Riemann surface. We start with a hyperelliptic Riemann surface Σ of genus g described by the equation

$$y^2 = \prod_{i=1}^{2g+2} (\lambda - \lambda_i). \tag{28}$$

We mark the two points ∞_+, ∞_-, which are permuted by the holomorphic involution $\tau(\lambda, y) = (\lambda, -y)$ exchanging the two sheets. We also choose a set of $g + 1$ distinct points $\mathcal{D}_0 = P_1 + \cdots + P_{g+1}$ placed in a generic position (a *non-special divisor*) and not containing $\infty_\pm$.

Let $\lambda(P) : \Sigma \to \mathbb{C} \cup \{\infty\}$ be the hyperelliptic projection to the Riemann sphere; this is a choice of meromorphic function on Σ whose pole divisor is $\infty_+ + \infty_-$ (i.e. whose poles are the preimage of ∞ via the map λ). In neighbourhoods of $\infty_\pm$, we choose the local parameters $k_\pm$ such that

$$(k_\pm)^{-1} = (\lambda(P))^{-1}$$

near $\infty_\pm$ respectively.

The main idea of Krichever is to construct a function $\psi(\lambda)$ on Σ which is uniquely defined by a prescribed behaviour at its singularities and which turns out a posteriori to be the simultaneous solutions of a pair of linear systems. The compatibility condition, i.e. the zero curvature representation of these two linear operators, is a completely integrable nonlinear equation for the coefficients. Thus, the construction of such a function provides one both with the nonlinear equation, an initial condition and its solution.

Definition 2.2. A associated to $(\Sigma, \mathcal{D}_0, \infty_\pm)$ is a function ψ on Σ which:

- is meromorphic everywhere except at $\infty_\pm$ and whose set of poles on $\Sigma \backslash \{\infty_\pm\}$ is $\mathcal{D}_0$,

- has essential singularities at $\infty_\pm$ that locally are of the form $\psi(k_\pm) \sim C e^{p(k_\pm)}$, where C is a constant and $p(k)$ an arbitrary polynomial with complex coefficients.

For our purposes, we recall the following result [31],

Proposition 2.3. *Suppose that the following technical condition holds:*

Condition 2.4. The divisor $P_1 + \cdots + P_{g+1} - \infty_+ - \infty_-$ is not linearly equivalent to a positive divisor.

Then, if $p(k) = ikx + iQ(k)t$, with x and t complex parameters with $|x|, |t|$ sufficiently small and $Q(k)$ a given polynomial, the linear vector space of Baker–Akhiezer functions associated to $(\Sigma, \mathcal{D}_0, k_\pm)$ is 2-dimensional and it has a unique basis ψ^1, ψ^2 with the following normalised expansions at $\infty_\pm$:

$$\psi^j(x, t; \lambda) = e^{ik_\pm x + iQ(k_\pm)t} \left(\sum_{n=0}^{\infty} \zeta_n^{j\pm}(x, t) k_\pm^{-n} \right), \quad j = 1, 2 \qquad (29)$$

where $\zeta_n^{j\pm}$ are functions of the parameters x and t and $\zeta_0^{1+} = 1$, $\zeta_0^{1-} = 0$, $\zeta_0^{2+} = 0$, $\zeta_0^{2-} = 1$.

Remarks. We make two remarks before proving the proposition.

1. For a non-special divisor of degree d (a formal integer linear combination of points on Σ counted with multiplicity) the Riemann–Roch formula (see for example [24]), states that the dimension $h^0(\mathcal{D})$ of the linear space of meromorphic functions on Σ whose pole divisor is $\mathcal{D}$ is

$$h^0(\mathcal{D}) = \left\{ \begin{array}{ll} 1, & d \le g, \\ d - g + 1, & d > g. \end{array} \right. \qquad (30)$$

2. Condition 2.4 can be rephrased as: there exists no non-constant meromorphic function with pole divisor $P_1 + \cdots + P_{g+1}$ which vanishes simultaneously at ∞_+ and at ∞_-.

Proof. Uniqueness: Suppose there are two functions ψ_1 and ψ_2 which satisfy the prescription; then their ratio ψ_1/ψ_2 is a meromorphic function (the essential singularities mutually cancel) whose poles are contained in the zero divisor of ψ_2. The condition of "non-speciality" assures that the dimension of the space of such functions is $2 (= g + 1 - g + 1$, according to the Riemann–Roch formula). As regards the normalisation, since $h^0(\mathcal{D}_0 - \infty_+) = h^0(\mathcal{D}_0 - \infty_-) = 1$, we can choose two such functions vanishing at ∞_+ and at ∞_- respectively; because $h^0(\mathcal{D}_0 - \infty_+ - \infty_-) = 0$ (Condition 2.4), they must be independent.

Existence: We follow an argument given in [46] and exhibit a pair of independent functions on Σ with the correct singularities by constructing them explicitly as ratios of Riemann Theta functions (a nice survey of the features which are relevant in this context can be found in [15]). Let

$$a_1, \ldots, a_g, \quad b_1, \ldots, b_g,$$

be a canonical homology basis for the Riemann surface Σ such that $a_i \cdot a_j = 0$, $b_i \cdot b_j = 0$, $a_i \cdot b_j = \delta_{ij}$, and

$$\omega_1, \ldots, \omega_g, \qquad \oint_{a_i} \omega_j = \delta_{ij}, \quad i, j = 1, \ldots, g$$

be g normalised holomorphic differentials. We introduce the period matrix B,

$$B_{ij} = \oint_{b_i} \omega_j, \qquad i, j = 1, \ldots, g$$

and construct the associated Riemann Theta function

$$\theta(z) = \sum_{n \in \mathcal{Z}^g} \exp i\pi \left(\langle n, Bn \rangle + 2\langle n, z \rangle \right), \quad z \in \mathbb{C}^g.$$

The essential behaviour at ∞ is introduced by means of the unique normalised differentials of the second kind η and ζ, which satisfy the following conditions:

1. η and ζ have a single pole at ∞ with local expansions

$$\eta \sim idk, \qquad \zeta \sim idQ(k),$$

   (these are dictated by the polynomial dependence of the exponent on the local parameter).

2. (normalisation)

$$\oint_{a_i} \eta = 0, \qquad \oint_{a_i} \zeta = 0.$$

We can now build the following function of $P \in \Sigma$ (where $P_0 \in \Sigma$ is a base point)

$$\tilde{\psi} = \exp\left(x \int_{P_0}^P \eta + t \int_{P_0}^P \zeta \right) \frac{\theta\big(\mathcal{A}(P) + Ux + Wt - K - \mathcal{A}(\mathcal{D})\big)}{\theta\big(\mathcal{A}(P) - K - \mathcal{A}(\mathcal{D})\big)}. \tag{31}$$

In this expression $\mathcal{D}$ is a divisor specified below and K is the Riemann constant (see [15]) chosen so that $\theta(\mathcal{A}(P) - K - \mathcal{A}(\mathcal{D}))$ has zeros precisely at $\mathcal{D}$, and

$$\mathcal{A}\left(\sum_{k=1}^g Q_k \right) := \sum_{k=1}^g \int_{P_0}^{Q_k} \omega$$

is the Abel map (ω is the vector of holomorphic differentials). The map $\mathcal{A}$ associates to a divisor $\sum_{k=1}^g Q_k$ on Σ a point of the complex torus $\mathbb{C}^g/\Lambda$ (the

Jacobian of the Riemann surface Jac(Σ)), where Λ is the $2g$-dimensional lattice spanned by the columns of the matrix $(I \,|\, B)$.

The vectors U and W have been introduced to make $\tilde{\psi}$ a well-defined function on Σ. The only indeterminacy is the path of integration which can be modified by adding any integer combination of homology cycles. Taking into account the vanishing condition for η and ζ, this produces the overall factor (the m_k's are integers)

$$\exp\left(\sum_{k=1}^{g}\left[m_k\left(x\oint_{b_k}\eta + t\oint_{b_k}\zeta\right) - m_k\left(xU_k + tW_k\right)\right]\right).$$

This is equal to 1 if we define the components of the "frequency vectors" U and W to be

$$U_k = \frac{1}{2\pi i}\oint_{b_k}\eta, \qquad \frac{1}{2\pi i}W_k = \oint_{b_k}\zeta.$$

At last we are left to choose the pole divisor $\mathcal{D}$ of degree g and to fix the normalisation. As discussed above, two independent functions are uniquely picked by requiring that one vanishes at ∞_+ and the other at ∞_-; for this purpose we introduce the following:

Definition 2.5. Let $\mathcal{D}_\pm$ be the unique positive divisors linearly equivalent to $\mathcal{D}_0 - \infty_\pm$.

The choices $\mathcal{D} = \mathcal{D}_+$ and $\mathcal{D} = \mathcal{D}_-$ in formula (31) produce two independent functions, whose poles are in $\mathcal{D}_+$ and $\mathcal{D}_-$ respectively. In order to make the pole divisor be $\mathcal{D}$ we multiply $\tilde{\psi}^\pm$ by a meromorphic function $g_\pm(P)$ whose zeros lie in $\mathcal{D}_\pm + \infty_\mp$ and whose poles lie in the original divisor $\mathcal{D}$. We finally obtain the correct Baker–Akhiezer functions

$$\psi^{1,2} = \exp\left[x\left(\int_{P_0}^{P}\eta - \eta_\pm^\infty\right) + t\left(\int_{P_0}^{P}\zeta - \zeta_\pm^\infty\right)\right] \times$$

$$\frac{\theta(\mathcal{A}(P) + Ux + Wt - K - \mathcal{A}(\mathcal{D}_\pm))}{\theta(\mathcal{A}(P) - K - \mathcal{A}(\mathcal{D}_\pm))} \times \tag{32}$$

$$\frac{\theta(\mathcal{A}(\infty_\pm) - K - \mathcal{A}(\mathcal{D}_\pm))}{\theta(\mathcal{A}(\infty_\pm) + Ux + Wt - K - \mathcal{A}(\mathcal{D}_\pm))} \times \frac{g_\pm(P)}{g_\pm(\infty_\pm)}.$$

The constant terms $\eta_\mp^\infty$ and $\zeta_\mp^\infty$ in the expansion of the argument of the exponential at $\infty_\pm$ have been subtracted to make the leading coefficient of the meromorphic part of the eigenfunction matrix be the identity. $\qquad\square$

2.3 The Gauge Transformation Revisited

Given the unique basis of the vector space of Baker–Akhiezer functions guaranteed in the previous theorem, we can build unambiguously a function and a corresponding pair of linear operators which will be identified with the Lax pair for the Continuous Heisenberg Model. We have the following result [8]:

Proposition 2.6. *If $Q(k) = 2k^2$ and $\Psi(x, t; P)$ is the matrix of Baker–Akhiezer functions constructed in Proposition 2.3*

$$\Psi(x, t; P) = \left(\begin{array}{cc} \psi^1(x, t; P) & \psi^1(x, t; \tau(P)) \\ \psi^2(x, t; P) & \psi^2(x, t; \tau(P)) \end{array} \right),$$

then the columns of the matrix

$$\Phi(x, t; \lambda) = \Psi(x, t; 0)^{-1} \Psi(x, t; \lambda) \tag{33}$$

are linearly independent simultaneous solutions of the following pair of linear systems

$$\frac{\partial \mathbf{F}}{\partial x} = i\lambda S \mathbf{F}$$

$$\frac{\partial \mathbf{F}}{\partial t} = (2i\lambda^2 S - \lambda S_x S)\mathbf{F},$$

where the matrix S is independent of λ and is given by

$$S(x, t) = \Psi(x, t; 0)^{-1} \sigma_3 \Psi(x, t; 0), \tag{34}$$

with $\sigma_3 = \left(\begin{array}{cc} 1 & 0 \\ 0 & -1 \end{array} \right)$.

The value of Ψ at $\lambda = 0$ defines the normalisation of the eigenfunction Φ at the essential singularity. We have in fact

$$\lim_{\lambda \to \infty} \Phi(x, t; \lambda) = \Psi(x, t; 0)^{-1}. \tag{35}$$

Moreover, the choice of $\lambda = 0$ as the point at which the HM Baker–Akhiezer function is normalised is arbitrary, provided it is not chosen to be at the essential singularity.

It is shown in [46] that the matrix $\Psi(x, t; \lambda)$ constructed in Proposition 2.3 when $Q(k) = 2k^2$ solves the Lax Pair (14) for the focusing nonlinear Schrödinger equation. These considerations lead to the following simple interpretation of the gauge transformation between the HM and the NLS equation:

Corollary 2.7. *Given the matrix $\Psi(x, t; \lambda)$ of Baker–Akhiezer functions for the NLS Lax Pair (14), the matrix obtained by renormalising Ψ to be the identity matrix at an arbitrary $\lambda_0 \neq \infty$ solves the HM Lax Pair at $\lambda - \lambda_0$. Moreover the potential S can be reconstructed by means of equation (35) with 0 replaced by λ_0.*

The proof of Proposition 2.6 is contained in the following lemma [8],

Lemma 2.8. *Given the Baker–Akhiezer matrix function $\Phi(x, t; \lambda)$ normalised as in Proposition 2.3, there exists a unique pair of matrix differential operators L_1 and L_2 of the following form*

$$L_1 = \sum_{\alpha=0}^{1} U_\alpha(x, t) \frac{\partial^\alpha}{\partial x^\alpha}, \qquad L_2 = \sum_{\alpha=0}^{2} V_\beta(x, t) \frac{\partial^\beta}{\partial x^\beta}, \tag{36}$$

such that $\Phi(x, t; \lambda)$ solves simultaneously the linear systems

$$L_1 \Phi = \lambda \Phi, \qquad L_2 \Phi = \frac{\partial \Phi}{\partial t}. \tag{37}$$

Proof. In order to determine the coefficients of the linear operators L_1 and L_2, we examine the asymptotic behaviour of both columns of Φ at $\infty_\pm$. We define $A(x,t) =: \Psi^{-1}(x,t,0)$ and look at an expansion for Φ in the global coordinate λ of the form

$$\Phi(x,t;\lambda) \sim \left(A(x,t) + \sum_{n=1}^{\infty} \frac{1}{\lambda^n} X_n(x,t) \right) \begin{pmatrix} e^{i(\lambda x + 2\lambda^2 t)} & 0 \\ 0 & e^{-i(\lambda x + 2\lambda^2 t)} \end{pmatrix}. \quad (38)$$

The operator L_1 is uniquely determined by the requirement

$$(L_1 - \lambda I)\Phi(x,t;\lambda) = \mathcal{O}\left(\frac{1}{\lambda}\right) \begin{pmatrix} e^{i(\lambda x + 2\lambda^2 t)} & 0 \\ 0 & e^{-i(\lambda x + 2\lambda^2 t)} \end{pmatrix}. \quad (39)$$

In fact, by substituting expression (38) in equation (39) and requiring that the $O(1)$ and $O(\lambda)$ terms are equal to zero, one finds the following recursive system of equations:

$$\begin{aligned} U_1 &= -iA\sigma_3 A^{-1} \\ U_0 &= X_1 A^{-1} - U_1 A_x A^{-1} - iU_1 X_1 \sigma_3 A^{-1} \end{aligned} \quad (40)$$

which determines U_0 and U_1 as functions of A and X_1. Since $\Phi(x,t;0) = I$, the multiplicative term U_0 in the expression of L_1 must vanish. We check this in the following way.

We consider the matrix of normalised Baker–Akhiezer functions $\Psi(x,t;\lambda)$ constructed in Proposition 2.3 and its asymptotic expansion at ∞,

$$\Psi(x,t;\lambda) \sim \left(I + \sum_{n=1}^{\infty} \frac{1}{\lambda^n} Z_n(x,t) \right) \begin{pmatrix} e^{i(\lambda x + 2\lambda^2 t)} & 0 \\ 0 & e^{-i(\lambda x + 2\lambda^2 t)} \end{pmatrix}.$$

Because Φ is obtained by normalising Ψ as in (34), then $Z_n = A^{-1} X_n$, $n = 1, \dots$. If we substitute the asymptotic expansion of Ψ into an equation of the same form as (39), we can show that Ψ satisfies the following linear eigenvalue problem,

$$-i\sigma_3 \frac{\partial \Psi}{\partial x} - (\sigma_3 Z_1 \sigma_3 - Z_1)\Psi = \lambda \Psi. \quad (41)$$

In particular, the coefficient $A^{-1}(x,t) = \Psi(x,t;0)$ satisfies (41) at $\lambda = 0$ which we rewrite as

$$\frac{dA^{-1}}{dx} - i[Z_1, \sigma_3]A^{-1} = 0. \quad (42)$$

We now solve for U_0 in the system (40), use equation (42) and obtain

$$U_0 = -A \left(\frac{dA^{-1}}{dx} - i[Z_1, \sigma_3]A^{-1} \right) = 0. \quad (43)$$

The coefficients of L_2 are determined in the same way. The requirement

$$(L_2 - \frac{\partial}{\partial t})\Phi(x,t;\lambda) = \mathcal{O}(\frac{1}{\lambda}) \begin{pmatrix} e^{i(\lambda x + 2\lambda^2 t)} & 0 \\ 0 & e^{-i(\lambda x + 2\lambda^2 t)} \end{pmatrix}, \quad (44)$$

together with the asymptotic expansion (38), determines the following recursive system of equations for the coefficients V_i:

$$
\begin{aligned}
V_2 &= -2iA\sigma_3 A^{-1} \\
V_1 &= 2X_1 A^{-1} - 2V_2 A_x A^{-1} - iV_2 X_1 \sigma_3 A^{-1} \\
V_0 &= A_t - V_1(A_x + iX_1\sigma_3) - V_2(A_{xx} + 2iX_{1x}\sigma_3 - X_2).
\end{aligned}
\tag{45}
$$

In order to simplify the expressions for V_1 and V_2 we use the equation for the time evolution of $\Psi(x,t;\lambda)$, which we can derive by substituting the asymptotic expansion of Ψ in the time linear system (36). We obtain the following λ-independent equation

$$
-2i\sigma_3 \frac{\partial^2 \Psi}{\partial x^2} + 2[Z_1, \sigma_3]\sigma_3 \frac{\partial \Psi}{\partial x} + 2i[Z_2, \sigma_3] - 2i[Z_1, \sigma_3]\sigma_3 Z_1 \sigma_3 - 2\sigma_3 Z_{1x}\sigma_3 = \frac{\partial \Psi}{\partial t}.
$$

Using the fact that $A^{-1}(x,t) = \Psi(x,t;0)$ satisfies this last equation we substitute the expression for A_t into (45) and obtain

$$
\begin{aligned}
V_2 &= -2iA\sigma_3 A^{-1} \\
V_1 &= -i(A\sigma_3 A^{-1})_x \\
V_0 &= 0.
\end{aligned}
$$

For this choice of U_i's and V_i's the right-hand side of the relation

$$
(L_j - \lambda\mathrm{I})\Phi(x,t;\lambda) = \mathcal{O}(\frac{1}{\lambda}) \begin{pmatrix} e^{i(\lambda x + 2\lambda^2 t)} & 0 \\ 0 & e^{-i(\lambda x + 2\lambda^2 t)} \end{pmatrix},
\tag{46}
$$

gives a pair of Baker–Akhiezer functions whose asymptotic expansions have leading coefficients vanishing at $\infty_\pm$. Because they are unique, they must be identically zero. Therefore $\Phi(x,t,\lambda)$ solves simultaneously the equations

$$
(L_j - \lambda\mathrm{Id})\Phi(x,t,\lambda) = 0, \quad j = 1,2
$$

which, by introducing the quantity $S = A\sigma_3 A^{-1}$, become the Lax Pair for the Continuous Heisenberg Model. $\qquad\square$

2.4 N-phase Curves

In the previous section, we built a basis of Baker–Akhiezer eigenfunctions for the linear problem of the Heisenberg Model and we expressed the potential S purely in terms of the leading coefficient of the expansion of its meromorphic part at $\lambda = \infty$. We now derive a formula for the corresponding N-phase curves. It is clear from what we have discussed so far that the Baker–Akhiezer function for the Heisenberg Model is specified by $g + 1$ poles, two essential singularities over ∞ and its normalisation at 0; it is also clear that normalising at a different point (so far as it does not coincide with one of the poles or the zeros) will affect neither the essential singularity nor the divisor of the meromorphic part. Moreover, if the pole divisor $\mathcal{D}$ is in general enough position, Condition 2.4 can be replaced by the following equivalent:

Condition 2.9. The divisor $P_1 + \cdots + P_{g+1} - 0_+ - 0_-$ is not linearly equivalent to a positive divisor.

Here 0_+ and $0_- = \tau(0_+)$ are the two points on Σ corresponding to $\lambda = 0$ which are exchanged by the hyperelliptic involution. Condition 2.9 and $\mathcal{D}$ being non-special assure that, among the meromorphic functions with poles in $\mathcal{D}$, there is only one which vanishes at 0_+ and just one which vanishes at 0_- (this assures the existence of two distinct functions with poles in $\mathcal{D}$), but there exists no non-constant function which vanishes at both points (this guarantees the independence of the two and therefore the ability to realise any normalisation at $\lambda = 0$). We introduce the following quantities:

a) $\mathcal{D}_+^0$ (resp. $\mathcal{D}_-^0$), the unique effective divisor which is linearly equivalent to $\mathcal{D} - 0_-$ (resp. $\mathcal{D} - 0_+$).

b) $h_\pm(P)$, a meromorphic function whose divisor is $(h_\pm) = \mathcal{D}_\pm^0 + 0_\mp - \mathcal{D}$.

By an argument identical to the one described in Proposition 2.3, we obtain the following result:

Proposition 2.10. *The linear problem associated to the Heisenberg Model is solved by a vector function with the following components*

$$
\phi^\pm = \exp\left[x\left(\int_{P_0}^P \eta - \eta_\pm^0\right) + t\left(\int_{P_0}^P \zeta - \zeta_\pm^0\right)\right] \times
$$
$$
\frac{\theta(\mathcal{A}(P) + Ux + Wt - K - \mathcal{A}(\mathcal{D}_\pm^0))}{\theta(\mathcal{A}(P) - K - \mathcal{A}(\mathcal{D}_\pm^0))} \times \tag{47}
$$
$$
\frac{\theta(\mathcal{A}(0_\pm) - K - \mathcal{A}(\mathcal{D}_\pm^0))}{\theta(\mathcal{A}(0_\pm) + Ux + Wt - K - \mathcal{A}(\mathcal{D}_\pm^0))} \times \frac{h_\pm(P)}{h_\pm(0_\pm)}.
$$

Moreover, the corresponding eigenfunction matrix

$$
\Phi(P) = \begin{pmatrix} \phi^+(P) & \phi^+(\tau(P)) \\ \phi^-(P) & \phi^-(\tau(P)) \end{pmatrix} \tag{48}
$$

is normalised to be the identity matrix at $\lambda = 0$.

In formula (47), $(\eta, \zeta, P_0, \boldsymbol{\omega})$ are the same as in Proposition 2.3. The terms $\eta_\pm^0 = \int_{P_0}^{0_\pm} \eta$ and $\zeta_\pm^0 = \int_{P_0}^{0_\pm} \zeta$ have been introduced to produce the correct normalisation at $\lambda = 0$.

Finally, we construct the N-phase curves. We use the reconstruction formula derived in section 2.1 (neglecting the additive constant associated to the normalization of Φ at $x = 0$)

$$
\Gamma(x,t) = -i \left.\frac{\partial \Phi(x,t,\lambda)}{\partial \lambda}\right|_{\lambda=0}.
$$

The derivatives of the entries of Φ can be expressed in terms of the following functions (and by the sheet information),

$$
\left.\frac{\partial \phi^\pm}{\partial \lambda}\right|_{\lambda=0} = a_\pm x + b_\pm t + \frac{\partial}{\partial P} \log \theta(\mathcal{A}(P) + Ux + Wt - K - \mathcal{A}(\mathcal{D}_\pm^0))\bigg|_{P=0_\pm} + c^\pm,
$$

where the $c^\pm$'s are constant in x and t depending on the Riemann surface data, $a_\pm = \pm\eta(0)$, and $b_\pm = \pm\zeta(0)$. We show that both coefficients of the linear terms in x and t are zero. Let λ be real for simplicity (we are just interested in what happens at $\lambda = 0$). In this case the fundamental solution Φ of the spatial linear problem is a unitary matrix. We introduce the *Transfer Matrix*

$$T(x, \lambda) = \Phi(x + 2\pi, \lambda)\Phi^{-1}(x, \lambda)$$

which takes a solution evaluated at a given x to its value after one spatial period. It is an easy exercise to show that its trace

$$\Delta(\lambda) = Tr\left[T(x, \lambda)\right]$$

is invariant both with respect to x and t. If we define the *Floquet eigenfunctions* by means of the following formula

$$\phi^{\pm}(x, \lambda) = e^{i\rho(\lambda)x} f^{\pm}(x, \lambda),$$

where $f^{\pm}(x, \lambda)$ are bounded, periodic functions, one can easily check that the following relation holds between the *Floquet exponent* $\rho(\lambda)$ and $\Delta(\lambda)$,

$$\frac{d\rho}{d\lambda} = \frac{1}{2\pi} \frac{\frac{d\Delta}{d\lambda}}{\sqrt{\Delta^2 - 4}}.$$

By comparing the expression of the Baker–Akhiezer eigenfunction with the one for the Floquet eigenfunction, we can identify the differential η with the derivative of the Floquet exponent,

$$\eta(\lambda) = \frac{d\rho}{d\lambda}(\lambda).$$

As a consequence of the form of the linear system we have the following symmetry

$$\Phi(x, -\lambda) = \Phi(-x, \lambda),$$

from which there follows

$$T(2\pi, -\lambda) = T^{-1}(2\pi, \lambda) = \bar{T}^T(2\pi, \lambda).$$

Since the trace of a unitary matrix is real, we have

$$\Delta(-\lambda) = \Delta(\lambda), \qquad \lambda \in \mathbb{R}.$$

Therefore $\Delta(\lambda)$ is an even function of λ and the Floquet exponent $\rho = \eta$ must vanish at 0. In order to show that the coefficient b vanishes as well, we use the symmetry $\frac{d\phi}{d\lambda}(x, -\lambda) = -\frac{d\phi}{d\lambda}(-x, \lambda)$, where ϕ is any of the components of the eigenmatrix. This formula must hold for all times and in particular in the limit as $t \sim \infty$. At leading order we have $bt = -bt$, it follows that $b = 0$.

Finally, we obtain the following compact formula for the components of the Hermitian matrix Γ (its left column is the vector $(\Gamma_+, \Gamma_-)^T$) representing the position vector of the curve,

$$\Gamma_{\pm} = -i \frac{\partial}{\partial P} \log \theta(\mathcal{A}(P) + Ux + Wt + \theta_0)\Big|_{P=0_{\pm}}, \qquad (49)$$

where we absorbed the information about the divisor and the Riemann constant in the initial condition θ_0.

Among the N-phase curves we find planar circles (1-phase solutions that are the analogues of plane wave solutions of the NLS equation) which will play an important role in section 5. Below, we show a pair of typical 2-phase curves, constrained extrema of the total squared curvature functional. Solutions to the associated variational problem in the case of unconstrained total torsion (the classical elastic curves) were completely classified by J. Langer and D. Singer [37], and shown to belong to certain classes of torus knots. The curves in Figure 2, a 5-2 and a 9-2 torus knot respectively, are critical points of the total squared curvature with constrained total torsion and total length (the most general 2-phase solutions), a vertical ribbon around each of them has been added to make the over and under crossings evident.

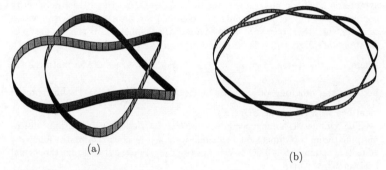

(a)

(b)

Figure 2: A 5-2 torus knot (a) and a 9-2 torus knot (b) are among the 2-phase solutions of the filament equation.

3   Natural Frames

If the vortex filament equation describes the evolution of an arclength parameterised curve in $\mathbb{R}^3$, the continuous Heisenberg Model describes the corresponding evolution of its unit tangent vector. In the course of this section, we will explain how, in order to understand the geometry of the solution space of HM, we need to consider not just the unit tangent vector, but orthonormal frames of the original curve.

The space of unit tangent vectors of space curves is the two-dimensional sphere of radius 1. The space of associated orthonormal frames can be identified with its unit tangent bundle $T_1 S^2$. Then, for a given curve in $\mathbb{R}^3$, choosing an orthonormal frame means choosing a way to lift the spherical curve described by the unit tangent vector into the bundle. In this section, we study the horizontal lifting associated to the canonical invariant connection on $T_1 S^2$ and introduce the notion of natural frame. We show that the horizontal lifting is the geometrical realisation of the gauge transformation between the HM and the NLS equation. Moreover, we use this concrete construction to characterise the differential of the gauge transformation in terms of the second symplectic operator for the NLS equation; we will show as a consequence that the HM and the NLS equation can be regarded as the same Hamiltonian system written with respect to different symplectic structures belonging to the same hierarchy of Hamiltonian flows.

3.1   The Circle Bundle of S^2

We start with recalling a few facts about the two-dimensional sphere and its associated circle bundle. The unit sphere

$$S^2 = \left\{ \mathbf{x} \in \mathbb{R}^3 \mid x_1^2 + x_2^2 + x_3^2 = 1 \right\}$$

is an example of a *symmetric homogeneous space*. Given a connected Lie group G and a nontrivial group automorphism σ such that $\sigma^2 = \mathrm{Id}$, the associated symmetric homogeneous space is the orbit space G/H, where H is the identity component of the set of elements $h \in G$ which are invariant under the action of σ, i.e. such that $\sigma(h) = h$. In the case of the 2-sphere, G is the Lie group of rotations $SO(3, \mathbb{R})$ and σ is "conjugation by T": $\sigma(g) = T^{-1}gT$, $g \in SO(3, \mathbb{R})$, where T is the matrix $T = \begin{pmatrix} I & \\ & -1 \end{pmatrix}$. For this choice of the matrix T, H becomes the subgroup of elements of the form $h = \begin{pmatrix} * & \\ & 1 \end{pmatrix}$ which can be identified with the group $SO(2, \mathbb{R})$.

The symmetric homogeneous space $SO(3, \mathbb{R})/SO(2, \mathbb{R})$ is naturally diffeomorphic to the two-dimensional sphere, we show it by constructing the following transitive action of $SO(3, \mathbb{R})$ on S^2. Let $\{e_1, e_2, e_3\}$ be the standard orthonormal basis in $\mathbb{R}^3$, the map

$$
\begin{array}{ccc}
SO(3, \mathbb{R}) & \overset{\delta}{\longrightarrow} & S^2 \\
g & \longrightarrow & g \cdot e_3
\end{array}
$$

induces a diffeomorphism between $SO(3, \mathbb{R})/SO(2, \mathbb{R})$ and S^2 since the subgroup H fixes the vector e_3. Moreover, given the canonical projection $p : SO(3, \mathbb{R}) \to SO(3, \mathbb{R})/SO(2, \mathbb{R}) \cong S^2$, the automorphism σ induces the involution $\sigma_o(px) = p\sigma(x)$, $x \in SO(3, \mathbb{R})$, for which the distinguished point $o = pe$ is an isolated fixed point (e denotes the identity element of $SO(3, \mathbb{R})$).

We now define the circle bundle of S^2 as the space of all unit tangent vectors to the sphere:

$$T_1 S^2 = \left\{ (x, v) \ : \ x \in S^2, \ v \in T_x S^2, \ \|v\| = 1 \right\},$$

($\| \cdot \|$ is the Euclidean norm in $\mathbb{R}^3$) together with the projection $\pi : T_1 S^2 \to S^2$, $\pi(x, v) = x$ with fibre $\pi^{-1}(x) = S^1$.

Remarks. 1. There is a natural action of the unit circle which "fixes" the base point x and rotates v in the tangent plane to S^2 at x described by the smooth map

$$
\begin{array}{ccc}
S^1 \times T_1 S^2 & \longrightarrow & T_1 S^2 \\
h \cdot (x, v) & \longrightarrow & (x, hv).
\end{array}
$$

This action is free, i.e. with no fixed points.

2. We can always construct a local cross-section, i.e. a smooth map $\phi^{-1} : \pi^{-1}(U) \to U \times S^1$ for some neighbourhood U of each point $x \in S^2$. To do so, we choose a smooth unit vector field $e(U)$ (for example, we can take the vector field $\partial_u / \|\partial_u\|$ in a local coordinate system (u, v)) and define $\phi(x, h) = (x, he(x))$.

Since the S^1-action is free and $e(x)$ is smooth and never zero in U, ϕ is smooth and invertible and so is ϕ^{-1}.

3. The properties described above define a principal fibre bundle with fibre S^1 and base S^2. The classical result that there does not exist a non-vanishing vector field on S^2 (and therefore a global cross-section) implies that $T_1 S^2$ is a non-trivial bundle.

4. The transitive, free action of $SO(3, \mathbb{R})$ on $T_1 S^2$ described by the map $g \cdot (x, v) = (gx, gv)$, $g \in SO(3, \mathbb{R})$, identifies the space of orthonormal frames with the circle bundle of S^2: $SO(3, \mathbb{R}) \cong T_1 S^2$.

3.2 The Canonical Connection

We summarise the notion of a connection in a principal fibre bundle and the construction of the canonical invariant connection for $T_1 S^2$. The content of this section can be found in various references, ([30] is comprehensive, [52] contains a simple treatment of the circle bundle of a two-dimensional Riemannian manifold).

As in Euclidean space, there is a natural way to parallel-translate and compare vectors at different points, likewise in a general manifold a choice of a connection prescribes a way of translating tangent vectors "parallel to themselves" and to define intrinsically a directional derivative. In the case of a principal bundle P with structure group G over a manifold M

$$
\begin{array}{ccc}
G & \longrightarrow & P \\
 & & \downarrow{\scriptstyle \pi} \\
 & & M
\end{array}
$$

we best explain the role of a connection when thinking of lifting a vector field $v \in TM$ to a vector field $\tilde{v} \in TP$ in a unique way. For each $p \in P$, let G_p be the subspace of $T_p P$ consisting of all the vectors tangent to the vertical fibre. The lifting of v will be unique if we require $\tilde{v}(p)$ to lie in a subspace of $T_p P$ complementary to G_p. A smooth and G-invariant choice of such a complementary subspace is called a *connection* on P. More precisely,

Definition 3.1. A connection on a principal bundle P is a smooth assignment of a subspace $H_p \subset T_p P$, $\forall p \in P$ such that:
1. $TP_p = G_p \oplus H_p$
2. $H_{gp} = (\mathcal{L}_g)_* H_p$, $\qquad \forall g \in G$ $\qquad$ ($\mathcal{L}_g$ is the left-translation in G).

Given a connection, the horizontal subspace H_p is mapped isomorphically by $d\pi$ onto $T_{\pi p} M$. Therefore the lifting of v is the unique horizontal $\tilde{v}$ which projects onto v. An equivalent way of assigning a connection is by means of a Lie algebra valued 1-form ϕ (the connection form). If $A \in \mathfrak{g}$ (the Lie algebra of G), let A^* be the vector field on P induced by the action of the 1-parameter subgroup e^{tA}. Since the action of G maps each fibre into itself, then A^* is tangent to the vertical fibre at each point. For each $X \in T_p P$, $\phi(X)$ is the unique $A \in \mathfrak{g}$ such that A^* is equal to the vertical component of X. It follows that $\phi(X) = 0$ if and only if X is horizontal.

Proposition 3.2. *A connection 1-form ϕ has the following properties:*
1. $\phi(A^*) = A$
2. $(\mathcal{L}_g)^* \phi = Ad_g \phi$, $\qquad \forall g \in G$ *(Ad is the adjoint representation of G).*

The first property follows immediately from the definition of connection 1-form; for a proof of the second property we refer to [30].

We are now ready to construct an invariant connection on $T_1 S^2$. The group involution σ defined in the previous section induces a Lie algebra automorphism on $\mathfrak{g} = \mathfrak{so}(3, R)$ (denoted with the same letter σ). The automorphism σ determines the following direct sum decomposition of $\mathfrak{g}$

$$\mathfrak{g} = \mathfrak{h} \oplus \mathfrak{k},$$

where $\mathfrak{h} = \{X \in \mathfrak{g} |\ \sigma X = X\}$ is the subalgebra of the invariant subgroup $H = SO(2, R)$ and $\mathfrak{k} = \{X \in \mathfrak{g} |\ \sigma X = -X\}$.

Let θ be the canonical 1-form of $SO(3, \mathbb{R})$, i.e. the left-invariant $\mathfrak{g}$-valued 1-form defined by

$$\theta(A) = A, \qquad A \in \mathfrak{g},$$

then

Theorem 3.3. *The $\mathfrak{h}$-component ϕ of the canonical 1-form θ of $SO(3, \mathbb{R})$ defines a left-invariant connection on $T_1 S^2$.*

Proof. Since θ is left-invariant, we restrict to left-invariant vector fields X on $T_1 S^2$. Then $\phi(X){=}0$ if and only if $X \in \mathfrak{k}$ and $\phi(Y) = Y$ if $Y \in \mathfrak{h}$. θ is left-invariant also with respect to the action of $H = SO(2, R) \cong S^1$; therefore ϕ satisfies both properties 1 and 2 of a connection form and it is invariant under the full action of $SO(3, \mathbb{R})$. $\square$

Structure Equations   Let (V, E_1, E_2) be the canonical basis for the Lie algebra $\mathfrak{so}(3, R)$ such that V spans the vertical space $\mathfrak{h}$ and (E_1, E_2) span the horizontal subspace $\mathfrak{k}$. We have

$$[V, E_1] = E_2, \quad [V, E_2] = -E_1, \quad [E_1, E_2] = V.$$

Setting $\theta = \phi V + \omega_1 E_1 + \omega_2 E_2$ and using the Maurer–Cartan equation

$$d\theta(X, Y) = -\frac{1}{2}\theta([X, Y]), \qquad X, Y \in \mathfrak{g},$$

we obtain the following structure equations for the dual basis $(\phi, \omega_1, \omega_2)$

$$d\phi = -\omega_1 \wedge \omega_2, \qquad d\omega_1 = \phi \wedge \omega_2, \qquad d\omega_2 = -\phi \wedge \omega_1.$$

Remarks.   1. If the Riemannian metric on S^2 is inherited from restricting the Euclidean metric on $\mathbb{R}^3$ to S^2, then the invariant connection constructed above coincides with the Riemannian connection on the frame bundle of S^2 (the unique connection which leaves the metric invariant and has zero torsion).

2. We can give an invariant definition of the 1-forms ω_1, ω_2 using the Riemannian structure, the left-invariance of the connection and the isomorphism $d\pi : H_{(x,v)} \rightarrow T_x S^2$ between the horizontal subspace at (x, v) and the tangent space to S^2 at x. We identify $T_o S^2$ with the horizontal subspace $\mathfrak{h}$. Then (E_1, E_2) may be taken to be the orthonormal basis (e_1, ie_1) of

T_oS^2, where ie_1 is obtained by a $90°$ rotation of e_1 within the tangent plane. Then for $t \in T_{(o,e_1)}\mathcal{T}_1S^2$, $\omega_1|_{(o,e_1)}(t)$ and $\omega_2|_{(o,e_1)}(t)$ are the components of the projection $d\pi(t)$ relative to the orthonormal basis (e_1, ie_1), we have for example $\omega_1|_{(o,e_1)}(t) = \langle d\pi(t), e_1 \rangle$. Since ω_1, ω_2 are invariant with respect to the S^1-action, we can define them on all tangent vectors at o of the form $v = he_1$, $h \in S^1$. The left-invariance of the metric under the full group action defines them everywhere:

$$\omega_1|_{(x,v)}(t) = \langle d\pi(t), v \rangle, \qquad \omega_2|_{(x,v)}(t) = \langle d\pi(t), iv \rangle, \qquad t \in T_{(x,v)}\mathcal{T}_1S^2.$$

3.3 Horizontal Lifting, Natural Curvatures, Gauge Transformation

Let $\gamma(s)$ be a smooth curve in $\mathbb{R}^3$ with nonvanishing curvature, $s \in [0, 2\pi]$ be its arclength parameter and $(\mathbf{T}, \mathbf{N}, \mathbf{B})$ be its Frenet frame. The unit tangent vector $\mathbf{T}$ describes a smooth curve $c(s)$ on the unit sphere. In this section, we construct the horizontal lifting of $c(s)$ in the bundle $\mathcal{T}_1S^2$, we introduce the notion of natural curvatures and discuss a geometric interpretation of the gauge transformation between the HM and NLS equations defined in section 1.5.

Given the curve $c(s) = \mathbf{T}(s)$ in S^2, we construct the unique horizontal lifting $\tilde{c}_o = (\mathbf{T}, v)$, with $v(0) = \mathbf{N}(0)$ (i.e. we require that the lifting to the Frenet frame and $\tilde{c}_o$ agree at $s = 0$). In order to compute the vector field tangent to $\tilde{c}_o$, we work locally in a coordinate patch $U \subset S^2$ and use the map $\phi : U \times S^1 \longrightarrow \pi^{-1}(U)$ to identify $\pi^{-1}(U)$ with the product space $U \times S^1$. In a local patch, there exist smooth real-valued functions $\beta(s)$, $\delta(s)$ for which $(\mathbf{T}, \mathbf{N}) \cong (\mathbf{T}, e^{i\beta(s)})$ and

$$\tilde{c}_o(s) \cong (\mathbf{T}(s), h(s)), \qquad \text{with } h(s) = e^{i(\beta(s)+\delta(s))} \in S^1.$$

If ∂_α is the unit tangent vector field to S^1 and $r : \mathbb{R} \to S^1$ is the exponential map $r(\alpha) = e^{i\alpha}$, then $\dfrac{dh}{ds} = \dfrac{d(\beta+\delta)}{ds}dr\left(\dfrac{d}{ds}\right) = \dfrac{d(\beta+\delta)}{ds}\partial_\alpha$. Therefore the velocity field of $\tilde{c}_o$ is

$$\frac{d\tilde{c}_o}{ds} = \left(k\mathbf{N}, \left(\frac{d\beta}{ds} + \frac{d\delta}{ds}\right)\partial_\alpha\right).$$

The lifting $\tilde{c}_o$ is horizontal if and only if the component of its tangent vector field along the vertical fibre vanishes, i.e. when $\dfrac{d\beta}{ds} = -\dfrac{d\delta}{ds}$. We will show in the next section (see equation (59)) that $\dfrac{d\beta}{ds} = \tau$; using this result and taking count of the initial condition, we require

$$\delta(s) = -\int_0^s \tau(u)du.$$

and we obtain the following expression for the corresponding horizontal lifting:

$$\tilde{c}_o(s) = \left(\mathbf{T}(s), e^{-i\int_0^s \tau(u)du}\mathbf{N}\right) = \left(\mathbf{T}, \cos(\int_0^s \tau du)\mathbf{N} - \sin(\int_0^s \tau du)\mathbf{B}\right) \quad (50)$$

(where we have introduced the binormal vector $\mathbf{B} = i\mathbf{N}$). Equation (50) defines the following orthonormal framing of the curve:

$$\mathbf{U} = \cos\left(\int_0^s \tau\, du\right)\mathbf{N} - \sin\left(\int_0^s \tau\, du\right)\mathbf{B}$$
$$\mathbf{V} = \sin\left(\int_0^s \tau\, du\right)\mathbf{N} + \cos\left(\int_0^s \tau\, du\right)\mathbf{B}. \tag{51}$$

This is known as the *natural frame* of the curve γ (see [4] for some history and discussion); it is indeed "natural" in the sense that it is the lifting of a parallel vector field along $\mathbf{T}(s)$ to the orthonormal frame bundle endowed with its canonical connection. The natural frame varies along the curve according to the following system of linear equations

$$\frac{d\mathbf{T}}{ds} = k\cos\left(\int_0^s \tau\, du\right)\mathbf{U} + k\sin\left(\int_0^s \tau\, du\right)\mathbf{V}$$
$$\frac{d\mathbf{U}}{ds} = -k\cos\left(\int_0^s \tau du\right)\mathbf{T} \tag{52}$$
$$\frac{d\mathbf{V}}{ds} = -k\sin\left(\int_0^s \tau du\right)\mathbf{T}.$$

Correspondingly, the components of the projection of the velocity field of the horizontal lifting with respect to $\mathbf{U}$ and $\mathbf{V}$ are called the *natural curvatures*, (we use the invariant expressions of ω_1, ω_2 derived at the end of the previous section)

$$k_{\mathbf{U}} = \omega_1|_{\tilde{c}_o}\left(\frac{d\tilde{c}_o}{ds}\right) = \left\langle k\mathbf{N}, \cos\left(\int_0^s \tau du\right)\mathbf{N} - \sin\left(\int_0^s \tau du\right)i\mathbf{N}\right\rangle = k\cos\left(\int_0^s \tau du\right)$$
$$k_{\mathbf{V}} = \omega_2|_{\tilde{c}_o}\left(\frac{d\tilde{c}_o}{ds}\right) = \left\langle k\mathbf{N}, \cos\left(\int_0^s \tau du\right)i\mathbf{N} + \sin\left(\int_0^s \tau du\right)\mathbf{N}\right\rangle = k\sin\left(\int_0^s \tau du\right).$$

We can now interpret the gauge transformation that relates HM and NLS in a geometric way: the procedure described above defines a map from curves in S^2 to curves in the complex plane

$$\mathcal{H} : \mathbf{T}(s) \longrightarrow \psi(s) = k(s)\exp\left(i\int_0^s \tau\, du\right).$$

The real and imaginary component of the complex function ψ are the components of the projection of the horizontal vector field onto $T_{\mathbf{T}}S^2$. If $\mathbf{T}$ satisfies the Heisenberg Model equation, then the complex function $\psi(s) = k(s)e^{i\int_0^s \tau\, du}$ is a solution of the cubic nonlinear Schrödinger equation. We summarise this discussion in the

Proposition 3.4. *For frozen t, the gauge transformation between the HM and NLS equations is the composition of the parallel transport of a vector $v(0)$ along the curve $c(s)$ described by $\mathbf{T}$ with its projection onto $T_c S^2$.*

Remarks.   1. If we parallel transport a different vector (there is a whole circle of choices) the change is reflected in a constant phase factor. The corresponding complex function has the form $\psi = k\exp\left[i\left(\int_0^s \tau\, du + \beta_0\right)\right]$

for some real constant β_0, which is still a solution of the NLS equation. Thus, choosing a particular initial tangent vector is equivalent to defining a map from the space of smooth curves in S^2 to the space of smooth complex-valued functions quotiented by the action of S^1.

2. A natural frame (a choice of a parallel vector field) is always defined, also when the curvature $k(s)$ vanishes, unlike the Frenet frame. In fact, once the vector $v(0)$ is chosen, its horizontal lifting is unique.

3. We also remark that while the Frenet lifting of a closed curve is always closed, the horizontal lifting need not be. Its holonomy is the element of the fibre S^1 which takes the initial value of the lifting to its value at $s = 2\pi$, therefore

$$\text{hol}(\tilde{c}_o) = \exp\left(i \oint \tau \, du\right).$$

Thus, the condition for a closed lifting (i.e. for trivial holonomy) is the following quantisation condition for the total torsion:

$$\oint \tau \, du = 2\pi j, \qquad \text{for } j \in \mathbb{Z}.$$

In particular, all closed liftings of parallel vectors along a solution of the HM are mapped to periodic solutions of the NLS equation, since $\psi(2\pi) = \psi(0) \exp\left(i \oint \tau \, du\right)$ while a general lifting is mapped to a quasi-periodic solution. A special case is when the curve γ itself lives on a sphere of radius r, then its total torsion vanishes [12] and its associated natural frame is closed.

3.4   Comparison between the HM and NLS Phase Spaces

In this section, we compare the phase spaces of the HM and the NLS equation, by computing the differential of the gauge transformation $\mathcal{H}$. Let $\mathcal{L}(S^2) = \{c : S^1 \to S^2\}$ be the space of smooth closed spherical curves, and let $\Psi = \{\psi : \mathbb{R} \to \mathbb{C}\}$ be the space of smooth complex-valued functions. We introduce the maps

$$\mathcal{L}(S^2) \xrightarrow{\mathcal{H}} \Psi$$
$$\Big\downarrow \pi$$
$$\Psi/S^1$$

where $\mathcal{H}(\mathbf{T}) = k \exp\left(i \int_0^s \tau du\right)$, and where π is the projection onto the quotient space Ψ/S^1. In this way, the composition $h = \pi \circ \mathcal{H}$ maps closed curves in S^2 to complex periodic functions (we factor out the holonomy of the horizontal lifting).

Next we compute the differential of $h = \pi \circ \mathcal{H}$ following a procedure used by J. Langer and R. Perline [34] in deriving a compact expression for the differential of the Hasimoto map. We express dh in terms the second symplectic operator for the NLS equation and relate it to the recursion operator for NLS. As a consequence, we show that the gauge transformation is a Poisson map which carries the Marsden–Weinstein Poisson structure for HM to the second Poisson structure for NLS inducing a corresponding shift in the hierarchy of Hamiltonian vector fields. Firstly, we recall a few facts about Poisson brackets.

Definition 3.5. A Poisson bracket on a manifold M is a bilinear skew-symmetric operation which endows the space of smooth functions on M with a Lie algebra structure; i.e. it is a bilinear map defined on smooth functions on M

$$\{\ ,\ \}:\ C^\infty(M) \times C^\infty(M) \longrightarrow M$$

satisfying the following properties:

$$\{f,g\} = -\{g,f\} \qquad \text{(Skew symmetry)}$$
$$\{f,gh\} = g\{f,h\} + \{f,g\}h \qquad \text{(Leibnitz rule)}$$
$$\{f,\{g,h\}\} + \{h,\{f,g\}\} + \{g,\{h,f\}\} = 0 \qquad \text{(Jacobi Identity)}$$

Remarks. 1. If M is a Riemannian manifold (i.e. if TM is endowed with an inner product $\langle\ ,\ \rangle_m$) then any skew-symmetric linear operator $J(m)$ on tangent vector fields induces a skew-symmetric bilinear map (not necessarily a Poisson bracket) of the form

$$\{f,g\}(m) = \langle J\nabla f, \nabla g\rangle(m), \qquad m \in M, \tag{53}$$

(in this case the Leibnitz rule is automatically satisfied).

2. If equation (53) defines a non-degenerate Poisson bracket, then the 2-form $\langle J\ ,\ \rangle$ is a symplectic form on M (the Jacobi identity is equivalent to such form being closed) and the operator J is called a symplectic operator. For example, the symplectic structure introduced on $\mathcal{L}(S^2)$ in section 1.2 is obtained in this way by using the inner product

$$\langle X, Y\rangle_{\mathcal{L}(S^2)}(\mathbf{T}) = \frac{1}{2\pi} \int_0^{2\pi} (X \cdot Y)(s)ds$$

and the skew-symmetric operator

$$J|_{\mathbf{T}} X = \mathbf{T} \times X \qquad \mathbf{T} \in S^2,\ \ X \in T_{\mathbf{T}}\mathcal{L}(S^2).$$

The resulting Poisson structure is known as the Marsden–Weinstein Poisson bracket

$$\{f,g\}_{\text{MW}}(\mathbf{T}) = \frac{1}{2\pi} \int_0^{2\pi} \nabla g \cdot (\mathbf{T} \times \nabla f)(s)ds.$$

If $f : M \to N$ is a differentiable map, we can obtain an explicit formula for its differential that is suitable for computations, by using the notion of directional derivative. Let $g : N \to \mathbb{R}$ be an arbitrary differentiable function on N, then the action of the vector field V on the real-valued function $g \circ f$ is described by $V_m(g \circ f) =: V[g \circ f](m)$, where the right-hand side is the directional derivative of $g \circ f$ along V at the point m. Thus, we write the following expression for $df : TM \to TN$:

$$df(V)[g] = V(g \circ f).$$

Before computing the differential of $\mathcal{H}$ we need the following variational formulas:

Lemma 3.6. *Consider a family of arclength parametrised curves* $\gamma(w, s)$: $(-\epsilon, \epsilon) \times S^1 \to \mathbb{R}^3$ *and let* k *and* τ *be the curvature and torsion of* $\gamma(0, s)$. *Let* $\mathbf{T}(w, s)$ *be the family of spherical curves described by the unit tangent vector of* γ *and denote with* $W = \mathbf{T}_w|_{(0,s)}$ *the variation vector field along* $\mathbf{T}$. *Then, the variation of* k *and* τ *along* W *are given by*

$$W(k) = \langle W_s, \mathbf{N} \rangle \tag{54}$$
$$W(\tau) = \langle \nabla_{\mathbf{N}} W, \mathbf{B} \rangle_s + k \langle W, \mathbf{B} \rangle. \tag{55}$$

In formulas (54) and (55) ∇ is the symbol of covariant differentiation and the subscript s denotes the derivative with respect to s (i.e. the covariant derivative $\nabla_{k\mathbf{N}}$ along the velocity vector of $\mathbf{T}$).

Proof. Since the velocity field $\mathbf{T}_s|_{w=0} = k\mathbf{N}$ of $\mathbf{T}$ commutes with the variation W, we have the sequence following identities

$$0 = [W, V] = \nabla_W(k\mathbf{N}) - \nabla_{k\mathbf{N}}W = W(k)\mathbf{N} + k\nabla_W\mathbf{N} - k\nabla_{\mathbf{N}}W. \tag{56}$$

Given $k^2 = \langle k\mathbf{N}, k\mathbf{N} \rangle$, we take the covariant derivative of both sides of this equation in the direction of W and use (56) to obtain

$$2kW(k) = 2\langle \nabla_W(k\mathbf{N}), k\mathbf{N} \rangle = 2k\langle \nabla_{k\mathbf{N}}W, \mathbf{N} \rangle = 2k^2\langle \nabla_{\mathbf{N}}W, \mathbf{N} \rangle.$$

Formula (54) follows directly together with the following useful relations,

$$[W, \mathbf{N}] = -\langle \nabla_{\mathbf{N}}W, \mathbf{N} \rangle \mathbf{N} \tag{57}$$
$$\nabla_W\mathbf{N} = \nabla_{\mathbf{N}}W - \langle \nabla_{\mathbf{N}}W, \mathbf{N} \rangle \mathbf{N} = \langle \nabla_{\mathbf{N}}W, \mathbf{B} \rangle \mathbf{B}. \tag{58}$$

As for $W(\tau)$, we write $\tau^2 = \langle \tau\mathbf{B}, \tau\mathbf{B} \rangle$, where $\tau\mathbf{B} = \nabla_{k\mathbf{N}}\mathbf{N}$. Using the chain rule and formula (56), we compute

$$2\tau W(\tau) = 2\tau\langle \nabla_W(\nabla_{k\mathbf{N}}\mathbf{N}), \mathbf{B} \rangle =$$
$$2\tau\langle \left(\nabla_{k\mathbf{N}}\nabla_W + \nabla_{[W, k\mathbf{N}]} + R(W, k\mathbf{N}) \right) \mathbf{N}, \mathbf{B} \rangle.$$

We have used the equation

$$\nabla_X\nabla_Y Z - \nabla_Y\nabla_X Z - \nabla_{[X,Y]}Z = R(X, Y)Z,$$

where $R(X, Y)$ is the Riemann curvature tensor. For the unit sphere we have $R(X, Y)Z = \langle Z, Y \rangle X - \langle Z, X \rangle Y$, there follows

$$W(\tau) = \langle \nabla_{k\mathbf{N}}\nabla_W\mathbf{N} + kW - \langle \mathbf{N}, W \rangle k\mathbf{N}, \mathbf{B} \rangle$$
$$= \langle \nabla_{k\mathbf{N}}\left(\langle \nabla_{\mathbf{N}}W, \mathbf{B} \rangle \mathbf{B} \right), \mathbf{B} \rangle + k\langle W, \mathbf{B} \rangle$$
$$= k\mathbf{N}(\langle \nabla_{\mathbf{N}}W, \mathbf{B} \rangle) + k\langle W, \mathbf{B} \rangle$$
$$= \langle \nabla_{\mathbf{N}}W, \mathbf{B} \rangle_s + k\langle W, \mathbf{B} \rangle. \qquad \square$$

Remark. In the set-up of the previous section, $\mathbf{N}$ belongs to the vertical fibre over $\mathbf{T}$ in $\mathcal{T}_1 S^2$. Then, in a local representation, $\mathbf{N}$ is identified with some element $e^{i\beta}$ of S^1, where β is a smooth real-valued function. Thus, the action of the vector field W on $\mathbf{N}$ can be written in terms of the unit tangent vector along the vertical fibre ∂_α as $\nabla_W\mathbf{N} = W(\beta)\partial_\alpha$. Comparing this expression with formula (58) and identifying $\mathbf{B}$ with ∂_α ($\mathbf{B}$ is a unit vector tangent to the fibre at $\mathbf{N}$), we compute $W(\beta) = \langle \nabla_{\mathbf{N}} W, \mathbf{B} \rangle$. In particular, if $W = k\mathbf{N}$, the velocity field of the curve, we obtain

$$\nabla_{k\mathbf{N}}(\beta) = \beta_s = \tau. \tag{59}$$

Next we compute the differential of the map $h = \pi \circ \mathcal{H}$ and re-express it in terms of the second Poisson structure for NLS. We first recall the following result which can be found in [40] and [20]:

Theorem 3.7. *There exist two compatible symplectic operators for the periodic NLS equation:*

$$\tilde{J}\phi \;=\; i\phi \tag{60}$$

$$\tilde{K}_\psi \phi \;=\; \phi_s + \frac{1}{4}\psi \left(\int_0^s + \int_{2\pi}^s \right) [\phi \bar{\psi} - \bar{\phi}\psi]\, du, \tag{61}$$

with respect to the inner product

$$\langle \phi_1, \phi_2 \rangle|_\psi = \frac{1}{2\pi} \int_0^{2\pi} (\phi_1 \bar{\phi}_2 + \bar{\phi}_1 \phi_2)\, du \qquad \phi_1, \phi_2 \in T_\psi \Psi. \tag{62}$$

The associated recursion operator $\tilde{\mathcal{R}} = \tilde{K}_\psi \tilde{J}^{-1}$ *generates the following infinite hierarchy of Poisson structures*

$$\{f,g\}_n = \langle \tilde{\mathcal{R}}^n \tilde{J} \nabla f, \nabla g \rangle, \quad f, g \in \Psi \tag{63}$$

and the sequence of Hamiltonian vector fields

$$X_n = \tilde{\mathcal{R}}^n X_0, \qquad X_0 = \psi_x$$

Remark. It is easy to check that the NLS equation can be written as a Hamiltonian system with respect to two different Hamiltonians using the operators $\tilde{J}$ and $\tilde{K}$. Whenever this happens and the two operators are compatible ($\tilde{J}$ and $\tilde{K}$ are compatible if $\tilde{J} + \tilde{K}$ defines a Poisson bracket) we say that $\tilde{J}$ and $\tilde{K}$ define a bi-Hamiltonian structure on Ψ. A bi-Hamiltonian formulation appears to be one of the unifying properties of soliton equations. Given a bi-Hamiltonian structure, one can formally construct an infinite sequence of constants of motion; it is still an open question though, whether every completely integrable equation admits a bi-Hamiltonian structure (this very question does not yet have a rigorous answer for the periodic HM itself!). The reader who is interested in this aspect of integrability is referred to [45, 40, 10].

Finally, we compute the differential of $\mathcal{H}(\mathbf{T}) = k \exp\left(i \int_0^s \tau\, du \right)$. Using the chain rule and Lemma 3.6, we write

$$d\mathcal{H}(W) = W(k) e^{i \int_0^s \tau du} + ik \left(\int_0^s W(\tau) du \right) e^{i \int_0^s \tau du} =$$

$$\left(\langle W_s, \mathbf{N} \rangle + i \langle W_s, \mathbf{B} \rangle \right) e^{i \int_0^s \tau\, du} + i\psi \int_0^s \langle W, k\mathbf{B} \rangle\, du + ic\psi,$$

c is a constant of integration, and since the kernel of $d\pi$ contains vector fields of the form $c\psi$, we can write equivalently

$$d\mathcal{H}(W) = \left(\langle W_s, \mathbf{N} \rangle + i \langle W_s, \mathbf{B} \rangle \right) e^{i \int_0^s \tau\, du} + \frac{i}{2}\psi \left(\int_0^s + \int_{2\pi}^s \right) \langle W, k\mathbf{B} \rangle\, du.$$

We rewrite $W = g\mathbf{U} + h\mathbf{V}$ on the basis of the natural frames and introduce the complex vector field $\eta(W) = g + ih$. Using the Darboux equations (52) for the components of the natural frame, we can rewrite the above expression as

$$d\mathcal{H}(W) = \frac{d}{ds}\eta(W) + \frac{1}{4}\psi\left(\int_0^s + \int_{2\pi}^s\right)[\eta(W)\bar{\psi} - \bar{\eta}(W)\psi]\,du. \qquad (64)$$

The right-hand side of (64) is nothing but the second Poisson operator K_ψ for the NLS equation acting on the tangent vector $\eta(W)$; we summarise this in the following

Proposition 3.8. *The differential of the composed map* $h : \mathcal{L}(S^2) \to \Psi/S^1$, $h = \pi \circ \mathcal{H}$ *is given by:*

$$dh = d\pi \circ \tilde{K}_{h(\mathbf{T})} \circ \eta. \qquad (65)$$

An important consequence of formula (65) is that h preserves Poisson structures, in particular it carries the Marsden–Weinstein Poisson bracket into the second Poisson bracket for the NLS equation. In order to show this, we need the following technical result:

Lemma 3.9.

$$\begin{aligned}
\nabla(f \circ h)|_{\mathbf{T}} &= dh^*(\nabla f)|_{h(\mathbf{T})}, \qquad f \in \Psi/S^1 \\
\eta \circ J \circ \eta^* &= \tilde{J}.
\end{aligned}$$

Here η^* is the adjoint of η with respect to the inner product (62), J is the symplectic operator $\mathbf{T}\times$ of HM and $\tilde{J}$ is simply multiplication by i. The proof of this lemma is a one-line computation (in a similar context see [34]) and it is left to the reader.

Corollary 3.10. *The map h is a Poisson map and the following relation holds*

$$\{f \circ h, g \circ h\}_{MW}(\mathbf{T}) = \{f, g\}_2(h(\mathbf{T})), \qquad (66)$$

for $f, g \in \Psi/S^1$.

Proof.

$$\begin{aligned}
\{f \circ h, g \circ h\}_{\mathrm{MW}}(\mathbf{T}) &= \langle J\nabla(f \circ h), \nabla(g \circ h)\rangle(\mathbf{T}) \\
&= \langle J \circ dh^*\nabla f, dh^*\nabla g\rangle(h(\mathbf{T})) \\
&= \langle dh \circ J \circ dh^*\nabla f, \nabla g\rangle(h(\mathbf{T})) \\
&= \langle d\pi \circ \tilde{K} \circ \eta \circ J \circ \eta^* \circ \tilde{K}^* \circ d\pi^*\nabla f, \nabla g\rangle(h(\mathbf{T})) \\
&= \langle \tilde{K} \circ \tilde{J} \circ \tilde{K}^* d\pi^*\nabla f, d\pi^*\nabla g\rangle(h(\mathbf{T})) \\
&= \langle \tilde{R}^2\tilde{J}\nabla(f \circ \pi), \nabla(g \circ \pi)\rangle(\mathcal{H}(\mathbf{T})) \\
&= \{f \circ \pi, g \circ \pi\}_2(\mathcal{H}(\mathbf{T})) = \{f, g\}_2(h(\mathbf{T})). \qquad \square
\end{aligned}$$

(We used the skew-symmetry of $\tilde{K}$ and the identity $\tilde{J}^{-1} = \tilde{J}$.) Therefore, we can view the NLS equation and the HM as the same Hamiltonian system written with respect to two different Poisson structures which belong to the same integrable hierarchy.

4   Bäcklund Transformation and Immersed Knots

A fundamental algebraic property of soliton equations is the fact that they arise as compatibility conditions for a pair of linear operators. In section 3.2 we discovered a deep relation among the infinite family of multi-phase solutions, the nonlinear equation and its associated Lax pair in the case of the NLS and HM equations. The zero-curvature formulation of integrable partial differential equations is at the basis of the inverse scattering method for finding exact solutions (see for example [20]). Moreover, the hierarchy of conserved quantities and the related conservation laws can be constructed in a purely algebraic way through the knowledge of the Lax pair, independently of global properties such as the boundary conditions and the differentiability class of the solutions. Another seemingly universal feature of soliton equations, also intimately connected to the presence of a Lax pair, is the existence of Bäcklund transformations. A Bäcklund formula produces more complicated solutions of the nonlinear equation from a simpler solution for which the pair of linear systems can be solved exactly. The close relation among inverse scattering, the infinite number of conservation laws and Bäcklund transformations was first discussed in 1975 in an elementary and intriguing paper by M. Wadati, H. Sanuki and K. Konno [55]. In their 1983 papers [21, 22, 23], H. Flaschka, A. Newell and T. Ratiu developed a general algebraic framework in which to interpret the unifying properties of soliton equations. The central idea is to view the hierarchy of integrable equations as a sequence of commuting flows on an infinite-dimensional loop algebra (a Kac–Moody Lie algebra, see the original papers or Chapter 5 in [44]), and to think of the solution as a function of an infinite number of independent variables, none of which play a distinguished role (x is now just one of the time variables).

In this section, we will simply focus on the use of Bäcklund transformations to construct new interesting solutions of soliton equations from simpler solutions. It is instructive to start with describing the Bäcklund formula for the nonlinear Schrödinger equation

$$iq_t = q_{xx} + 2|q|^2 q, \tag{67}$$

to be considered together with its associated Lax pair

$$\frac{\partial}{\partial x} \mathbf{F}_{\mathrm{NS}} = \left[i\lambda\sigma_3 + i \begin{pmatrix} 0 & \bar{q} \\ q & 0 \end{pmatrix} \right] \mathbf{F}_{\mathrm{NS}}$$

$$\frac{\partial}{\partial t} \mathbf{F}_{\mathrm{NS}} = \left[(2i\lambda^2 - i|q|^2)\sigma_3 + \begin{pmatrix} 0 & 2i\lambda\bar{q} + \bar{q}_x \\ 2i\lambda q - q_x & 0 \end{pmatrix} \right] \mathbf{F}_{\mathrm{NS}}. \tag{68}$$

The Bäcklund transformation for equation (67) is obtained in the following way (the proof is a direct verification, see [50] for its derivation in the context of gauge transformations). Let (ψ_+, ψ_-) be two independent solutions of the linear system (68) at (q, ν). We construct the following quantities (where $c_\pm$ are complex arbitrary constants):

$$\boldsymbol{\psi} = c_+\boldsymbol{\psi}_+ + c_-\boldsymbol{\psi}_-$$

$$N_{\mathrm{NS}} = \begin{pmatrix} \psi_1 & -\bar{\psi}_2 \\ \psi_2 & \bar{\psi}_1 \end{pmatrix}$$

$$G_{\mathrm{NS}} = N_{\mathrm{NS}} \begin{pmatrix} \lambda - \nu & 0 \\ 0 & \lambda - \bar{\nu} \end{pmatrix} N_{\mathrm{NS}}^{\;-1}.$$

Then

$$\mathbf{F}_{\mathrm{NS}}^{(1)}(x, t; \lambda, \nu) = G_{\mathrm{NS}}(\lambda, \nu, \boldsymbol{\psi})\mathbf{F}_{\mathrm{NS}}(x, t; \lambda) \tag{69}$$

solves equation (68) at $(q^{(1)}, \lambda)$, where

$$q^{(1)} = q + 2(\nu - \bar{\nu})\frac{\psi_1 \bar{\psi}_2}{|\psi_1|^2 + |\psi_2|^2} \tag{70}$$

is the corresponding new solution of the NLS equation.

Bäcklund transformations can be used to generate homoclinic orbits of a given solution. The importance of understanding the homoclinic manifolds of N-phase solutions that possess nonlinear instabilities is discussed in the work of N. Ercolani, G. Forest and D. McLaughlin [16, 17, 18]. Unstable N-phase solutions, just like saddle points and periodic orbits of saddle-type in finite-dimensional integrable systems, reside on singular level sets in phase space. A complete topological description of the singularities of the foliation of the space of N-phase solutions can be achieved by means of Bäcklund transformations. Bäcklund formulas produce explicitly the homoclinic manifolds and thus label the underlying level sets. Moreover, using Bäcklund transformations, one is able to construct both the tangent and normal vector fields to the level set of a given solution and to generate families of solutions residing on the same level set. In fact [17], a finite number of iterated Bäcklund transformations generates the entire level set of a given N-phase solution.

These ideas are hereby applied to curves in space, a context in which the relevant structures can be more easily visualised. We will use Bäcklund formulas in a two-fold way. Firstly, we will construct curves which are homoclinic to a given curve and conjecture that such homoclinic curves play a role in separating different classes of knots. Secondly, we will use the theory of Bäcklund transformations to understand the symmetries of a given curve. All the computations will be carried out in the simplest case of planar circles, where interesting results already arise.

4.1   The Bäcklund Formula for the Heisenberg Model

We first derive the Bäcklund transformation for the HM equation and the expression for the corresponding transformed curve [8]. To this end, we use the Bäcklund formula (69) and the expression of the gauge transformation between the HM and NLS.

Proposition 4.1 (HM Bäcklund transformation). *Let* $\phi = c_+\phi_+ + c_-\phi_-$ *be a complex linear combination of linearly independent solutions of the Lax pair (24) at* (S, ν). *We construct the matrix of gauge transformation*

$$G(\lambda, \nu, \phi) = N \begin{pmatrix} \frac{\nu - \lambda}{\nu} & 0 \\ 0 & \frac{\bar{\nu} - \lambda}{\bar{\nu}} \end{pmatrix} N^{-1} \tag{71}$$

with

$$N = \begin{pmatrix} \phi_1 & -\bar{\phi}_2 \\ \phi_2 & \bar{\phi}_1 \end{pmatrix}. \tag{72}$$

Then, if $\mathbf{F}$ solves the linear system (24) at (S, λ), the vector

$$\mathbf{F}^{(1)}(x, t; \lambda, \nu) = G(\lambda, \nu, \phi)\mathbf{F}(x, t; \lambda) \tag{73}$$

solves (24) at $(S^{(1)}, \lambda)$ and the new solution of the HM equation is given by the Bäcklund formula

$$S^{(1)}(x, t) = N \begin{pmatrix} e^{-i\theta} & 0 \\ 0 & e^{i\theta} \end{pmatrix} N^{-1} S(x, t) N \begin{pmatrix} e^{i\theta} & 0 \\ 0 & e^{-i\theta} \end{pmatrix} N^{-1} \tag{74}$$

with $e^{i\theta} = \dfrac{\nu}{|\nu|}$.

Remarks. 1. The matrix $U = N \begin{pmatrix} e^{i\theta} & 0 \\ 0 & e^{-i\theta} \end{pmatrix} N^{-1}$ is unitary, hence the solution $S^{(1)}$ is Hermitian and has zero trace (and its determinant is 1). The corresponding vector $\mathbf{T}^{(1)}$ is therefore obtained by a rotation of the tangent vector $\mathbf{T}$ of the original curve by a constant angle $\theta = \frac{\nu}{|\nu|}$ depending only on the eigenvalue parameter ν, around the instantaneous axis of rotation

$$\left(\frac{2\,\mathrm{Re}(\phi_1\bar{\phi}_2)}{|\phi_1|^2 + |\phi_2|^2}, -\frac{2\,\mathrm{Im}(\phi_1\bar{\phi}_2)}{|\phi_1|^2 + |\phi_2|^2}, \frac{|\phi_1|^2 - |\phi_2|^2}{|\phi_1|^2 + |\phi_2|^2} \right)^T.$$

2. If $\nu \in \mathbb{R}$, then $S^{(1)} = S$ and the Bäcklund transformation is the identity. If ν is chosen such that the corresponding eigenfunction ϕ is periodic or antiperiodic in the x-variable (such ν is called a periodic/antiperiodic eigenvalue), then the corresponding solution is periodic (i.e. the transformed curve is closed). For any other choice of $\nu \in \mathbb{C}$, formula (74) produces quasi-periodic solutions of the nonlinear equation.

Proof. [8] The gauge formula allows us to write the eigenfunction of the NLS Lax pair at (q, λ) in terms of the solution of the HM linear system at (S, λ) as $\mathbf{F}_{\mathrm{NS}} = V\mathbf{F}$, where V is the gauge matrix introduced in section 1.5. The eigenvector $\mathbf{F}_{\mathrm{NS}}^{(1)}$ of the NLS Lax pair at $(q^{(1)}, \lambda)$ can similarly be expressed as $\mathbf{F}_{\mathrm{NS}}^{(1)} = V^{(1)}\mathbf{F}^{(1)}$, where $V^{(1)}$ is the new gauge matrix. Since $V^{(1)}$ solves the linear system for NLS at $(q^{(1)}, 0)$, it can be written as $V^{(1)} = G_{\mathrm{NS}}|_{\lambda=0} V$ and we can write the new eigenfunction for the HM

$$\mathbf{F}^{(1)} = V^{-1} \left(G_{\mathrm{NS}}|_{\lambda=0} \right)^{-1} G_{\mathrm{NS}} V \mathbf{F}. \tag{75}$$

We make the important observation that the expression of $\mathbf{F}^{(1)}$ is independent of the gauge transformation, and therefore it does not depend on the NLS eigenfunctions. In order to show this, we use the fact that V is a unitary matrix with determinant 1, and compute

$$V^{-1}N_{\mathrm{NS}} = V^{-1} \begin{pmatrix} \psi_1 & -\bar{\psi}_2 \\ \psi_2 & \bar{\psi}_1 \end{pmatrix} = \begin{pmatrix} \phi_1 & -\bar{\phi}_2 \\ \phi_2 & \bar{\phi}_1 \end{pmatrix} = N, \tag{76}$$

with $\psi = V\phi$ and $\phi = c_+\phi_+ + c_-\phi_-$ a complex linear combination of independent eigenfunctions of the linear system (24). Using (76) to simplify equation (75), we obtain expression (73) for the new eigenfunction. Substituting the expression for $\mathbf{F}^{(1)}$ in the equation

$$\frac{\partial}{\partial x}\mathbf{F}^{(1)} = i\lambda S^{(1)}\mathbf{F}^{(1)}, \tag{77}$$

we derive the following formula for the transformed potential

$$S^{(1)} = \frac{1}{i\lambda}\frac{\partial G}{\partial x}G^{-1} + GSG^{-1}. \tag{78}$$

The dependence on λ is only apparent, by computing the x-derivative of G and rearranging terms we can reduce (78) to the form of equation (74). $\quad\square$

In order to derive the formula for the corresponding curve in space, we use the representation of the curve in terms of the fundamental solution of the linear problem.

$$\Gamma_{\text{new}} = -i\left.\frac{\partial}{\partial\lambda}\right|_{\lambda=0}\Phi_{\text{new}}\Phi_{\text{new}}^{-1}|_{\lambda=0} + C, \tag{79}$$

where C is a Hermitian x-independent. Using expression (73) for Φ_{new}, we find (up to an x-independent translation)

Corollary 4.2. *The Bäcklund transformation for the filament equation is described by the following formula (in matrix representation):*

$$\Gamma_{\text{new}} = \Gamma + \frac{\text{Im}(\nu)}{|\nu|^2}V \tag{80}$$

with

$$V = N\sigma_3 N^{-1} \tag{81}$$

Formula (80) reduces to the identity when the eigenvalue ν is real, and it produces bounded solutions of the filament equation (not necessarily periodic in the space variable x) for any complex ν. In fact, directly from the formula for the new curve we deduce the following confinement result:

Proposition 4.3.

$$\|\Gamma\| - \frac{|\text{Im}(\nu)|}{|\nu|^2} \leq \|\Gamma_{\text{new}}\| \leq \|\Gamma\| + \frac{|\text{Im}(\nu)|}{|\nu|^2} \tag{82}$$

($\|\cdot\|$ is the standard Euclidean norm in $\mathbb{R}^3$). The transformed curve is confined to the interior of a sphere if $\dfrac{|\text{Im}(\nu)|}{|\nu|^2} \geq \max_x \|\Gamma(x)\|$ or to the interior of a torus if $\dfrac{|\text{Im}(\nu)|}{|\nu|^2} < \max_x \|\Gamma(x)\|$.

Cieslinski et al. [11] obtained this same confinement result (which is of interest in the context of vortex dynamics) by reconstructing the curve from a basis of eigenfunctions of the NLS Lax pair and by using directly the NLS Bäcklund transformation.

4.2   An Application to Planar Circles

The simplest solutions which possess homoclinic instabilities are planar circles. Their tangent vector fields are fixed points of the Continuous Heisenberg Model, therefore the expression of the associated curve can be chosen to be time-independent. We consider k copies of a circle lying in the (x, y)-plane with non-zero curvature $k \in \mathbb{Z}$ (an integral curvature is required for the circle to be closed). If x denotes the arclength parameter, the Hermitian matrix which represents the unit tangent vector is

$$S_0(x, t) = \begin{pmatrix} 0 & e^{-ikx} \\ e^{ikx} & 0 \end{pmatrix}. \tag{83}$$

Following [16, 17, 18], we make use of the Floquet spectrum of the linear operator

$$\mathcal{L}(S, \lambda) = -\begin{pmatrix} 1 & 0 \\ 0 & -1 \end{pmatrix} \frac{d}{dx} + i\lambda S \tag{84}$$

(the spatial part of the HM Lax pair) in order to characterise the level set on which S_0 resides. The fundamental solution matrix $\Phi(x, y; S, \lambda)$ is defined by the following properties

$$\mathcal{L}(S, \lambda)\Phi = 0,$$
$$\Phi(x, x; S, \lambda) = \begin{pmatrix} 1 & 0 \\ 0 & 1 \end{pmatrix}. \tag{85}$$

We introduce the *transfer matrix* $\Phi(x + 2\pi, x; S, \lambda)$ as the transformation which takes a solution of (84) evaluated at x to its value after a period 2π. It is easy to check that the transfer matrix has determinant 1, as a consequence its eigenvalues are completely characterised in terms of its trace

$$\Delta(S, \lambda) = \text{Trace}\left[\Phi(x + 2\pi, x; S, \lambda)\right]. \tag{86}$$

The quantity $\Delta(S, \lambda)$ is called the Floquet discriminant of the linear operator $\mathcal{L}$. Recalling that the spectrum of a linear operator is the set of eigenvalues for which the corresponding eigenfunction is bounded in x, then the spectrum $\sigma(\mathcal{L})$ of $\mathcal{L}$ is the set of λ's for which the eigenvalues of the transfer matrix lie on the unit circle. In terms of the trace of the Floquet discriminant,

$$\sigma(\mathcal{L}) = \{\lambda \in \mathbb{C} \mid \Delta(S, \lambda) \in \mathbb{R} \text{ and } -2 \leq \Delta(S, \lambda) \leq 2\}.$$

The discriminant has the following properties (property 1 follows from the analyticity of the fundamental solution of the linear system, the verification of property 2 is a simple computation):

Theorem 4.4. *If S solves the Continuous Heisenberg Model then,*

1. *$\Delta(S, \lambda)$ is an analytic function of λ.*

2. *$\Delta(S, \lambda)$ is invariant under the evolution, i.e.*

$$\frac{d}{dt}\Delta(S(t), \lambda) = 0. \tag{87}$$

Remark. Since the discriminant is a constant of motion which depends analytically on the complex parameter λ, Δ provides us with a 1-complex parameter family of invariants. In fact, one can extract the usual hierarchy of constants of motion from the coefficients of the asymptotic expansion of the discriminant at a distinguished value of λ [20]. Moreover, we can characterise the level set on which a given solution S_0 resides as its isospectral set (the set of all potentials which have the same spectrum as S_0).

We introduce the following distinguished points of $\sigma(\mathcal{L})$:

critical points λ_c:

$$\frac{d}{d\lambda}\Delta(S,\lambda)\bigg|_{\lambda=\lambda_c} = 0,$$

periodic (antiperiodic) points $\lambda_\pm$:

$$\Delta(S,\lambda)|_{\lambda=\lambda_\pm} = \pm 2,$$

multiple points λ_m:

$$\Delta(S,\lambda)|_{\lambda=\lambda_m} = \pm 2$$
$$\frac{d}{d\lambda}\Delta(S,\lambda)\bigg|_{\lambda=\lambda_m} = 0.$$

The periodic (antiperiodic) points are associated with periodic (antiperiodic) eigenfunctions. The Bäcklund transformation at one such point produces a solution which is periodic in x. The existence of complex multiple points (we will consider only double points, i.e. critical points of multiplicity 2) is related to the presence of linear instabilities. The level set of a solution whose spectrum contains complex double points is saddle-like and the corresponding homoclinic orbits can be constructed by means of Bäcklund transformations. We refer the reader to references [16, 17, 18] for the discussion of these general results. We will instead infer these conclusions for the simplest example of planar circles and their associated Bäcklund transformations.

The HM Lax pair can be easily solved when S is the matrix S_0 representing a multiple covered circle as in (83). The associated Floquet discriminant is

$$\Delta(S_0,\lambda) = \Delta(k,\lambda) = (-1)^k 2\cos(\delta\pi) \quad \text{with} \quad \delta = \sqrt{k^2 + 4\lambda^2},$$

and the corresponding spectrum is (see Figure 3)

$$\sigma(S_0,\lambda) = \{\lambda \in \mathbb{C} \mid \cos(\delta\pi) \in \mathbb{R}\} = \mathbb{R} \cup \{i\mu \mid \mu \in \mathbb{R} , \ -k/2 \le \mu \le k/2\}.$$

The critical points are the set of solutions of

$$\frac{d}{d\lambda}\Delta = -\sin(\delta\pi)\frac{4\lambda}{\sqrt{k^2+4\lambda^2}} = 0.$$

All of them are double points except for $\lambda = 0$ which is a critical point of multiplicity 4. There is a finite number of complex double points (in this case they are purely imaginary) given by

$$\lambda_n = \pm\frac{i}{2}\sqrt{k^2 - n^2} \qquad n = 1, 2, \ldots, k-1.$$

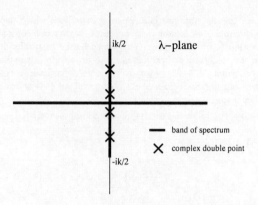

Figure 3: The spectral configuration of S_0.

We now implement the Bäcklund transformation derived in the previous section and construct the corresponding new solutions of the HM. We choose two linearly independent solutions of the linear problem at (S_0, ν) as follows,

$$\phi_- = e^{-\frac{i}{2}\delta(x+2\nu t)} \begin{pmatrix} -2\frac{\nu}{k+\delta}e^{-\frac{i}{2}kx} \\ e^{\frac{i}{2}kx} \end{pmatrix}, \quad \phi_+ = e^{\frac{i}{2}\delta(x+2\nu t)} \begin{pmatrix} e^{-\frac{i}{2}kx} \\ \frac{\delta-k}{2}e^{\frac{i}{2}kx} \end{pmatrix}.$$

ϕ_+ and ϕ_- are called Baker eigenfunctions and are uniquely characterised by their expression in terms of an exponential phase factor multiplied by a bounded (periodic) vector function.

We distinguish two classes of Bäcklund transformations which generate qualitatively different solutions in the level set of S_0:

Type I Bäcklund Transformation. If we take ϕ in formula (80) to coincide with one of the Baker eigenfunctions at the eigenvalue ν, we obtain a periodic solution for every choice of $\nu \in \mathbb{C}$. Moreover such solution belongs to the same class S_0; in particular, for $\phi = \phi_+$ we obtain the following family of planar circles

$$\Gamma_1(x,\nu) = \begin{pmatrix} 0 & \frac{i}{k}e^{-ikx} \\ -\frac{i}{k}e^{ikx} & 0 \end{pmatrix} + \frac{\mathrm{Im}(\nu)}{|\nu|^2(4|\nu|^2+|\delta-k|^2)} \begin{pmatrix} 4|\nu|^2-|\delta-k|^2 & 4\nu(\bar{\delta}-k)e^{-ikx} \\ 4\bar{\nu}(\delta-k)e^{ikx} & 4|\nu|^2-|\delta-k|^2 \end{pmatrix} \tag{88}$$

parametrised by the complex parameter ν.

Remark. In order to interpret the expression of the new curve as the effect of the action of a symmetry, we compute an infinitesimal Bäcklund transformation as $\mathrm{Im}(\nu) \to 0$. In this limit, $\Gamma_1 - \Gamma_0$ approximates a family of vector fields along the original curve Γ_0 of the form

$$\mathbf{V}(x,\nu) = c_1(\nu)(0,0,1)^T + c_2(\nu)\mathbf{S}_0, \tag{89}$$

where c_i's are constants depending on ν. Formula (89) describes a family of Killing vector fields (infinitesimal isometries) for the the original circle: the sum of a constant translation field and a rotation. Already in this simple example we see that a type I Bäcklund transformation realises the symmetries of the level set of the solution S_0. Such symmetries manifest themselves at the curve level as infinitesimal isometries.

Type II Bäcklund Transformation We now take $\phi = c_- \phi_- + c_+ \phi_+$ to be a general complex linear combination of Baker eigenfunctions. In this case, the transformed curve is closed if ν is a periodic/antiperiodic eigenvalue. When ν is chosen to be one of the complex double points $\nu = i\mu_n = \pm\frac{i}{2}\sqrt{k^2 - n^2}$, $n = 1, \ldots, k - 1$, setting $c_+/c_- = \rho e^{i\theta}$, we obtain the following formula:

$$\Gamma_2(x, t; \mu_n) = \begin{pmatrix} 0 & \frac{i}{k}e^{-ikx} \\ -\frac{i}{k}e^{ikx} & 0 \end{pmatrix} +$$
$$\frac{1}{\mu_n}\begin{pmatrix} f(x,t) - f(0,t) & \overline{g(x,t) - g(0,t)} \\ g(x,t) - g(0,t) & -(f(x,t) - f(0,t)) \end{pmatrix},$$

where

$$f(x,t) = -\frac{1}{2\mu_n}\frac{h'(t)}{kh(t) + 8\rho\mu_n^2\sin(nx + \theta)},$$

$$g(x,t) = -2i\frac{\mu_n}{k}e^{-ikx} - 4\rho n\mu_n\frac{\cos(nx + \theta) + i\frac{n}{k}\sin(nx + \theta)}{kh(t) + 8\rho\mu_n^2\sin(nx + \theta)}e^{-ikx},$$

$$h(t) = [k + n + (k - n)\rho^2]\sinh(n\mu t) + [k + n - (k - n)\rho^2]\cosh(n\mu t).$$

Since $\lim_{t\to\pm\infty} \Gamma_2(x, t) = \Gamma_0(x)$, the new curve Γ_2 is a homoclinic orbit for the original solution (this also implies that multiple covers of circles are unstable solutions of the filament equation). Moreover, we conclude that Γ_2 resides on the same level set as Γ_0, indeed spectral configurations such as the one depicted in Figure 3 describe both the target solution and its homoclinic orbits.

Figure 4 shows a sequence of time frames of the evolution of the curve Γ_2 homoclinic to a 6-fold circle. The new curve possesses five points of self-intersection which persist throughout the dynamics. It appears to be a general feature of these singular curves that the number of such "stable" self-intersections equals the ordering of the complex double point.

An interesting question is whether the presence of self-intersections has a topological meaning. A more general related question is whether the curves produced by a type II Bäcklund transformation play a role in distinguishing different knot classes of multiphase solutions, and whether the Floquet spectrum can be used to characterise topological invariants. Figures 5b, 5c show typical spectral configurations of initial data close to a planar circle with a single complex double point (Figure 5a). The spectra can be computed using standard perturbation analysis (we consider only spatially even solutions, the general case being far more complicated [19]). The degenerate double point splits in a "cross" (b) or a "gap" (c) configuration characteristic of different 3-phase solutions which in turn correspond to complicated, likely knotted curves in space. Figure 5 illustrates the role of the homoclinic orbit to the planar circle as a separatrix between different types of 3-phase solutions. How this should be interpreted at the curve level and whether this fact carries a topological

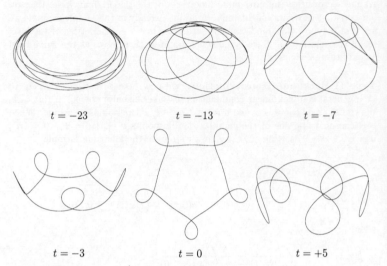

$t = -23$ $t = -13$ $t = -7$

$t = -3$ $t = 0$ $t = +5$

Figure 4: Evolution of the Bäcklund transformation of a 6-fold planar circle. $k = 6$, $n = 5$, $\rho = 1$, $\theta = 0$.

significance is an open problem being currently investigated by the author. A positive answer would support the use of the Floquet spectrum in the topological classification of N-phase curves.

5   Future Directions

We conclude with a description of current research activities and a list of open questions.

To my knowledge, the following groups have been working on related topics: A. Sym et al. (see e.g. [11, 54, 53]) have worked on applications of soliton theory to vortex filament dynamics and on the characterisation of soliton surfaces generated by the evolution of vortex filaments in the shape of solitons. H.K. Moffat, R.L. Ricca and collaborators (at Cambridge and University College, London) have focussed on applications of knot theory to fluid mechanics (see e.g. [49, 47, 48]). The work of J.P. Keener and J.J. Tyson [29, 28] on scroll waves in excitable media and on vortex filaments in ideal fluids has important points of contacts with the author's work. The group at Case Western Reserve University including J. Langer, D. Singer, T. Ivey and myself, together with collaborator R. Perline at Drexel University, have worked on higher-dimensional generalisations of the filament equation hierarchy and on the topological implications of related Bäcklund transformations. A. Doliwa and P.M. Santini have addressed similar questions about integrable curve evolutions [13] on n-dimensional spheres and have also discussed the discrete problem in the context of polygonal spherical curves [14]. On the same theme of polygonal curves,

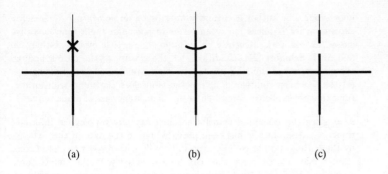

(a)   (b)   (c)

Figure 5: Typical spectral configurations of initial data close to a planar circle with a single complex double point.

J. Millson at the University of Maryland and M. Kapovich at Utah State University [27] have studied the topology of the space of polygonal knots, its symplectic structure and an associated integrable evolution. The continuous counterpart has been studied by J. Millson and B. Zombro [43], with a focus on the symplectic geometry of smooth curves in Euclidean space and on the bi-Hamiltonian structure of the periodic Heisenberg Model.

There are several open directions of research.

1. The formulation and study of more realistic models of vortex filament dynamics which take account of the finite structure of the vortex core, the longitudinal stretching of the filament, nonlocal interactions and the effect of finite viscosity.

2. Discrete models of integrable curve dynamics (i.e. evolution equations of polygonal curves): discrete soliton equations such as the Ablowitz–Ladik model or integrable discretisations of the Heisenberg Chain are among them.

3. Other integrable hierarchies are associated to curve dynamics, among them the KdV, mKdV and the sine-Gordon equations: a natural question is whether any soliton equation can be associated with a curve evolution.

4. Closely related to the previous question is the construction of a general framework in which to describe integrable hierarchies of curve dynamics in a space of arbitrary dimension; a suitable set-up appears to be the Kac–Moody Lie algebra framework where a soliton equation is regarded as a dynamical system on an infinite-dimensional loop algebra. Since this work was written, Langer and Perline have shown that dynamical systems associated with Hermitian symmetric Lie algebras can be used to model the evolution of arclength parametrized curves in Euclidean and other spaces, see [36] for details.

5. In the context of higher phase solutions of the filament equation, what do the higher N-phase curves look like? Is it possible to use the Floquet spectrum of the spatial part of the associated Lax pair to classify their

knot type? The author is currently working on implementing the general expression for N-phase curves in terms of Riemann theta functions discussed in this work. Recently, in an interesting article by P.G. Grinevich and M.U. Schmidt [25], conditions for the closure of the corresponding curves in terms of the NLS spectral problem are derived (generalizing the relation between complex double points and close Bäcklund transformations that we explicitly computed in the elementary case of planar circles).

6. Also, does the Bäcklund transformation carry any topological information? A partial answer has been provided (since the date of this article) by the author and her collaborator T. Ivey [9] with respect to closed constant torsion elastic rods, that can be associated with the sine-Gordon equation. Bäcklund transformations are shown to provide a variety of topological phenomena such as knotting, unknotting and self-intersections. Questions about the mechanisms of topological changes are gaining attention in a variety of fields (e.g. [41]).

7. Moffat and Ricca posed an analogous question for the sequence of global geometrical invariants: whether they restrict the possible knot types of the corresponding curves and whether they themselves have a topological interpretation.

8. The Marsden–Weinstein Poisson structure is realised naturally at the level of curves as a 90-degree rotation of any normal vector field to the curve and it is associated to the almost-complex structure on the loop space of $\mathbb{R}^3$: an open question is whether the higher Poisson structures have a natural geometrical realisation, and whether the zero-curvature formulation of the sequence of NLS flows (and of general soliton equations) can be given a direct geometric interpretation.

Acknowledgements

The original results in this article are based on my dissertation work at the University of Arizona. Many thanks go to my advisor Nick Ercolani for hours and hours of stimulating discussions and exciting mathematics. I am also grateful to my collaborators in the "geometry group" at Case Western Reserve University: Joel Langer, David Singer and Tom Ivey.

References

[1] M.J. Ablowitz and H. Segur, *Solitons and the Inverse Scattering Transform*, SIAM, Philadelphia, 1981.

[2] V.I. Arnold, *Mathematical Methods of Classical Mechanics*, Springer-Verlag, New York, 1980.

[3] G.K. Batchelor, *An Introduction to Fluid Dynamics*, Cambridge University Press, 1967.

[4] A. Bishop, There is more than one way to frame a curve, *Amer. Math. Monthly* **82**, 246–251, 1975.

[5] J-L. Brylinski, *Loop Spaces, Characteristic Classes and Geometric Quantization*, Birkhäuser, Basel, 1992.

[6] J.L. Burchnall and T.W. Chaundy, Commutative ordinary differential operators, *Proc. London Math. Soc.* **211**, 420–440, 1923.

[7] J.L. Burchnall and T.W. Chaundy, Commutative ordinary differential operators, *Proc. Roy. Soc. London Ser. A* **118**, 557–583, 1928.

[8] A. Calini, *Integrable Curve Dynamics*, PhD thesis, University of Arizona, 1994.

[9] A. Calini and T. Ivey, Bäcklund transformations and knots of constant torsion, *J. Knot Theory Ramifications* **7**, 719–746, 1998.

[10] P. Casati, F. Magri, and M. Pedroni, Bihamiltonian manifolds and Sato's equations, In *Integrable Systems: the Verdier Memorial Conference*, volume 115 of *Progress in Mathematics*. Birkhäuser, 1993.

[11] J. Cieslinski, P. Gragert, and A. Sym, Exact solutions to localized approximation equation modelling smoke ring evolution, *Phys. Rev. Lett. A* **57**, 1507, 1986.

[12] M. do Carmo, *Differential Geometry of Curves and Surfaces*, Prentice Hall, 1976.

[13] A. Doliwa and P.M. Santini, An elementary characterization of the integrable motions of a curve, *Phys. Lett. A* **185**, 373–384, 1994.

[14] A. Doliwa and P.M. Santini, Integrable dynamics of a discrete curve and the Ablowitz–Ladik hierarchy, *J. Math. Physics* **36**, 1259–1273, 1995.

[15] B.A. Dubrovin, Theta functions and nonlinear equations, *Russian Math. Surveys* **36**, 11–92, 1981.

[16] N.M. Ercolani, M.G. Forest, and D.W. McLaughlin, Geometry of the modulational instability, I: Local analysis, University of Arizona preprint, 1987.

[17] N.M. Ercolani, M.G. Forest, and D.W. McLaughlin, Geometry of the modulational instability, II: Global analysis, University of Arizona preprint, 1987.

[18] N.M. Ercolani, M.G. Forest, and D.W. McLaughlin, Geometry of the modulational instability, III: Homoclinic orbits for the periodic sine-Gordon equation, *Physica D* **43**, 349–384, 1990.

[19] N.M. Ercolani and D.W. McLaughlin, Towards a topological classification of integrable PDE's, In *The Geometry of Hamiltonian Systems*. Springer-Verlag, 1991.

[20] L.D. Faddeev and L.A. Takhtajain, *Hamiltonian Methods in the Theory of Solitons*, Springer-Verlag, 1980.

[21] H. Flaschka, A.C. Newell, and T. Ratiu, Kac–Moody Lie algebras and soliton equations II. Lax equations associated with $A_1^{(1)}$, *Physica D* **9**, 303–323, 1983.

[22] H. Flaschka, A.C. Newell, and T. Ratiu, Kac–Moody Lie algebras and soliton equations III. Stationary equations associated with $A_1^{(1)}$, *Physica D* **9**, 324–332, 1983.

[23] H. Flaschka, A.C. Newell, and T. Ratiu, Kac–Moody Lie algebras and soliton equations IV, *Physica D* **9**, 333–345, 1983.

[24] P. Griffith and J. Harris, *Principles of Algebraic Geometry*, Wiley Interscience, New York, 1978.

[25] P.G. Grinevich and M.U. Schmidt, Closed curves in $\mathbb{R}^3$: a characterization in terms of curvature and torsion, the Hasimoto map and periodic solutions of the filament equation, 1997, Los Alamos preprint no. dg-ga/9703020.

[26] R. Hasimoto, A soliton on a vortex filament, *J. Fluid Mech.* **51**, 477–485, 1972.

[27] M. Kapovich and J.J. Millson, The symplectic geometry of polygons in the euclidean plane, *J. Diff. Geom.* **44**, 479–513, 1996.

[28] J.P. Keener, Knotted vortex filaments in an ideal fluid, *J. Fluid Mech.* **211**, 629–651, 1990.

[29] J.P. Keener and J.J. Tyson, The dynamics of scroll waves in excitable media, *SIAM Review* **34**, 1–39, 1992.

[30] S. Kobayashi and K. Nomizu, *Foundations of Differential Geometry*, Interscience Publishers, New York, 1963.

[31] I.M. Krichever, Methods of algebraic geometry in the theory of non-linear equations, *Russian Math. Surveys* **32**, 185–213, 1977.

[32] M. Lakshmanan, Continuum spin systems as an exactly solvable dynamical system, *Phys. Lett. A* **61**, 53–54, 1977.

[33] G.L. Lamb, *Elements of Soliton Theory*, Wiley Interscience, New York, 1980.

[34] J. Langer and R. Perline, Poisson geometry of the filament equation, *J. Nonlinear Sci.* **1**, 71–93, 1991.

[35] J. Langer and R. Perline, The localized induction equation, the Heisenberg chain, and the non-linear Schrödinger equation, In *Fields Institute Communications Vol. 7*. American Math. Soc., 1996.

[36] J. Langer and R. Perline, Geometric realizations of Fordy–Kulish integrable systems, Sym–Pohlmeyer curves and related variation formulas, *Pacific J. Maths.*, 1999, to appear.

[37] J. Langer and D. Singer, Knotted elastic curves in $\mathbb{R}^3$, *J. London Math. Soc.* **30**, 512–520, 1984.

[38] P. Lax, Integrals of nonlinear equations of evolution and solitary waves, *Comm. Pure and App. Math.* **21**, 467–490, 1968.

[39] P.D. Lax, Periodic solutions of the KdV equation, *Comm. Pure and App. Math.* **28**, 141–188, 1975.

[40] F. Magri, A simple model of the integrable Hamiltonian equation, *J. Math. Phys.* **19**, 1156, 1978.

[41] A. Malevanets and R. Kapral, Links, knots and knotted labyrinths in bistable systems, *Phys. Rev. Lett.* **77**, 767–770, 1996.

[42] J. Marsden and A. Weinstein, Coadjoint orbits, vortices, Clebsh variables for incompressible fluids, *Physica D* **7**, 305–323, 1983.

[43] J.J. Millson and B. Zombro, The symplectic geometry of smooth closed curves in euclidean space, University of Maryland preprint, 1996.

[44] A.C. Newell, *Solitons in Mathematics and Physics*, SIAM, 1985.

[45] P.J. Olver, *Applications of Lie Groups to Differential Equations*, Springer-Verlag, New York, 2nd edition, 1993.

[46] E. Previato, Hyperelliptic quasi-periodic and soliton solutions of the nonlinear Schrödinger equation, *Duke Mathematical Journal* **52**, 329–377, 1985.

[47] R.L. Ricca, Torus knots and polynomial invariants for a class of soliton equations, *Chaos* **3**, 83–91, 1993.

[48] R.L. Ricca, Geometric and topological aspects of vortex filament dynamics under LIA, *Lecture Notes in Physics* **462**, 99–104, 1995.

[49] R.L. Ricca and H.K. Moffat, The helicity of a knotted vortex filament, In H.K Moffat, editor, *Topological aspects of the dynamics of fluids and plasmas*. Kluwer Academic Publishers, Dordrecht, 1992.

[50] D.H. Sattinger and V.D. Zurkowski, Gauge theory of Bäcklund transformations, *Physica D* **26**, 225–250, 1987.

[51] A. Seeger, H. Donth, and A. Kochendörfer, Theorie der Versetzungen in eindimension den Atomreiher III Versetzungen, Eigenbewegungen und ihre Wechselworkung, *Zeit. Physik* **134**, 173–193, 1953.

[52] I.M. Singer and J.A. Thorpe, *Lecture Notes on Elementary Topology and Geometry*, Springer-Verlag, 1967.

[53] A. Sym, Soliton surfaces I-V, *Lettere al Nuovo Cimento* **33, 36, 39, 40, 42**, 394–400, 1982-84.

[54] A. Sym, Vortex filament motion in terms of Jacobi Theta functions, *Fluid Dynamics Research* **3**, 151–156, 1988.

[55] M. Wadati, H. Sanuki, and K. Konno, Relationships among inverse method, Bäcklund transformation and an infinite number of conservation laws, *Prog. Theor. Phys.* **53**, 419–436, 1975.

An Elementary Introduction to Exterior Differential Systems

NIKY KAMRAN

Keywords: Exterior differential systems, Frobenius theorem, Cartan–Kähler theorem

1 Introduction

The theory of exterior differential systems was founded by Élie Cartan. It provides a geometric and coordinate-free approach to the study of differential equations, and is particularly well suited to the investigation of problems arising in local differential geometry. Well-known examples include the local isometric imbedding problem, the classical deformation problem for surfaces, and the local equivalence problem for G-structures, also known as the Cartan equivalence problem.

This chapter is meant to serve as a brief and elementary introduction to the theory of exterior differential systems. Our aim will be to review some of the main existence theorems for local integral manifolds and to illustrate them on simple examples. These theorems fall within two general categories. On the one hand, we have those theorems which are based on normal form results, such as the Frobenius theorem, in which the expression of the integral manifolds and their degree of generality are manifest. These theorems are typically valid for exterior differential systems of class C^∞ since their proofs rely essentially on the fundamental theorems for the existence, uniqueness and smooth dependence on the initial conditions of the solutions of ordinary differential equations. On the other hand, we have the Cartan–Kähler existence theorem, which solves the fundamental problem of proving that a subspace of the tangent space at a point of the underlying manifold, which is annihilated by the differential system and which is sufficiently generic, is tangent to an integral manifold of maximal dimension. The Cartan–Kähler theorem is proved by successive applications of the Cauchy–Kovalevskaia theorem and thus generally only valid for systems of class C^ω.

Most of the exterior differential systems arising in applications are endowed with an independence condition which expresses a prescribed choice of independent variables for the problem under consideration. By applying the Cartan–Kähler theorem, one obtains a very important and useful criterion, known as E. Cartan's involutivity test, for determining whether an exterior differential system with an independence condition admits integral manifolds which are compatible with the independence condition. Finally, we should emphasize that all the existence theorems that will be reviewed in these notes are concerned with *local* integral manifolds. The global converses to Lie's fundamental theorems provide important examples of global existence theorems for integral manifolds, [Che].

It is inevitable that any attempt to summarize the main results of such a rich and extensive subject within the bounds of a short introductory chapter will be superficial and incomplete. In particular, no proofs will be given for any of the theorems stated and many important topics, such as the Cartan–Kuranishi prolongation theorem, will not be covered. The interested reader is therefore urged to study the fundamental treatise of E. Cartan [Ca], *Les systèmes différentiels extérieurs et leurs applications géométriques*, and its worthy successor, the monograph entitled *Exterior Differential Systems* by Bryant, Chern, Gardner, Goldschmidt and Griffiths [BC3G]. All the results that appear in the present notes are borrowed from these two essential references and also from [BCG], which covers part of the material appearing in [BC3G].

2 Exterior Differential Systems and Integral Manifolds

In this section and in Section 3, all the manifolds, maps, differential forms, vector fields and other geometric structures will be assumed to be of class C^∞.

Definition 2.1. An exterior differential system on an n-dimensional manifold M_n is a differential ideal $\mathcal{I}$ of the ring $\Omega^*(M_n)$ of exterior differential forms on M_n, that is an ideal of $\Omega^*(M_n)$ which is closed under exterior differentiation. A Pfaffian system is an exterior differential system generated by 1-forms as a differential ideal.

We will assume throughout these notes that all the exterior differential systems considered are *finitely generated* . We shall use the notation

$$\mathcal{I} = \{\omega^1, ..., \omega^p\}$$

to denote the algebraic ideal generated by p differential forms $\omega^1, \ldots, \omega^p$ in $\Omega^*(M_n)$. If $\mathcal{I} = \{\theta^1, \ldots, \theta^s, d\theta^1, \ldots, d\theta^s\}$ is a Pfaffian system generated by s linearly independent 1-forms $\theta^1, \ldots, \theta^s$, then the subbundle of $T^*(M_n)$ generated by $\theta^1, \ldots, \theta^s$ will be denoted by I. We will use the same notation I for the $C^\infty(M_n)$-module of sections of I.

Definition 2.2. An m-dimensional integral manifold of $\mathcal{I}$ is an m-dimensional immersed submanifold $h : W_m \to M_n$ such that

$$h^*\omega = 0,$$

for all $\omega \in \mathcal{I}_m$.

Example 2.1. On $\mathbf{R}^3$ with coordinates (x, y, p), consider the Pfaffian system

$$\mathcal{I} = \{dy - p\,dx,\ dp \wedge dx\}.$$

The one-dimensional integral manifolds of $\mathcal{I}$ are the curves $\big(x(t), y(t), p(t)\big)$, such that $y' - px' = 0$. Note that the curves for which $x' \neq 0$ may be reparametrized in the form $\big(x, f(x), f'(x)\big)$, where f is an arbitrary function. The one-dimensional integral manifolds of $\mathcal{I}$ thus depend on one arbitrary function of one variable.

Example 2.2. On $\mathbf{R}^5$ with coordinates (x, y, u, p, q), consider a function $F :$ $\mathbf{R}^5 \to \mathbf{R}$ such that $F_p \neq 0, F_q \neq 0$. Consider the exterior differential system

$$\mathcal{I} = \{F, \, du - p\,dx - q\,dy, \, dF, \, dx \wedge dp + dy \wedge dq\}.$$

The two-dimensional integral manifolds of $\mathcal{I}$ are surfaces

$$\big(x(s,t), y(s,t), u(s,t), p(s,t), p(s,t)\big)$$

such that

$$F\big(x(s,t), y(s,t), u(s,t), p(s,t), p(s,t)\big) = 0,$$

$$u_s - px_s - qy_s = 0, \quad u_t - px_t - qy_t = 0.$$

Note that if $\left| \frac{\partial(x,y)}{\partial(s,t)} \right| \neq 0$, then the integral surfaces may be reparametrized in the form $\big(x, y, u(x,y), p(x,y), q(x,y)\big)$, where $p = u_x, q = u_y$ and $u(x,y)$ is a solution of the first-order partial differential equation

$$F(x, y, u, u_x, u_y) = 0.$$

Thus we see that in local coordinates, the condition that an immersion defines an integral manifold of an exterior differential system is expressed as a system of differential equations on the component maps of the immersion in the corresponding local coordinate charts.

Let $\mathcal{I}_p$ denote the homogeneous component of degree p in $\mathcal{I}$. We assume that the equations $\omega = 0$, $\omega \in \mathcal{I}_0$, define locally a submanifold of M_n. Unless otherwise indicated, we shall restrict $\mathcal{I}$ to this submanifold, and assume that $\mathcal{I}_0$ is empty.

3   Pfaffian Systems

Our purpose in this section is to state some basic existence theorems for local integral manifolds of Pfaffian systems. The general principle behind these existence theorems is to establish local normal forms in which the integral manifolds are manifest. The proofs of these theorems are based on the fundamental theorems for the existence, uniqueness and smooth dependence on the initial conditions of solutions of ordinary differential equations.

Consider a Pfaffian system $\mathcal{I}$ generated as a differential ideal by $s \leq n$ linearly independent 1-forms

$$\omega^a = \sum_{i=1}^{n} A_i^a(x^1, \ldots, x^n)\, dx^i, \qquad 1 \leq a \leq s. \tag{1}$$

The integer s will be called the *dimension* of the Pfaffian system.

An integral manifold of $\mathcal{I}$ is thus an immersion $f : U \to M_n$ such that

$$f^*\omega^a = 0, \quad 1 \leq a \leq s. \tag{2}$$

When written out in local coordinates, the integral manifolds of a Pfaffian system $\mathcal{I}$ correspond to the solutions of a system of partial differential equations. Indeed, if the immersion f is given locally in a domain U of $\mathbf{R}^p$ with coordinates

u^α, $1 \le \alpha \le p$, by an n-tuple of functions $\left(f^1(u^1, \ldots, u^p), \ldots, f^n(u^1, \ldots, u^p) \right)$, satisfying rank $\left(\frac{\partial f^i}{\partial u^\alpha} \right) = p$ in U, then (2) reads

$$\sum_{i=1}^{n} \sum_{\alpha=1}^{p} A_i^a \left(f^1(u^1, \ldots, u^p), \ldots, f^n(u^1, \ldots, u^p) \right) \frac{\partial f^i}{\partial u^\alpha} = 0, \quad 1 \le a \le s. \quad (3)$$

The simplest existence theorem for integral manifolds of a Pfaffian system is the *Frobenius theorem*:

Theorem 3.1. *Let $\mathcal{I}$ be a Pfaffian system generated by linearly independent 1-forms ω^a, $1 \le a \le s$, satisfying*

$$d\omega^a \wedge \omega^1 \wedge \cdots \wedge \omega^s = 0, \quad 1 \le a \le s. \quad (4)$$

There exist local coordinates $(u^1, \ldots, u^n)$ such that $\mathcal{I}$ is generated by $du^1, \ldots, du^s$.

If $\mathcal{I}$ satisfies the Frobenius conditions (4), then there exist, at least locally, $(n - s)$-dimensional integral manifolds of $\mathcal{I}$ given by $u^1 = c^1, \ldots, u^s = c^s$, where $c^1, \ldots, c^s$ are arbitrary real constants. These integral manifolds are of maximal dimension.

Example 3.1. On $\mathbf{R}^4$ with local coordinates (x, y, z, u), minus the locus $x + z = 0$, $y = 0$, $u = 0$, consider the Pfaffian system $\mathcal{I} = \{\omega^1, \omega^2, d\omega^1, d\omega^2\}$, where

$$\omega^1 = u^2 (x + z)(dx + dz) + u^2 (dy + u\, du), \quad \omega^2 = y^4 (dy + u du).$$

The integrability conditions (4) are satisfied and we have

$$\mathcal{I} = \left\{ d(x + z), d(2y + u^2) \right\}$$

so that the two-dimensional integral manifolds are the surfaces obtained by taking the intersection of the 3-planes $x + z = c_1$ with the parabolic 3-cylinders $2y + u^2 = c_2$, where c_1 and c_2 are arbitrary constants.

The Frobenius conditions (4) thus amount to saying that the algebraic ideal generated by ω^a, $1 \le a \le s$ is a differential ideal. A Pfaffian system generated as a differential ideal by linearly independent 1-forms $\omega^1, \ldots, \omega^s$ is said to be *completely integrable* if it satisfies the Frobenius conditions. The functions $u^1, \ldots, u^s$ are called *first integrals* of the completely integrable system $\mathcal{I}$. There are classical topological obstructions to the global existence of a completely integrable Pfaffian system on a manifold M_n in terms of the Pontryagin classes of the quotient bundle $T(M_n)/I^\perp$, where I denotes the subbundle of $T^*(M_n)$ generated by ω^i, $1 \le i \le s$, [Bott].

Let $\mathcal{I}$ be a Pfaffian system and let I be the corresponding $C^\infty(M, \mathbf{R})$-module of sections of $T^*(M_n)$. Let $\{I\}$ denote the algebraic ideal generated by I in $\Omega^*(M_n)$. The exterior derivative $d : I \to \Omega^2(M_n)$ induces a $C^\infty(M_n; \mathbf{R})$-linear map $\delta : I \to \Omega^2(M_n)/\{I\}$. We define the first derived system $I^{(1)}$ of I to be the kernel of δ, which we assume to have constant dimension. We have therefore a short exact sequence of $C^\infty(M; \mathbf{R})$-modules

$$0 \to I^{(1)} \to I \to dI/\{I\} \to 0.$$

We see that $I^{(1)} = I$ if and only if $\mathcal{I}$ is completely integrable. The quotient $I/I^{(1)}$ can thus be thought of as a measure of amount by which $\mathcal{I}$ fails to be completely integrable. By iterating this construction, we obtain a flag of $C^\infty(M_n, \mathbf{R})$-modules

$$\cdots \subset I^{(k)} \subset \cdots \subset I^{(1)} \subset I, \tag{5}$$

where $I^{(k)}$ is defined recursively by

$$I^{(k)} := (I^{(k-1)})^{(1)}.$$

The flag (5) is called the *derived flag* of $\mathcal{I}$. It stabilizes at a $C^\infty(M_n, \mathbf{R})$-module $I^{(k)}$ associated to the maximal completely integrable subsystem $\mathcal{I}^{(k)}$ of $\mathcal{I}$.

There is a classical construction, due to É. Cartan, for obtaining a minimal set of coordinates in which to express the generators of an exterior differential system $\mathcal{I}$. Let $\mathrm{Char}(I)$ denote the system of *Cauchy characteristic vector fields* of $\mathcal{I}$, defined by

$$\mathrm{Char}(\mathcal{I}) = \{X \mid X \in \mathcal{I}^\perp,\ X \,\rfloor\, d\mathcal{I} \subset \mathcal{I}\}. \tag{6}$$

The *Cartan system* of $\mathcal{I}$, denoted by $C(\mathcal{I})$, is the differential ideal generated by the module $\mathrm{Char}(\mathcal{I})^\perp$ of 1-forms which are annihilated by the Cauchy characteristic vector fields.

The *class* of a Pfaffian system is, by definition, the dimension of its Cartan system. It is easy to prove that the Cartan system $C(\mathcal{I})$ of any Pfaffian system $\mathcal{I}$ is always completely integrable. The first integrals of $C(\mathcal{I})$ provide the required minimal set of coordinates:

Theorem 3.2. *Let $\mathcal{I}$ be a Pfaffian system of class r and let $\{w^1, \ldots, w^r\}$ denote a set of first integrals of the Cartan system $C(\mathcal{I})$. There exists a local coordinate chart $(w^1, \ldots, w^r;\ y^{r+1}, \ldots, y^n)$ such that $\mathcal{I}$ is generated by 1-forms in $w^1, \ldots, w^r$ and their differentials.*

The *codimension* of a Pfaffian system is defined to be its class minus its dimension.

We have seen that the Frobenius theorem leads to a normal form in which the integral manifolds of a completely integrable Pfaffian system are manifest. There are similar normal form results which apply to Pfaffian systems which are not completely integrable, but whose structure equations are of a special type. We begin with the simplest such result, known as the solution to the *Pfaff problem*. Let $\mathcal{I}$ be a Pfaffian system generated as a differential ideal by a single 1-form ω. The *rank* of $\mathcal{I} = \{\omega, d\omega\}$ is the integer r defined by

$$(d\omega)^r \wedge \omega \neq 0, \quad (d\omega)^{r+1} \wedge \omega = 0. \tag{7}$$

In what follows, we shall assume r to be constant.

Theorem 3.3. *If $\mathrm{rank}\,\{\omega, d\omega\} = r$, then class $\{\omega\} = 2r + 1$ and there exist local coordinates $(z, p_1, \ldots, p_r, x^1, \ldots, x^r, u^{2r+2}, \ldots, u^n)$ such that*

$$\{\omega, d\omega\} = \Big\{ dz - \sum_{i=1}^{r} p_i dx^i,\ \sum_{i=1}^{r} dp_i \wedge dx^i \Big\}. \tag{8}$$

In contrast with the case in which $\mathcal{I}$ is completely integrable, the integral manifolds of $\mathcal{I} = \{\omega, d\omega\}$ depend now on one arbitrary *function* of r variables. They can be generically put in the form $z = f(x^1, \ldots, x^n)$, $p_i = \frac{\partial f}{\partial x^i}$, $1 \le i \le r$.

Example 3.2. On $\mathbf{R}^4$ with coordinates (x, y, z, u), minus the locus $y(x+y^2) = 0$, consider the Pfaffian system $\mathcal{I} = \{\omega, d\omega\}$, where

$$\omega = (x + y^2)y^2 \, dz - y\left(yz + u^2(x + y^2)^2\right) dx + \left(u^2 x (x + y^2)^2 - 2y^3 z\right) dy.$$

We have $d\omega \wedge d\omega \wedge \omega = 0$ and $\mathcal{I} = \{dZ - P\,dX, dP \wedge dX\}$, where

$$X = \frac{x}{y}, \quad Z = \frac{z}{x + y^2}, \quad P = u^2.$$

The one-dimensional integral manifolds of $\mathcal{I}$ thus depend on one arbitrary function of one variable and are given by

$$\frac{z}{x + y^2} = f\left(\frac{x}{y}\right), \quad u^2 = f'\left(\frac{x}{y}\right).$$

Actually, one can prove a slightly stronger result which can be thought of as the solution to a relative version of the Pfaff problem.

Theorem 3.4. *Let $\mathcal{I}$ be an $(s+1)$-dimensional Pfaffian system of the form*

$$\mathcal{I} = \{\theta^1, \ldots, \theta^s, d\theta^1, \ldots, d\theta^s, \omega, d\omega\},$$

where $\mathcal{J} = \{\theta^1, \ldots, \theta^s, d\theta^1, \ldots, d\theta^s\}$ is completely integrable and the rank r of $\mathcal{I} = \{\omega, d\omega\}$ relative to $\mathcal{J}$, defined by

$$(d\omega)^r \wedge \omega \not\equiv 0 \mod \mathcal{J}, \quad (d\omega)^{r+1} \wedge \omega \equiv 0 \mod \mathcal{J}, \tag{9}$$

is constant.

There exist local coordinates

$$\left(z, p_1, \ldots, p_r, x^1, \ldots, x^r, w^1, \ldots, w^s, u^{2r+s+1}, \ldots, u^n\right)$$

in which

$$\mathcal{I} = \left\{ dz - \sum_{i=1}^{r} p_i \, dx^i, \ \sum_{i=1}^{r} dp_i \wedge dx^i, \ dw^1, \ldots, \ dw^s \right\}. \tag{10}$$

The local integral manifolds of $\mathcal{I}$ now depend on one arbitrary function of r variables and s arbitrary constants. They are parametrized by $z = f(x^1, \ldots, x^n)$, $p_i = \frac{\partial f}{\partial x^i}$, $w^i = c^i$, $1 \leq i \leq r$, $1 \leq \alpha \leq s$.

Example 3.3. The solution of the relative Pfaff problem has interesting applications to the classification and normal forms problem for partial differential equations [GaK]. We consider a system

$$F\left(x, y, z, w, \frac{\partial z}{\partial x}, \frac{\partial z}{\partial y}, \frac{\partial w}{\partial x}, \frac{\partial w}{\partial y}\right) = 0,$$
$$G\left(x, y, z, w, \frac{\partial z}{\partial x}, \frac{\partial z}{\partial y}, \frac{\partial w}{\partial x}, \frac{\partial w}{\partial y}\right) = 0, \tag{11}$$

of two first-order partial differential equations for maps $(z, w) : \mathbf{R}^2 \to \mathbf{R}^2$.

The system (11) is said be parabolic if and only if the matrix

$$\begin{pmatrix} \frac{\partial(F,G)}{\partial(m,n)} & \frac{1}{2}\left(\frac{\partial(F,G)}{\partial(q,m)} - \frac{\partial(F,G)}{\partial(p,n)}\right) \\ \frac{1}{2}\left(\frac{\partial(F,G)}{\partial(q,m)} - \frac{\partial(F,G)}{\partial(p,n)}\right) & \frac{\partial(F,G)}{\partial(p,q)} \end{pmatrix} \tag{12}$$

has rank 1. Our first step is to formulate the p.d.e. system (11) geometrically as a Pfaffian system. We consider the bundle $J^1(\mathbf{R}^2, \mathbf{R}^2)$ of 1-jets of maps from $\mathbf{R}^2$ to $\mathbf{R}^2$ and use local coordinates (x, y, z, w, p, q, m, n) in which the Pfaffian system $\Omega^1_{\text{cont}}(\mathbf{R}^2, \mathbf{R}^2)$ of contact 1-forms is the differential ideal generated by

$$\theta^1 = dz - p\,dx - q\,dy, \quad \theta^2 = dw - m\,dx - n\,dy. \tag{13}$$

Thus if $f : \mathbf{R}^2 \to J^0(\mathbf{R}^2, \mathbf{R}^2) : (x, y) \mapsto \big(x, y, z(x,y), w(x,y)\big)$ is a section, then

$$p(j^1 f) = \frac{\partial z}{\partial x}, \quad q(j^1 f) = \frac{\partial z}{\partial y}, \quad m(j^1 f) = \frac{\partial w}{\partial x}, \quad n(j^1 f) = \frac{\partial w}{\partial y}.$$

We assume that the equations

$$F(x, y, z, w, p, q, m, n) = 0, \quad G(x, y, z, w, p, q, m, n) = 0, \tag{14}$$

corresponding to the p.d.e.s (3.11) give rise to a six-dimensional submanifold $i : \Sigma_6 \to J^1(\mathbf{R}^2, \mathbf{R}^2)$.

It can be shown [GaK] that the local sections $f : \mathbf{R}^2 \to \Sigma_6$ which are the integral manifolds of the Pfaffian system $\mathcal{I}$ obtained by pulling back the contact system $\Omega^1_{\text{cont}}(\mathbf{R}^2, \mathbf{R}^2)$ to Σ_6 are in one-to-one correspondence with the solutions of (3.11). It can be shown moreover [GaK] that if (3.11) is parabolic and $\mathcal{I}$ has a one-dimensional first derived system, then there exist generators π^1, π^2 such that $\mathcal{I} = \{\pi^1, \pi^2, d\pi^1, d\pi^2\}$ and such that the following structure equations are valid,

$$\begin{aligned}
d\pi^1 &\equiv 0, \\
d\pi^2 &\equiv \omega^3 \wedge \omega^5 + \omega^4 \wedge \omega^6, \qquad \mod\{\pi^1, \pi^2\}.
\end{aligned} \tag{15}$$

These structure equations are obtained by putting the symmetric matrix (12) in Witt normal form.

Theorem 3.5. *Every p.d.e. system (3.11) of parabolic type whose associated Pfaffian system I has a one-dimensional first derived system can be locally transformed to the normal form*

$$\frac{\partial z}{\partial x} = 0, \quad \frac{\partial z}{\partial y} = 0,$$

by a contact transformation.

Proof. The hypotheses of the theorem allow us to apply the Cartan–Von Weber theorem [Ga] to conclude that $I^{(1)} = \{\pi^1\}$ must be completely integrable. We can therefore apply the solution to the relative Pfaff problem to argue that there exist local coordinates (x, y, z, w, m, n) such that

$$\{\pi^1, \pi^2\} = \{dw - m\,dx - n\,dy, \ dz\} \,. \qquad \square$$

Another normal form result which we will use is the *Goursat normal form*, of which we present directly the relative version.

Theorem 3.6. *Let $\mathcal{I}$ be an $(r+s)$-dimensional Pfaffian system of codimension two, given by*

$$\mathcal{I} = \{\omega^1, \ldots, \omega^r, \theta^1, \ldots, \theta^s, d\omega^1, \ldots, d\omega^r, d\theta^1, \ldots, d\theta^s\} \qquad (16)$$

where $\mathcal{J} := \{\theta^1, \ldots, \theta^s, d\theta^1, \ldots, d\theta^s\}$ is completely integrable. Suppose that there exist 1-forms α and π such that $\alpha \not\equiv 0$, $\pi \not\equiv 0$, mod $\mathcal{I}$ and such that the following structure equations are valid

$$d\omega^1 \equiv \omega^2 \wedge \pi \qquad \mod \{\omega^1, \theta^1, \ldots, \theta^s\}$$
$$\vdots$$
$$d\omega^i \equiv \omega^{i+1} \wedge \pi \qquad \mod \{\omega^1, \ldots, \omega^i, \theta^1, \ldots, \theta^s\}, \quad 1 \le i \le r-1 \qquad (17)$$
$$\vdots$$
$$d\omega^r \equiv \alpha \wedge \pi \qquad \mod \{\omega^1, \ldots, \omega^r, \theta^1, \ldots, \theta^s\} = I.$$

There exist local coordinates $\left(x, y, y', \ldots, y^{(r)}, w^1, \ldots, w^s, u^{r+s+3}, \ldots, u^n\right)$ such that

$$\mathcal{I} = \left\{ dy - y'\, dx, \ldots, dy^{(r-1)} - y^{(r)}\, dx,\ dy' \wedge dx, \ldots, dy^{(r)} \wedge dx,\ dw^1, \ldots, dw^s \right\}. \tag{18}$$

The local integral manifolds of $\mathcal{I}$ depend on one arbitrary function of one variable and r arbitrary constants. They can be generically parametrized as

$$y = f(x), \ y' = f'(x), \ \ldots, \ y^{(s)} = f^{(s)}(x), \ u^1 = c^1, \ \ldots, \ u^r = c^r.$$

Example 3.4. Consider the under-determined equation,

$$\frac{dz}{dx} = F\left(x, y, z, \frac{dy}{dx}, \frac{d^2y}{dx^2}\right), \tag{19}$$

which has been studied by É. Cartan in the context of exterior differential systems. This equation can be expressed as a Pfaffian system $\mathcal{I}$, on $\mathbf{R}^5$, with coordinates (x, y, z, y', y''), where $\mathcal{I}$ is the differential ideal generated by

$$dy - y'\, dx, \quad dy' - y''\, dx, \quad dz - F(x, y, z, y', y'')\, dx.$$

The hypotheses appearing in the relative Goursat normal form are satisfied (with $\mathcal{J} = 0$ and $s = 2$) if and only if F is linear in y''. The solutions of (19) thus depend on one arbitrary function of one variable and its derivatives up to order 2. An example given by Cartan is

$$\frac{dz}{dx} = y^m \frac{d^2y}{dx^2},$$

whose solutions in parametric form are

$$x = -2tf''(t) - f'(t),$$
$$y^{m+1} = (m+1)^2 t^3 (f''(t))^2,$$
$$z = (m-1)t^2 f''(t) - mtf'(t) + mf(t).$$

4   The Cartan–Kähler Theorem and E. Cartan's Involutivity Test

Throughout this section, we shall assume that all the manifolds, maps, differential forms, vector fields and other geometric data are *real-analytic*.

A natural approach to the task of constructing integral manifolds of exterior differential systems is to proceed one dimension at a time, by extending p-dimensional manifolds into $p+1$-dimensional ones. This is the starting point of the Cartan–Kähler Theorem. Let us first recall the usual pairing between the exterior algebras $\Lambda^*(V)$ and $\Lambda^*(V^*)$ of a finite-dimensional vector space V and its dual V^*. Let $\{e_1, \dots, e_n\}$ be a basis of V and let $\{\omega^1, \dots, \omega^n\}$ be the dual basis of V^*. The standard pairing between $\Lambda^*(V)$ and $\Lambda^*(V^*)$ is defined by

$$\langle v, \phi \rangle = 0,$$

if $v \in \Lambda^p(V), \phi \in \Lambda^q(V^*)$, with $p \neq q$, and

$$\langle v, \phi \rangle = \frac{1}{p!} \sum_{i_1 \cdots i_p}^{n} v^{i_1 \cdots i_p} \phi_{i_1 \cdots i_p}$$

if

$$v = \frac{1}{p!} \sum_{i_1 \cdots i_p}^{n} v^{i_1 \cdots i_p} e_{i_1} \wedge \cdots \wedge e_{i_p}, \quad \phi = \frac{1}{p!} \sum_{i_1 \cdots i_p}^{n} \phi_{i_1 \cdots i_p} \omega^{i_1} \wedge \cdots \wedge \omega^{i_p}.$$

An indication on how to proceed follows from the elementary observation that if W_m is an integral manifold of $\mathcal{I}$, then

$$\langle h_* T_u(W_m), \omega \rangle = 0$$

for all $u \in W_m$ and for all $\omega \in \mathcal{I}_m$.

Definition 4.1. Let $x \in M_n$ and let E^p denote a p-dimensional subspace of $T_x(M_n)$. We say that (x, E^p) is a p-dimensional integral element of $\mathcal{I}$ if $\langle E^p, \omega \rangle = 0$ for all $\omega \in \mathcal{I}_x$.

Note that every subspace of an integral element of $\mathcal{I}$ is also an integral element of $\mathcal{I}$. The basic strategy is to construct flags of integral elements and to show that integral elements which are suitably generic are tangent to integral manifolds. The latter step is the substance of the Cartan–Kähler theorem. It requires the solution of a sequence of initial value problems which can be solved by applying the Cauchy–Kovalevskaia Theorem. Thus, let (x, E^p) be a p-dimensional integral element of $\mathcal{I}$ and let ξ be a tangent vector in $T_x(M_n)$. To E^p we associate a decomposable p-vector $[E^p]$ in $\Lambda^p(T_x(M_n))$ such that a tangent vector $v \in T_x(M_n)$ lies in E^p if and only if $[E^p] \wedge v = 0$. This p-vector is, of course, unique up to a non-vanishing scalar multiple. We may ignore this ambiguity and identify subspaces with their multi-vector representatives for the purposes of contructing integral elements. Indeed, the condition that $(x, [E^p] \wedge v)$ be an integral element is given by

$$\langle [E^p] \wedge v, \omega \rangle = 0,$$

for all $\omega \in \mathcal{I}_p$. This is a homogeneous system of linear equations in v. Its solution space is called the *polar space* of E^p and denoted by $H(E^p)$. Note that we have the obvious inclusion $E^p \subseteq H(E^p)$. The dimension of the polar space $H(E^p)$ is denoted by $p + 1 + \sigma_{p+1}$, with $\sigma_{p+1} \geq -1$ in view of the above inclusion.

In order to define what is meant by a "sufficiently generic" integral element, it is natural to consider the Grassmann bundles $G_k(M_n)$, $0 \leq k \leq n$, determined by the tangent bundle $T(M_n)$. The fiber at $x \in M_n$ of $G_k(M_n)$ is thus the Grassmannian of k-dimensional subspaces of the tangent space $T_x(M_n)$. The p-dimensional integral elements of $\mathcal{I}$ form an analytic variety $V_p(\mathcal{I})$ in $G_p(M_n)$.

Example 4.1. This example is borrowed from [BC3G]. Let $\mathcal{I}$ be the differential ideal generated on $\mathbf{R}^5$ by the two 1-forms

$$\omega^1 = dx^1 + (x^3 - x^4 x^5) \, dx^4, \quad \omega^2 = dx^2 + (x^3 + x^4 x^5) \, dx^5$$

We thus have

$$\mathcal{I} = \left\{ \omega^1, \, \omega^2, \, d\omega^1 = \omega^3 \wedge dx^4, \, d\omega^2 = \omega^3 \wedge dx^5 \right\}$$

where

$$\omega^3 = dx^3 + x^5 \, dx^4 - x^4 \, dx^5.$$

A one-dimensional subspace E^1 of $T_x(\mathbf{R}^5)$ gives rise to an integral element (x, E^1) of $\mathcal{I}$ if and only if

$$E^1 \subset K_x := \left\{ v \in T_p(\mathbf{R}^5) \mid \omega^1(v) = \omega^2(v) = 0 \right\} \subset T_x(\mathbf{R}^5).$$

It follows that

$$V_1(\mathcal{I}) = \mathbf{P}(\mathcal{K}),$$

where $\mathbf{P}(\mathcal{K})$ denotes the projectivization of the distribution $\mathcal{K} \subset T(M)$ defined by the field of 3-planes K_x, $x \in \mathbf{R}^5$. On the other hand, there is a unique two-dimensional integral element (x, E^2) through every $x \in \mathbf{R}^5$, where

$$E^2 := \left\{ v \in T_p(\mathbf{R}^5) \mid \omega^1(v) = \omega^2(v) = \omega^3(v) = 0 \right\},$$

so that $V_2(\mathcal{I}) = \mathbf{R}^5$. It is easy to see that the integral elements (x, E^2) are of maximal dimension.

Definition 4.2. An integral element (x, E^p) is said to be Kähler-regular if there exist s independent p-forms $\beta^1, \ldots, \beta^s$ such that the variety $V_p(\mathcal{I})$ is defined in a neighborhood of (x, E^p) by the equations $\langle [E^p], \beta^1 \rangle = 0, \ldots, \langle [E^p], \beta^s \rangle = 0$, and the rank of the polar equations of a p-dimensional integral element is constant in a neighborhood of (x, E^p). A *Kähler-regular integral manifold* is submanifold of M_n all of whose tangent spaces are Kähler-regular.

Thus we see that the rank of the polar equations of the tangent space to any connected Kähler-regular integral manifold W_m is the same at every point of W_m.

In most of the applications of the theory of exterior differential systems, there is a set of variables which plays a role very similar to that of the independent variables in a system of partial differential equations. Thus we define an *exterior differential system with independence condition* $(\mathcal{I}, \Omega)$ to consist of an exterior differential system $\mathcal{I}$ together with a decomposable p-form Ω which is well defined and non-zero modulo $\mathcal{I}$. An *admissible integral element* of $(\mathcal{I}, \Omega)$ is a p-dimensional integral element of $(\mathcal{I}, \Omega)$ on which Ω is non-zero and an *admissible integral manifold* of $(\mathcal{I}, \Omega)$ is an integral manifold $h : W_m \to M_n$ such that $h_*\big(T_u(W_m)\big)$ is an admissible integral element for all $u \in W_m$.

Example 4.2. On $\mathbf{R}^5$ with coordinates (x, y, z, p, q), consider the exterior differential system

$$\mathcal{I} = \big\{ dz - pdx - qdy, \ dx \wedge dp + dy \wedge dq, \ dx \wedge dq + dy \wedge dp \big\},$$

with independence condition given by $\Omega = dx \wedge dy$. The admissible two-dimensional integral manifolds of $(\mathcal{I}, \Omega)$ can be locally parametrized as surfaces $\big(x, y, f(x,y), f_x(x,y), f_y(x,y)\big)$, where f satisfies Laplace's equation

$$f_{xx} + f_{yy} = 0.$$

An admissible integral element (x, E^p) of $(\mathcal{I}, \Omega)$ is said to be *ordinary* if there exists a flag

$$E^1 \subset E^2 \subset \cdots \subset E^p,$$

such that each integral element (x, E^k), $1 \leq k \leq p - 1$, is a Kähler-regular integral element of $\mathcal{I}$.

Example 4.3. Consider the Pfaffian system of Example 4.1, endowed with the independence condition given by $\Omega := dx^4 \wedge dx^5$. Every 2-plane in $G_2\big(T(\mathbf{R}^5), \Omega\big)$ has a basis of tangent vectors of the form

$$X_4 = \tfrac{\partial}{\partial x^4} + p_4^1 \tfrac{\partial}{\partial x^1} + p_4^2 \tfrac{\partial}{\partial x^2} + p_4^3 \tfrac{\partial}{\partial x^3},$$

$$X_5 = \tfrac{\partial}{\partial x^5} + p_5^1 \tfrac{\partial}{\partial x^1} + p_5^2 \tfrac{\partial}{\partial x^2} + p_5^3 \tfrac{\partial}{\partial x^3}.$$

An element $\big(x, [X_4 \wedge X_5]\big) \in G_2\big(T(\mathbf{R}^5), \Omega\big)$ will be an admissible integral element if and only if

$$p_5^1 = p_4^2 = p_4^1 - x^3 + x^4 x^5 = p_5^2 + x^3 + x^4 x^5 = p_5^3 - x^4 = p_4^3 - x^5 = 0.$$

The left-hand sides of the above equations are linearly independent and their joint zero sets cut out exactly the variety $V_2(\mathcal{I}, \Omega)$; all the integral elements $\big(x, [X_4 \wedge X_5]\big)$ are Kähler-regular. It is likewise easily verified that they are ordinary.

We may now state the Cartan–Kähler theorem.

Theorem 4.3 (Cartan–Kähler). *Let $\mathcal{I}$ be an analytic exterior differential system on M_n. Let $W_p \subset M_n$ be a connected, p-dimensional Kähler-regular integral manifold of $\mathcal{I}$. Suppose that the dimension $p + 1 + \sigma_{p+1}$ of the polar space of $\big(x, T_x(W_p)\big)$ is greater than or equal to $p + 1$. Let $V_{n-\sigma_{p+1}}$ be an $n - \sigma_{p+1}$-dimensional submanifold of M_n containing W_p. Suppose that $T_x(V_{n-\sigma_{p+1}})$ and $H\big(T_x(W_p)\big)$ are transverse in $T_x(M_n)$ for all $x \in W_p$. Then there exists locally a unique $(p+1)$-dimensional connected integral manifold N_{p+1} of $\mathcal{I}$ which satisfies $W_p \subset N_{p+1} \subset V_{n-\sigma_{p+1}}$.*

The assumption of real-analyticity is essential since the proof of the Cartan–Kähler Theorem relies on successive applications of the Cauchy–Kovalevskaia Theorem.

Definition 4.4. An exterior differential system with independence condition $(\mathcal{I}, \Omega)$ is in involution if there exists an ordinary integral element (x, E^p) for every $x \in M_n$.

The importance of the notion of involutivity is underscored by the following theorem, which is a direct consequence of the Cartan–Kähler Theorem.

Theorem 4.5. *If $(\mathcal{I}, \Omega)$ is in involution and (x, E^p) is an admissible integral element, then there exists an admissible local integral manifold $W_p \subset M_n$ through x such that $T_x(W_p) = E^p$.*

There is a very important criterion, due to E. Cartan, for determining whether an exterior differential system $(\mathcal{I}, \Omega)$ is in involution.

Let thus $(\mathcal{I}, \Omega)$ be an exterior differential system with independence condition. Let X be a Zariski-open set in an irreducible component of the variety $V_{p,\Omega}(\mathcal{I})$ of admissible p-dimensional integral elements, which we assume of course to be non-empty. Let $G_k(X)$ denote the subset of points $(x, E^k) \in G_k(M_n)$ such that $E^k \subset E^p$ for some admissible integral element $(x, E^p) \in X$. We define $S_k(X)$ be the maximum rank of the polar equations of a k-dimensional integral element (x, E^k) close to a point $(x_0, E_0^k) \in G_k(X)$ and let $\sigma_k(X) = S_{k-1}(X) - (n - k)$. Also let $S_k'(X)$ be the maximum rank of the polar equations of $(x_0, E_0^k) \in G_k(X)$ and let $\sigma_k'(X) = S_{k-1}'(X) - (n - k)$. We have the obvious inequalities

$$S_k'(X) \leq S_k(X).$$

Proposition 4.6. *The open set $X \subseteq V_{p,\Omega}(\mathcal{I})$ contains an ordinary integral element if and only if $S_k'(X) = S_k(X)$.*

The integers $\sigma_k(X)$ and $\sigma_k'(X)$ are called called the *Cartan characters* and the *reduced Cartan characters* of $(\mathcal{I}, \Omega)$ respectively, with respect to the open set X. The Cartan characters of an involutive system $(\mathcal{I}, \Omega)$ give a measure of the degree of generality of the admissible integral manifolds. If $\sigma_q' = l$ is the highest non-vanishing character, then the admissible integral manifolds may be shown to depend on l arbitrary functions of q variables, [Ca], [BC3G].

Theorem 4.7 (E. Cartan's involutivity test). *Let $(\mathcal{I}, \Omega)$ be an exterior differential system with independence condition on an n-manifold M_n and let X be an irreducible component of the variety $V_{p,\Omega}(\mathcal{I})$ of admissible integral elements. We have*

$$\dim X \leq n + \sum_{i=1}^{p} i \, \sigma_i'(X),$$

with equality holding if and only if a general element $(x, E^p) \in X$ is a Cartan-ordinary integral element, or equivalently $(\mathcal{I}, \Omega)$ is in involution.

In general, the structure of the variety $V_{p,\Omega}(\mathcal{I})$ can be quite complicated. However, the variety $V_{p,\Omega}(\mathcal{I})$ has an affine bundle structure for the class of *quasilinear* exterior differential systems, so that the task of applying the involutivity test becomes then relatively straightforward. These systems arise in many important geometric applications, such as the local equivalence problem (see D. Hartley's chapter). Let $(\mathcal{I}, \Omega)$ be an exterior differential system generated (as a differential ideal) by s linearly independent 1-forms $\theta^1, \ldots, \theta^s$, with independence condition given by $\Omega = \omega^1 \wedge \cdots \wedge \omega^p$. Let $\pi^1, \ldots, \pi^l$, where $l = n - s - p$ be linearly independent 1-forms on M_n such that $(\theta^1, \ldots, \theta^s, \omega^1, \ldots, \omega^p, \pi^1, \ldots, \pi^l)$ forms a coframe on M_n. Thus we have

$$d\theta^i \equiv \sum_{\alpha,a=1}^{l,p} A^i_{\alpha a}\pi^\alpha \wedge \omega^a + \frac{1}{2}\sum_{a,b=1}^{p} B^i_{ab}\omega^a \wedge \omega^b + \frac{1}{2}\sum_{\alpha,\beta=1}^{l} C^i_{\alpha\beta}\pi^\alpha \wedge \pi^\beta, \quad \mathrm{mod}\,\mathcal{I}$$

$$(20).$$

We say of an exterior differential system $(\mathcal{I}, \Omega)$ generated by linearly independent 1-forms $\theta^1, \ldots, \theta^s$ that it is *quasilinear* if the condition

$$C^i_{\alpha\beta} = 0, \quad 1 \leq i \leq s,\, 1 \leq \alpha \leq l, \tag{21}$$

is satisfied. The condition of quasilinearity is clearly invariant under local automorphisms of $(\mathcal{I}, \Omega)$, that is local diffeomorphisms $f : M_n \to M_n$ which satisfy $f^*\mathcal{I} = \mathcal{I}$, $f^*\Omega \equiv w\Omega$, $\mathrm{mod}\,\mathcal{I}$, for some positive function $w : M_n \to \mathbf{R}$.

For quasilinear systems, the admissible integral elements of $(\mathcal{I}, \Omega)$ are given by linear equations,

$$\pi^\alpha - \sum_{a=1}^{p} t^\alpha{}_a\omega^a = 0, \quad 1 \leq \alpha \leq l, \tag{22}$$

where the coefficients $t^\alpha{}_a$ satisfy the system of inhomogeneous linear equations given by

$$\sum_{\alpha=1}^{t} A^i_{\alpha a}t^\alpha{}_b - \sum_{\alpha=1}^{t} A^i_{\alpha b}t^\alpha{}_a + B^i_{ab} = 0, \quad 1 \leq i \leq s,\, 1 \leq a, b \leq p.$$

It thus follows that if $(\mathcal{I}, \Omega)$ is quasilinear, then $V_{p,\Omega}(\mathcal{I})$ is an affine bundle over M_n. We shall drop the reference to the component X and also assume that the dimension of the space of solutions $\mathcal{S}$ of (22) does not vary with the choice of base point $x \in M_n$. Cartan's involutivity test takes now a simpler form.

Theorem 4.8. *Let $(\mathcal{I}, \Omega)$ be a quasilinear exterior differential system with independence condition. Let $\mathcal{S}(A := A^i_{\alpha a})$ denote the space of solutions of the polar equations (4.5). We have*

$$\dim \mathcal{S}(A) \leq \sum_{i=1}^{p} i\,\sigma'_i$$

with equality holding if and only if $(\mathcal{I}, \Omega)$ is in involution.

Following the standard terminology, we will say that $(A^i_{\alpha a})$ forms an *involutive tableau* if

$$\dim \mathcal{S}(A) = \sum_{i=1}^{p} i\,\sigma'_i.$$

There is a useful inductive formula for computing the reduced characters σ'_i, $1 \leq i \leq p$, of a quasilinear system.

Proposition 4.9. *Let* $(\mathcal{I}, \Omega)$ *be a quasilinear system with structure equations given by* (20) *and* (21). *We have*

$$
\sigma_1' + \cdots + \sigma_q' = \max_{v_1, \ldots, v_q \in \mathbf{R}^l} \operatorname{rank} \begin{pmatrix} \sum_{\alpha=1}^{l} v_1^\alpha A_{\alpha a}^i \\ \vdots \\ \sum_{\alpha=1}^{l} v_q^\alpha A_{\alpha a}^i \end{pmatrix} \tag{23}
$$

Example 4.4. We consider a second-order partial differential equation

$$
F\left(x^i, u, \frac{\partial u}{\partial x^i}, \frac{\partial^2 u}{\partial x^i \partial x^j}\right) = 0, \qquad 1 \leq i, j, \leq n.
$$

To this partial differential equation corresponds a Pfaffian system with independence condition $(\mathcal{I}, \Omega := dx^1 \wedge \cdots \wedge dx^n)$ on $\mathbf{R}^{\frac{n(n+5)}{2}+1}$ with local coordinates (x^i, u, u_i, u_{ij}), where $\mathcal{I}$ is the differential ideal generated by

$$
F(x^i, u, u_i, u_{ij}), \ \theta_0 := du - \sum_{i=1}^{n} p_i \, dx^i, \ \theta_i = du_i - \sum_{j=1}^{n} u_{ij} \, dx^i, \quad 1 \leq i, j, \leq n.
$$

We assume that the symmetric matrix $\left(\frac{\partial F}{\partial u_{ij}}\right)$, $1 \leq i, j, \leq n$, has maximal rank on the hypersurface of $\mathbf{R}^{\frac{n(n+5)}{2}+1}$ defined by the equation $F(x^i, u, u_i, u_{ij}) = 0$. The structure equations of $(\mathcal{I}, \Omega)$ are given by

$$
d\theta_0 \equiv 0, \ d\theta_i \equiv -\sum_{j=1}^{n} du_{ij} \wedge dx^j, \quad \bmod\{\theta_0, \theta_1, \ldots, \theta_n\}, \quad 1 \leq i, j, \leq n.
$$

The computation of the reduced characters is greatly simplified if one performs an admissible transformation from the given coframe ($\omega^i := dx^i, \theta_0, \theta_i$, $\pi_{ij} := du_{ij}$), $1 \leq i, j, \leq n$, to a coframe ($\bar{\omega}^i, \bar{\theta}_0, \bar{\theta}_i, \bar{\pi}_{ij}$), $1 \leq i, j, \leq n$, in which the symmetric matrix $\left(\frac{\partial F}{\partial u_{ij}}\right)$, $1 \leq i, j, \leq n$ assumes the algebraic normal form $\operatorname{diag}(\epsilon^1, \ldots, \epsilon^n)$. After such a transformation, we have structure equations of the form

$$
d\bar{\theta} \equiv 0, \ d\bar{\theta}_i \equiv -\sum_{j=1}^{n} \pi_{ij} \wedge dx^j; \quad \bmod\{\bar{\theta}_0, \bar{\theta}_1, \ldots, \bar{\theta}_n\}, \qquad 1 \leq i, j, \leq n,
$$

where

$$
\sum_{i=1}^{n} \epsilon^i \pi_{ii} \equiv 0, \ \pi_{ij} \equiv \pi_{ji}, \quad \bmod\{\bar{\theta}_0, \bar{\theta}_1, \ldots, \bar{\theta}_n\}.
$$

The exterior differential system $(\mathcal{I}, \Omega)$ is thus quasilinear and the structure equations (20) take the form

$$
d\bar{\theta}_{i'} \equiv \sum_{\alpha, i=1}^{n} A_{\alpha i}^{i'} \pi^\alpha \wedge \omega^j, \quad \bmod\{\theta_0, \ldots, \theta_n\}, \quad 0 \leq i' \leq n
$$

with

$$\sum_{\alpha=1}^{l:=n+1} A^i_{\alpha a}\pi^\alpha = \begin{pmatrix} 0 & \cdots & 0 \\ \pi_{11} & \cdots & \pi_{1n} \\ \vdots & \vdots & \vdots \\ \pi_{n1} & \cdots & \pi_{nn} \end{pmatrix}.$$

It is easy to compute the reduced characters of $(\mathcal{I}, \Omega)$ using the inductive formula (23). We obtain

$$\sigma'_1 = n, \ \sigma'_2 = n-1, \ \ldots, \ \sigma'_{n-1} = 2, \ \sigma_n = 0,$$

so that

$$\sum_{i=1}^{n} i\,\sigma_i = \frac{n(n+1)(n+2)}{6} - n.$$

On the other hand, the admissible integral elements take the form

$$\pi_{ij} - \sum_{k=1}^{n} L_{ijk}\omega^k = 0,$$

where

$$L_{ijk} = L_{jik} = L_{ikj}, \quad \sum_{i=1}^{n} \epsilon^i L_{iik} = 0.$$

So the space of admissible p-dimensional integral elements has dimension

$$\dim \mathcal{S}(A) = \binom{n+2}{n+1} - n = \frac{n(n+1)(n+2)}{6} - n$$

so that the system $(\mathcal{I}, \Omega)$ is indeed involutive and its solutions are parametrized by two functions of $n-1$ variables.

Example 4.5. We now review an example from [BC3G] of an exterior differential system with independence condition which is not involutive. On an open set $U \subset \mathbf{R}^6$, consider the Pfaffian system with independence condition $(\mathcal{I}, \omega^1 \wedge \omega^2)$, where $\mathcal{I} = \{\theta^1, \theta^2, d\theta^1, d\theta^2\}$, with structure equations given by

$$d\theta^1 \equiv \pi^1 \wedge \omega^1, \quad d\theta^2 \equiv \pi^1 \wedge \omega^2, \quad \mod\{\theta^1, \theta^2\},$$

and here $\{\theta^1, \theta^2, \omega^1, \omega^2, \pi^1, \pi^2\}$ are linearly independent 1-forms on U.

Let $\{D_{\theta^1}, D_{\theta^2}, D_{\omega^1}, D_{\omega^2}, D_{\pi^1}, D_{\pi^2}\}$ denote the dual basis of tangent vector fields on U. Let

$$V = v^1 D_{\omega^1} + v^2 D_{\omega^2} + v^3 D_{\pi^1} + v^4 D_{\pi^2},$$

$$W = w^1 D_{\omega^1} + w^2 D_{\omega^2} + w^3 D_{\pi^1} + w^4 D_{\pi^2},$$

denote a pair of linearly independent tangent vectors in the 4-plane $\theta^1 = 0, \theta^2 = 0$. The polar equations of W are given by

$$v^3 w^1 - v^1 w^3 = 0, v^3 w^2 - v^2 w^3 = 0.$$

Let us choose any point $x \in U$. The equations of a 2-plane $E^2 \subset T_x(U)$ on which $\theta^1 = \theta^2 = 0$ can be put in the form

$$\pi^1 + a\omega^1 + b\omega^2 = 0, \pi^2 + a'\omega^1 + b'\omega^2 = 0.$$

Thus (x, E^2) will be an integral element if and only if $a = b = 0$. It follows that any one-dimensional integral element (x, E^1) such that E^1 is contained in E^2 will be spanned by a tangent vector $V' = v'^1 D_{\omega^1} + v'^2 D_{\omega^2} + v'^4 D_{\pi^2}$. The rank of the polar equations of (x, E^1) is equal to 1 and therefore $(\mathcal{I}, \omega^1 \wedge \omega^2)$ is not involutive since it has no ordinary admissible integral elements at any $x \in U$.

Acknowledgements

The author wishes to thank Dr. Peter Vassiliou and the School of Mathematics and Statistics at the University of Canberra for their warm hospitality.

References

[BCG] R. Bryant, S.S. Chern and P.A. Griffiths, Exterior differential systems, in *Proceedings of the Beijing Symposium on Partial Differential Equations and Geometry*, China Scientific Press, 1982.

[BC3G] R. Bryant, S.S. Chern, R.B. Gardner, H. Goldschmidt and P.A. Griffiths, *Exterior Differential Systems*, MSRI Publications Vol. 18, Springer-Verlag, 1991.

[Bott] R. Bott, On a topological obstruction to integrability, in *Proc. Symposia Pure Mathematics, Vol. 16*, 127–131, American Mathematical Society, 1970.

[Ca] É. Cartan, *Les systèmes différentiels extérieurs et leurs applications géométriques*, Hermann, 1971.

[Che] C. Chevalley, *The Theory of Lie Groups, Vol.1*, Princeton University Press, 1946.

[Ga] R.B. Gardner, Invariants of Pfaffian systems, *Trans. Amer. Math. Soc.*, **126**, 514–533, 1967.

[GaK] R.B. Gardner and N. Kamran, Normal forms and focal systems for determined systems of two first-order partial differential equations in the plane, *Indiana University Mathematics Journal*, **44**, 1127–1162, 1995.

Cartan Structure of Infinite Lie Pseudogroups

IAN G. LISLE AND GREGORY J. REID

Keywords: Lie pseudogroups, Lie symmetries, Cartan structure

1   Introduction

Since Chevalley's seminal work [12], the definition of Lie group has been universally agreed. Namely, a Lie group G is an analytic manifold G on which is defined an analytic group operation $*$ with analytic inverse. The historical evolution of this definition was not direct. In the 1870s when Lie began his work on "continuous groups of transformations", the notion of abstract group was in its infancy [42]. Lie developed his group theory in the context of transformation properties of differential equations, seeking a 'Galois theory' of DEs.

Example 1.1. The Lie (point) symmetry group of the differential equation

$$y'' = y'\left(\frac{y'}{y} + \frac{2}{x}\right)$$

consists of the two-parameter family $\mathcal{G}$ of transformations

$$X = \frac{x}{b - ax}, \qquad Y = by, \qquad b \neq 0. \tag{1}$$

That is, if one performs this change of variables, then (X, Y) obey the same differential equation as (x, y). Given two transformations $T_{(a_1, b_1)}$ and $T_{(a_2, b_2)}$ in $\mathcal{G}$ (1), their composition is still in $\mathcal{G}$, and is given by

$$T_{(a_2, b_2)} \circ T_{(a_1, b_1)} = T_{(b_2 b_1, a_2 + b_2 a_1)}.$$

However, this is only true in a formal sense. Any 'transformation' in $\mathcal{G}$ (1) with $a \neq 0$ fails to be defined on the whole of $\mathbb{R}^2(x, y)$. Moreover, if $P = (x_0, y_0)$ with $x_0 \neq 0$ is some point, there are transformations in $\mathcal{G}$ which are undefined at P. Hence the composition above is valid only on a subset of $\mathbb{R}^2$, and the subset depends upon the values (a, b) under consideration.

Thus Lie's transformation "groups" make sense only as pseudogroups of *local* diffeomorphisms. The pejorative "pseudo" is earned because composition is defined only if no attention is paid to the domains involved. Nevertheless, Lie recognized that underlying these transformation "groups" is an abstract group G, where the parameters (a, b) live, and whose structure is independent of the nature of the space (x, y) on which it acts. Gradually, this abstract group G became our modern Lie group, the "parameters" became coordinates on a chart of G, and Lie's transformation group became a realization of G. However, additional problems arise in seeking the symmetries of differential equations.

Example 1.2. The symmetry "group" of Liouville's equation

$$u_{xy} = e^u$$

consists of local diffeomorphisms of the form

$$X = f(x), \qquad Y = g(y), \qquad U = u + \log\big(f'(x)g'(y)\big)$$

where f, g are arbitrary smooth functions such that $f'(x) \neq 0$, $g'(y) \neq 0$.

Example 1.3. The symmetries of the heat equation

$$u_t - u_{xx} = 0 \tag{2}$$

include superpositions of the form

$$u \mapsto u + f(x, t)$$

where f is any solution of (2). Indeed any linear equation $\mathcal{L}[u] = 0$ admits symmetries $u \mapsto u + f$ where $\mathcal{L}f = 0$. This "group" is the basis of methods [19] which detect whether a nonlinear PDE can be mapped to a linear one.

The symmetry pseudogroups above have no finite parametrization, and are not Lie groups in the modern sense. The analogues of the "group parameters" are the arbitrary functions f, g, and seem to demand explicit appearance of the coordinates (x, y) of the space on which the "group" is to act. Indeed it is very difficult to see how one could find an abstract group underlying these objects.

Lie proposed a novel way out of this impasse, by altering the definition of "continuous group of transformations". Starting in about 1890, Lie [22] developed a theory of what are now called *Lie pseudogroups*, which are defined as a "group" $\mathcal{G}$ of transformations τ which are the solutions of a *defining system* of partial differential equations. At first sight, this definition seems arbitrary, but Lie pointed out

1. That every finite-parameter Lie transformation group has a defining system. For this, start with a Lie group $\mathcal{G}$ and construct its left-invariant Maurer–Cartan forms ω^i. A map $\tau : \mathcal{G} \to \mathcal{G}$ is left action by an element of $\mathcal{G}$ if and only if $\tau^* \omega^i = \omega^i$. Treating these as conditions on τ yields (in coordinates) a system of first order partial differential equations for the components of τ.

2. That symmetries of differential equations (or other geometric objects) automatically are specified in terms of a defining system.

Example 1.4. The defining system for the 2-parameter pseudogroup (1) is

$$\begin{aligned} X_x &= \frac{X^2 Y}{x^2 y} & X_y &= 0 \\ Y_x &= 0 & Y_y &= Y/y. \end{aligned} \tag{3}$$

The early theory of Lie pseudogroups was established by Lie [22] and his student Tresse [38]. However, they did not provide a structure theory, and it was left to Vessiot [39, 40] (another of Lie's students) and especially to Élie Cartan [5, 6, 7, 8] to fill this gap. Vessiot's approach was to seek a result of

Jordan–Hölder type for the Lie pseudogroup via a decomposition of the defining system. Much later Guillemin [14] proved that a Jordan–Hölder decomposition really does exist for infinite Lie pseudogroups. But it was Cartan who made decisive advances in the theory. His main tool was his own theory of exterior differential systems in involution, developed in [5] specifically for this purpose. Cartan [5, 8] proved that infinite Lie pseudogroups obey an analogue of the Maurer–Cartan equations for Lie groups, and used these to define structure coefficients $a_{i\rho}^k$, c_{ij}^k which generalize the structure constants of a Lie group. Cartan succeeded [7] in classifying the simple infinite groups, and also provided [6, 9] a mechanism (method of equivalence) for deciding whether two Lie pseudogroups are isomorphic.

In this expository paper, we provide a general introduction to Cartan's structure equations for Lie pseudogroups. We establish one of the fundamental theorems, and point out some of the difficulties arising in the intransitive case. Our orientation is towards explicit computation, with examples for motivation. An alternative method for calculating Cartan's structure coefficients is described, based on the (linear) infinitesimal defining system.

2   Infinite Lie Pseudogroups

2.1   Pseudogroup

The modern definition of pseudogroup [20, 21, 37] is as follows:

Definition 2.1 (Pseudogroup). Let M be a (real) manifold, and $\mathcal{G}$ be a collection of diffeomorphisms of open subsets of M into M. Then $\mathcal{G}$ is a *pseudogroup* if

   i. $\mathcal{G}$ is closed under restriction: if $\tau : U \to M$ is in $\mathcal{G}$, then so is $\tau|_V$ for any open $V \subseteq U$.

  ii. If $U \subseteq M$ is an open set with $U = \bigcup_s U_s$, and $\tau : U \to M$ is a diffeomorphism with $\tau|_{U_s} \in \mathcal{G}$, then $\tau \in \mathcal{G}$.

 iii. $\mathcal{G}$ is closed under composition: if $\tau : U \to M$, and $\sigma : V \to M$ are any two members of $\mathcal{G}$, then $\sigma \circ \tau \in \mathcal{G}$ also, wherever this composition is defined.

  iv. $\mathcal{G}$ contains the identity diffeomorphism of M.

   v. $\mathcal{G}$ is closed under inverse: if $\tau : U \to M$ is in $\mathcal{G}$, then τ^{-1} (the domain of which is $\tau(U)$) is also in $\mathcal{G}$.

To paraphrase, a pseudogroup $\mathcal{G}$ is a collection of local diffeomorphisms of a manifold M, which is closed under composition *when defined*, contains an identity and is closed under inverse. Because the diffeomorphisms are local, composition $\tau_2 \circ \tau_1$ is only defined if the domain of τ_2 contains the range of τ_1. Hence the pejorative "pseudo": for a true group, $\tau_2 \circ \tau_1$ would have to work for all τ_1, τ_2.

2.2   Lie Pseudogroup

A pseudogroup is a very general object. We restrict ourselves to Lie pseudogroups:

Definition 2.2 (Lie pseudogroup). A *Lie pseudogroup* is a pseudogroup with diffeomorphisms defined as the local analytic solutions of an analytic system $\mathcal{R}$ of defining partial differential equations.

Thus we have an analytic variety $\mathcal{R} \subseteq J^k(M, M)$ in k-th order jet space. If x are coordinates on M, and X a copy of this coordinate system, the defining system $\mathcal{R}$ is a system of PDEs with the x as independent variables and X as dependent variables. It is locally described by equations of the form

$$F(x, X, X^{(1)}, \ldots, X^{(k)}) = 0.$$

Cartan's structure theory is greatly simplified if the defining system is of *first order*, and henceforth we *assume that this is so*. Reduction to first order can be achieved via a method described in [23].

The demand of analyticity can be relaxed to C^∞ in certain parts of the theory. However, a theory based on defining PDEs requires an existence-uniqueness result so that the size of the solution space can be predicted. For analytic systems, Cauchy–Kovalevskaya or related theories based on Taylor series (Cartan–Kähler, Riquier–Janet) serve in this role. Such results do not generalize to the C^∞ case except in special cases.

2.2.1 Involutivity

We demand that the defining system be put into a form where analytic existence-uniqueness results can be applied. All the *compatibility conditions* of the equations must be satisfied identically before we can use E-U results. This is a serious issue for defining systems, since they are typically *overdetermined* (more equations than unknowns). It is possible (and usual) that a collection of equations of order k or less can imply other equations of order k or less, obtained by differentiation and algebraic combination.

Example 2.1. We seek point symmetries of the ordinary DE $y'' = -x/y$, as vector fields

$$\xi(x, y)\partial_x + \eta(x, y)\partial_y.$$

This gives the infinitesimal defining system

$$\xi_{xx} = 2\eta_{xy} + 3\frac{x}{y}\xi_y$$

$$\xi_{xy} = \frac{1}{2}\eta_{yy}$$

$$\xi_{yy} = 0$$

$$\eta_{xx} = -2\frac{x}{y}\xi_x + \frac{x}{y}\eta_y - \frac{1}{y}\xi + \frac{x}{y^2}\eta.$$

Let a point (x_0, y_0) be chosen, along with values of $\xi, \eta, \xi_x, \xi_y, \eta_x, \eta_y, \xi_{xx}, \xi_{xy}, \xi_{yy},$ $\eta_{xx}, \eta_{xy}, \eta_{yy}$ at (x_0, y_0) consistent with the d.e.s. Then there is no guarantee that a solution can be built on these initial values (i.e. the system is not "locally solvable"). To see why not, apply the algorithm of Reid [31]. Any analytic solution must satisfy the identities $\xi_{xxy} = \xi_{xyx}$ and $\xi_{xyy} = \xi_{yyx}$. Enforcing these compatibility conditions on the defining system, one deduces that

$$\eta_{yyy} = 0, \qquad \eta_{xyy} = 2\frac{x}{y^2}\xi_y.$$

Computing further compatibility conditions gives $\xi_y = 0$, and eventually the system collapses to:

$$\xi = \frac{2}{3}\frac{x}{y}\eta, \qquad \eta_x = 0, \qquad \eta_y = \frac{1}{y}\eta.$$

In this form, compatibility conditions yield only algebraic consequences of the original equations. A formal power series about some initial point (x_0, y_0) can then be constructed to any desired order according to methods described in [31]. Here one chooses the value $\eta(x_0, y_0)$ arbitrarily: the defining system furnishes values for all other derivatives $\xi, \xi_x, \xi_y, \ldots$ and $\eta_x, \eta_y, \ldots$ at (x_0, y_0). According to Riquier's existence theorem [18], the resulting Taylor series has nonzero radius of convergence, and is the unique solution of the problem with the given initial data.

There are actually several related canonical forms for PDE systems: differential Gröbner basis, standard form, involutive form. This is a topic for religious war; the reader is referred to [17, 18, 27, 30, 31, 34, 35, 36] for further information. The point is that by systematically appending compatibility conditions, any linear PDE system[1] can be brought to a canonical form where local existence results can be applied. Implementations of several reduction algorithms are available [25, 31, 36] and are successful even on systems containing hundreds of linear PDE. Henceforth, we assume WLOG that any infinitesimal defining system has been put into one of these reduced forms.

The example above is linear, while the defining system for a pseudogroup is typically nonlinear. For polynomial nonlinearity, implementations of reduction methods are available [26, 34, 41]. However, nonlinearity is very computationally demanding, and such programs often fail due to an explosion in memory requirements.

2.2.2   Finite vs. Infinite

A Lie pseudogroup is *finite* if the general solution of its defining system depends on a finite number of arbitrary constants. Otherwise it is *infinite*. An infinite Lie pseudogroup contains arbitrary *functions* in the general solution of its defining system.

Assume the defining system of a Lie pseudogroup is first order involutive. If the pseudogroup is finite, this defining system consists of 0-th order (algebraic) equations, plus a full complement of n^2 first order equations:

$$\frac{\partial X^i}{\partial x^j} = f^i_j(x, X), \qquad i = 1, \ldots, n, \quad j = 1, \ldots, n$$

We say that all the $\frac{\partial X^i}{\partial x^j}$ are *principal* derivatives .

An *infinite* Lie pseudogroup has a defining system with 0-th and first order equations, but now the first order equations are less than n^2 in number, so that some first order derivatives $\frac{\partial X^i}{\partial x^j}$ do not occur on the left-hand side of any DE. Those which do occur on the LHS of a DE are principal, those which do not are called *parametric* first order derivatives.

---

[1] In fact the coefficients in the PDE must come from a computable differential field such as $\mathbb{Q}(x)$ for the reduction to be algorithmically achieved.

2.3   Examples of Lie Pseudogroups

First a finite Lie pseudogroup:

Example 2.2. Take $\mathcal{G}$ as the collection of local transformations of $\mathbb{R}^2$ of the form

$$X = \frac{x}{ax+b}, \qquad Y = by, \qquad Z = z/b \qquad (4)$$

where $b \neq 0$. Differentiating and eliminating a, b, we obtain the pseudogroup defining system

$$X_x = \frac{X^2 Y}{x^2 y} \qquad X_y = 0 \qquad X_z = 0 \qquad Z = \frac{yz}{Y}$$
$$Y_x = 0 \qquad Y_y = Y/y \qquad Y_z = 0. \qquad (5)$$

Note that this is first order and that all the first order derivatives occur on the left-hand side (equations for Z_x, Z_y and Z_z are implicitly present, though we do not trouble to write them since they are simply deducible from the 0-th order equation). That is, every first order derivative is principal, indicating that the pseudogroup is finite. Indeed, since the only parametric derivatives are X, Y, the pseudogroup has a general solution near (x_0, y_0) depending on two parameters: the values $X(x_0, y_0) = a$, $Y(x_0, y_0) = b$.

Now some infinite pseudogroups:

Example 2.3 (Volume preserving maps). Take $\mathcal{G}$ as the collection of transformations $(x, y) \mapsto (X, Y)$ preserving the volume form $dx \wedge dy$ on $\mathbb{R}^2$. A local diffeomorphism is in $\mathcal{G}$ if and only if $X_x Y_y - X_y Y_x = 1$. This is a nonlinear analytic partial differential equation, and $\mathcal{G}$ is a Lie pseudogroup. There is an obvious generalization to $\mathbb{R}^n$.

Example 2.4 (Conformal maps). The set of local transformations mapping (x, y) to (X, Y) and satisfying the Cauchy–Riemann equations:

$$X_x = Y_y, \qquad X_y = -Y_x$$

constitute the infinite pseudogroup of conformal maps of the plane. In the form we have written it here, the derivatives X_x, X_y are principal, while Y_x, Y_y are parametric, along with X, Y.

The transformations of Example 2.4 are intimately associated with classical complex analysis, while Example 2.3 is central to incompressible flow theory. Similarly, Hamiltonian dynamics studies maps which preserve a symplectic 2-form and which constitute an infinite Lie pseudogroup.

However, it is easy to construct pseudogroups not of Lie type. The following examples are due to Lie and Cartan:

Example 2.5 (Isotropy subgroup). The isotropy subgroup $\mathcal{G}_{x_1}$ of a Lie pseudogroup $\mathcal{G}$

$$\mathcal{G}_{x_1} = \{\tau \in \mathcal{G} \mid \tau(x_1) = x_1\}$$

is a pseudogroup, but not generally of Lie type, since the condition that τ fix x_1 is not a DE.

Example 2.6. The pseudogroup

$$X = f(x), \qquad Y = f(y)$$

(n.b. same f!) is not of Lie type .

The significance of the defining system is that it provides a fully *local* criterion for membership of a pseudogroup $\mathcal{G}$. Given a local diffeomorphism τ on some neighbourhood of P, the membership question $\tau \in \mathcal{G}$ is decided by checking finitely many derivatives of τ in a neighbourhood of P. In contrast, in Example 2.5 we cannot check membership $\tau \in \mathcal{G}$ without reference to a remote point x_1. Similar comments hold for Example 2.6.

2.4   Symmetries of Defining System

Our approach here follows Olver [28, §2.2]. Suppose we have a system $\mathcal{R}$ of differential equations, specifying a variety in $J^1(M, M)$. Let the independent variables be x and dependent variables X. Let $\tau \colon U \to M$ be a local diffeomorphism of M. Let $V \subseteq M$, and let $f \colon V \to N$ be some map. The *graph* of f is the set $\Gamma_f \subseteq M \times N$ given by

$$\Gamma_f = \big\{ \big(x, f(x)\big) \mid x \in V \big\}.$$

If $V \subseteq U$, then τ acts on Γ_f by

$$\begin{aligned} \tau \cdot \Gamma_f &= \big\{ \big(\tau(x), f(x)\big) \mid x \in V \big\} \\ &= \big\{ \big(x, f \circ \tau^{-1}(x)\big) \mid x \in \tau(V) \big\} \end{aligned}$$

and hence $\tau \cdot \Gamma_f$ is the graph of the function $f \circ (\tau|_V)^{-1}$. This motivates the following:

Definition 2.3 (Symmetry of d.e.). Let $\mathcal{R}$ be a system of differential equations with independent variables $x \in M$ and dependent variables $X \in N$. Let $U \subseteq M$ be an open set, and $\tau \colon U \to M$ be a local diffeomorphism. Then τ is a *symmetry* of $\mathcal{R}$ if for every local solution $f \colon V \to N$ defined on an open subset $V \subseteq U$, we have that $f \circ (\tau|_V)^{-1}$ is also a local solution of $\mathcal{R}$.

The following lemma is of central importance in Lie pseudogroup theory.

Lemma 2.4. *Let $\mathcal{G}$ be a Lie pseudogroup on M. Let $\mathcal{G}$ have involutive defining system $\mathcal{R}$ with independent variables x and dependent variables X. Then a local diffeomorphism τ of M is in $\mathcal{G}$ if and only if τ is a symmetry of $\mathcal{R}$.*

That is, a defining system for $\mathcal{G}$ admits $\mathcal{G}$ as symmetries acting on its independent variables x only.

Proof. Let $\tau \in \mathcal{G}$ be a local diffeomorphism $\tau \colon U \to M$ of M. Let f be a local solution $f \colon V \to M$ of $\mathcal{R}$, with $V \subseteq U$. By Definition 2.2 of Lie pseudogroup, $f \in \mathcal{G}$. Definition 2.1 (i, v, iii) of pseudogroup then imply in turn that $\tau|_V$, $(\tau|_V)^{-1}$ and $f \circ (\tau|_V)^{-1}$ are in $\mathcal{G}$. But by Definition 2.2 of Lie pseudogroup, we then have $f \circ (\tau|_V)^{-1}$ is a solution of $\mathcal{R}$. So by Definition 2.3, τ is a symmetry of $\mathcal{R}$.

For the converse, suppose that τ is a symmetry of $\mathcal{R}$. Thus $f \circ (\tau|_V)^{-1}$ is a local solution of $\mathcal{R}$, and hence is in $\mathcal{G}$ by Definition 2.2. Premultiplying by

f^{-1} (which is in $\mathcal{G}$ by Definitions 2.2 and 2.1(v)), we find that $(\tau|_V)^{-1}$, and hence (by Definition 2.1(v)) $\tau|_V$ are in $\mathcal{G}$. But then $\tau|_V$ is a solution of $\mathcal{R}$ (by Definition 2.2) for all open $V \subseteq U$, and hence by Definitions 2.1(ii) and 2.2, τ is a solution of $\mathcal{R}$. □

Actually it is equally true that $\mathcal{R}$ admits $\mathcal{G}$ acting on the dependent variables: $x' = x$, $X' = \tau(x)$.

If a PDE system is involutive it is locally solvable. That is, at any point P on the surface $f(x, X, X^{(1)}) = 0$, there is a graph of a local solution passing through P. Olver [28, p.165] shows that a symmetry (which maps solutions to solutions of $\mathcal{R}$) maps the variety $\mathcal{R}$ to itself if it is locally solvable. For involutive systems, this implies that the subset $\mathcal{R}_0$ of 0-th order equations in $\mathcal{R}$ is mapped to itself by a symmetry of $\mathcal{R}$.

This gives us a method for finding scalar invariants of a Lie pseudogroup. Let $\mathcal{R}_0$ be given by

$$f^i(x, X) = 0, \qquad i = 1, \ldots, l.$$

Let X_0 be a point where $\left[\frac{\partial f^i}{\partial X^j}\right]$ is of full rank l. The implicit function theorem guarantees that we can write $\mathcal{R}_0$ in the form

$$Z^i = g^i(x, Y), \qquad i = 1, \ldots, l$$

where $X = (Y, Z)$, Y are the parametric 0-th order derivatives, and Z the principal.

Proposition 2.5. *The functions $J^i(x) = g^i(x, Y_0)$ are a basis for the scalar invariants of $\mathcal{G}$ in a neighbourhood of $X_0 = (Y_0, Z_0)$.*

Example 2.7. Consider the defining system (5), which has 0-th order equation in solved form

$$Z = yz/Y.$$

Choosing a convenient value of Y_0 (say $Y_0 = 1$) the proposition shows that $J = yz$ is an invariant of the pseudogroup defined by (5), as can be seen directly.

3 Cartan Structure Theory

Lie [22] and his student Tresse [38] defined infinite Lie pseudogroups and established some of their properties. However, they did not exhibit a structure theory for infinite groups: Vessiot [39, 40] and Cartan [5, 8] did this later. Vessiot worked with the defining system of the Lie pseudogroup, aiming at a Jordan–Hölder decomposition of the pseudogroup into a sequence of normal subgroups with simple quotients. (See [30] for a modern exposition.) It took until 1966 before Guillemin [14] rigorously established a Jordan–Hölder theorem for infinite Lie pseudogroups.

Unfortunately the Vessiot theory provides no analogue of the structure constants c_{ij}^k so central to the structure of finite Lie groups. However, Cartan [5] developed an alternative structure theory for infinite Lie pseudogroups, based on a generalization of the Maurer–Cartan structure equations, and which does provide analogues of the c_{ij}^k. Cartan developed his structure theory in a remarkable sequence of papers collected in [11]. There are three interwoven strands. First is Cartan's theory of integrability of exterior differential systems in involution, which he developed in a paper of 1905 *Sur la structure des groupes infinis de transformations* [5], and described in [10]. The second strand is the Cartan structure equations, which appear in the same paper. The third strand is the method of equivalence, developed in a long paper of 1908 called *Les sous-groupes des groupes continus de transformations* [6]. The fact that both the integrability theory and the method of equivalence appear in papers about infinite groups indicates the central role they played in Cartan's thinking at this time. Cartan reviewed his structure theory in a very readable seminar paper from 1937 called *La structure des groupes infinis* [8]. Cartan used Lie's definition of pseudogroup essentially unchanged. However, where Lie worked with the defining system as differential equations, Cartan worked with defining system as 1-forms.

3.1 Isomorphism

For an algebraic object defined axiomatically, a "structural" feature is one which is preserved under isomorphism. Infinite Lie pseudogroups are defined via an action on some manifold M, so it is not obvious what is meant by "isomorphism" here. Certainly we demand that the structure of a pseudogroup should be independent of coordinates:

Definition 3.1 (Similarity). A pseudogroup $\mathcal{G}_2$ on manifold M_2 is similar to $\mathcal{G}_1$ on M_1, if there is a diffeomorphism $\phi : M_2 \to M_1$ such that $\mathcal{G}_2 = \phi^{-1} \circ \mathcal{G}_1 \circ \phi$.

However, more is needed, since the "same" pseudogroup can act on different manifolds.

Example 3.1. The pseudogroup of local diffeomorphisms of $\mathbb{R}$ $X = f(x)$ and the pseudogroup on $\mathbb{R}^2$

$$X = f(x), \qquad Y = f'(x)y$$

share the same law of composition on f. The first pseudogroup sits inside the second one, and can be obtained from it by projection.

Definition 3.2 (Prolongation). Let $\mathcal{G}$ be a pseudogroup acting on M. Let $\pi : \hat{M} \to M$ be fibred over M and let $\hat{\mathcal{G}}$ be a pseudogroup acting on $\hat{M}$. If every $\hat{\tau} \in \hat{\mathcal{G}}$ projects to $\tau \in \mathcal{G}$ via $\pi \circ \hat{\tau} = \tau \circ \pi$, then we call $\hat{\mathcal{G}}$ a *prolongation* of $\mathcal{G}$.

Example 3.2. The pseudogroup

$$X = f(x), \qquad Y = g(x, y)$$

acting on $\mathbb{R}^2$ is a prolongation of $X = f(x)$ acting on $\mathbb{R}$.

Projection of $\hat{\mathcal{G}}$ constitutes a Lie pseudogroup homomorphism. We are most interested in prolongations where the kernel of this homomorphism is trivial:

Definition 3.3 (Isomorphic prolongation). A prolongation $\hat{\mathcal{G}}$ on $\hat{M}$ of $\mathcal{G}$ on M is called *isomorphic* if the only diffeomorphism in $\hat{\mathcal{G}}$ that projects to id_M is $\mathrm{id}_{\hat{M}}$.

The prolongation of Example 3.1 is isomorphic, while that of Example 3.2 is not.

Definition 3.4 (Isomorphism). Two Lie pseudogroups $\mathcal{G}_1$, $\mathcal{G}_2$ acting on manifolds M_1, M_2 are called isomorphic if there exist isomorphic prolongations $\hat{\mathcal{G}}_1$, $\hat{\mathcal{G}}_2$ such that $\hat{\mathcal{G}}_1$ and $\hat{\mathcal{G}}_2$ are similar.

Example 3.3. Consider the realizations

$$
\text{(a)} \ \left\{ \begin{array}{l} P = p + a \\ Q = q + b \end{array} \right. \qquad \text{(b)} \ \left\{ \begin{array}{l} X = x \\ Y = y \\ U = u + ax + by \end{array} \right. \tag{6}
$$

on $\mathbb{R}^2$, $\mathbb{R}^3$ of a two-dimensional Abelian Lie group. Note that the orbit dimensions also differ (2 vs. 1). The pseudogroups are isomorphic, since they both prolong isomorphically to the pseudogroup on $\mathbb{R}^5$:

$$
X = x, \quad Y = y, \quad U = u + ax + by, \quad P = p + a, \quad Q = q + b. \tag{7}
$$

The usual way to prolong a space is where the coordinates represent independent and dependent variables in a PDE, and we lift to an action on derivatives. For instance, if we take $J^0(\mathbb{R}^2, \mathbb{R})$ with coordinates (x, y, u), prolongation means constructing action on (x, y, u, p, q) (where $p \equiv u_x$, $q \equiv u_y$) via the standard methods [28]. Start with a diffeomorphism $\tau : (x, y, u) \mapsto (X, Y, U)$. Lift it to $\hat{\tau} : (x, y, u, p, q) \mapsto (X, Y, U, P, Q)$ by requiring that $\hat{\tau}$ preserve the contact form

$$
\Omega_1^1 = du - p\, dx - q\, dy.
$$

For instance, this process lifts the action (6b) to (7).

3.2 Construction of Invariant 1-forms

Cartan's structure equations are closely analogous to the Maurer–Cartan equations for a Lie group. The method is still to construct invariant 1-forms ω^i, but now these forms do not live on the manifold M on which $\mathcal{G}$ acts. Instead ω^i are invariant under an isomorphic prolongation $\hat{\mathcal{G}}$. We now show how Cartan constructs this prolongation and hence how to obtain the forms ω^i. See [4, 10] for a thorough discussion of exterior differential systems, and [1, 15] for algorithmic implementations.

It is supposed that the defining system $\mathcal{R}$ of the pseudogroup $\mathcal{G}$ is available, and that $\mathcal{R}$ is involutive. For simplicity assume that $\mathcal{R}$ is of first order and contains no equations of order 0. Then $\mathcal{R}$ may be taken as locally having the form

$$
X_{x^j}^i = f_j^i(x, X, u)
$$

where the principal derivatives $X_{x^j}^i$ are some subset of all the first order derivatives, and u denotes the parametric first order derivatives. Begin with the contact ideal

$$
\Omega^1(\mathbb{R}^n, \mathbb{R}^n) = \{\Omega^i, i = 1, \dots, n\}
$$

where

$$\Omega^i = dX^i - u^i_j dx^j.$$

Pulling it back to $\mathcal{R}$, we obtain an exterior differential system in involution, which is the defining system recast in terms of 1-forms.

Example 3.4. Consider the defining system (5) for the finite pseudogroup (4). Pull back the contact forms

$$dX - X_x\,dx - X_y\,dy - X_z\,dz$$
$$dY - Y_x\,dx - Y_y\,dy - Y_z\,dz$$
$$dZ - Z_x\,dx - Z_y\,dy - Z_z\,dz$$

to get the involutive Pfaff system

$$\theta^1 = dX - \left(\frac{X}{x}\right)^2 \frac{Y}{y}\,dx,$$

$$\theta^2 = dY - \frac{Y}{y}\,dy,$$

$$\theta^3 = dZ - \frac{y}{Y}\,dz. \tag{8}$$

Since there is also a 0-th order equation $Z = yz/Y$, adjoin this to the Pfaff system. The resulting system can be written in the pregnant form

$$YZ = yz, \qquad \frac{dX}{X^2Y} = \frac{dx}{x^2y}, \qquad \frac{dY}{Y} = \frac{dy}{y}, \qquad Y\,dZ = y\,dz$$

expressing invariance of the scalar yz and the 1-forms

$$\omega^1 = \frac{1}{x^2y}dx, \qquad \omega^2 = \frac{1}{y}dy, \qquad \omega^3 = z\,dy + y\,dz.$$

Cartan showed that this rewriting of $\mathcal{R}$ as expressing invariance of scalars and 1-forms is always possible. First, note that the 1-forms of the restricted contact system are

$$\theta^i = dX^i - g^i_j(x, X, u)\,dx^j, \qquad i = 1, \ldots, n \tag{9}$$

where g^i_j are some functions. Start with a diffeomorphism $x \mapsto \tau(x)$ from the pseudogroup $\mathcal{G}$, then prolong it to $M \times M$ by $x \mapsto \tau(x)$, $X \mapsto X$, and then prolong to $\hat{\tau}$ on $J^1(M, M)$ by the usual prolongation process. According to Lemma 2.4, the variety $\mathcal{R}$ in $J^1(M, M)$ specified by the defining system is invariant under the action of this prolonged diffeomorphism $\hat{\tau}$. Hence $\hat{\tau}$ restricts to the variety $\mathcal{R}$. It also follows that the ideal $\mathcal{I}$ generated by these forms is invariant under pullback by a diffeomorphism from the pseudogroup obtained by restricting $\hat{\mathcal{G}}$ to $\mathcal{R}$. Because of the way that dX^i occur in θ^i (9), invariance of the ideal $\mathcal{I}$ implies that each θ^i is invariant. But since $\hat{\tau}$ has trivial action on X^i, it follows that each dX^i is invariant, and hence the forms $g^i_j(x, X, u)\,dx^j$ are invariant. Moreover, evaluating at $X = X_0$ some regular value of X, we have the invariant 1-forms

$$\omega^i = g^i_j(x, X_0, u)\,dx^j, \qquad i = 1, \ldots, n$$

Example 3.5. Discarding the dX^i from (8), and setting $X = 1$, $Y = 1$ (or any other value $X \neq 0$, $Y \neq 0$), we obtain the invariant 1-forms

$$\omega^1 = \frac{1}{x^2 y} dx, \qquad \omega^2 = \frac{1}{y} dy, \qquad \omega^3 = z \, dy + y \, dz.$$

as before.

In the above example the pseudogroup is finite so there are no u's (parametric first order derivatives) present.

Example 3.6. Start with the first order involutive defining system for a pseudogroup $\mathcal{G}$ on $\mathbb{R}^2$:

$$X_x = 1, \qquad X_y = 0, \qquad Y_y = 1 \tag{10}$$

which can be recast as 1-forms

$$dX - dx, \qquad dY - dy - u \, dx$$

(where u represents the parametric derivative Y_x), yielding the invariant forms

$$\omega^1 = dx, \qquad \omega^2 = dy + u \, dx. \tag{11}$$

It is worth explicitly constructing the prolongation. Starting with an explicit transformation $\tau \in \mathcal{G}$ on $\mathbb{R}^2$:

$$x \mapsto x + a, \qquad y \mapsto y + f(x) \tag{12}$$

we prolong to $J^0(\mathbb{R}^2, \mathbb{R}^2)$ (x, y, X, Y) by adjoining $X \mapsto X$, $Y \mapsto Y$, then to $J^1(\mathbb{R}^2, \mathbb{R}^2)$ $(x, y, X, Y, X_x, X_y, Y_x, Y_y)$ by adjoining

$$X_x \mapsto X_x - f'(x) X_y, \qquad X_y \mapsto X_y, \qquad Y_x \mapsto Y_x - f'(x) Y_y, \qquad Y_y \mapsto Y_y.$$

Finally, restricting to the variety of the defining system (10), we obtain the prolonged pseudogroup $\hat{\mathcal{G}}$:

$$x \mapsto x + a, \qquad y \mapsto y + f(x), \qquad u \mapsto u - f'(x) \tag{13}$$

where $u \equiv Y_x$. It is readily confirmed that the forms ω^1, ω^2 (11) are invariant under $\hat{\mathcal{G}}$.

3.3 Structure Equations

Cartan developed his structure equations in [11, pp.594–624,1335–1351]. We note that the n 1-forms ω^i are

$$\omega^k = A_i^k(x, u) dx^i$$

where the matrix A_i^k is of full rank. Hence on taking the exterior derivative one finds

$$d\omega^k = \theta_i^k \wedge \omega^i$$

for some forms θ_j^k. Also, since $d\omega^k$ are invariant, it follows that the θ_i^k are invariant mod $\{\omega^1, \ldots, \omega^n\}$. Resolving θ_i^k with respect to a basis $\{\omega^1, \ldots, \omega^n, \pi^1, \ldots, \pi^r\}$, we get

$$d\omega^k = a_{i\rho}^k \pi^\rho \wedge \omega^i - \tfrac{1}{2} c_{ij}^k \omega^i \wedge \omega^j \tag{14}$$

where $a_{i\rho}^k$ and c_{ij}^k are scalar invariants and the factor of $1/2$ is for convenience. The forms π^ρ are invariant mod ω^k. Equations (14) are called the (first) Cartan structure equations. Note also that if the pseudogroup is finite (with first order involutive defining system), no prolongation is necessary, the π^ρ are absent, and the Maurer–Cartan equations are recovered.

Example 3.7. Hitting ω^1, ω^2 (11), we find

$$d\omega^1 = 0, \qquad d\omega^2 = du \wedge dx = \pi^1 \wedge \omega^1 \tag{15}$$

where $\pi^1 = du$ (or indeed $du + $ (any multiple of ω^1)). Observe that acting on π^1 with the prolonged pseudogroup (13), we find

$$\pi^1 \mapsto \pi^1 - f''(x)\omega^1$$

confirming that π^1 is invariant mod ω^i.

Example 3.8. In Table 1 we present the defining systems, invariant differential 1-forms, and Cartan structure of Cartan's list [11, pp.620–624] of infinite pseudogroups on $\mathbb{R}^2$ with first order defining system. The labelling in Table 1 follows Cartan:

(a) 1 arbitrary function   of 1 variable.
(b) 2 arbitrary functions   of 1 variable.
(c) 1 arbitrary function   of 2 variables.
(d) 1 arbitrary function   of 2 variables + 1 function of 1 variable.
(e) 2 arbitrary functions   of 2 variables.

The pseudogroups are transitive except as noted.

3.3.1 Linear Isotropy Group

For later use we now define the linear isotropy group. Let M be the manifold (with coordinates x) on which a pseudogroup $\mathcal{G}$ acts. Let $\tau \in \mathcal{G}$ be a local diffeomorphism of M; it maps a point $P \in M$ to the point $\tau(P) \in M$. The *isotropy* (or stabilizer) subgroup for a particular point $P \in M$ is the set

$$\mathcal{G}_P = \{\tau \in \mathcal{G} : \tau(P) = P\}$$

of transformations leaving the point P fixed. Let $T_P M$ be the tangent space of M at P. Any $\tau \in \mathcal{G}$ induces a map $\tau_* : TM \to TM$, which when written out in coordinates encodes the Jacobian matrix B of τ. That is, if $V_P \in T_P M$ is a tangent vector at P, then the transformation τ_* carries V_P to a vector $W_{\tau(P)}$ at the image point $\tau(P)$. This is a linear map $T_P M \to T_{\tau(P)} M$. In particular if $\tau \in \mathcal{G}_P$ leaves P fixed then τ_* induces a linear map $T_P M \to T_P M$ specified by this Jacobian matrix B. Since $\mathcal{G}_P$ is a pseudogroup, it follows that the collection of all such linear maps

$$G_P = \{\tau_{P*} : T_P M \to T_P M, \tau \in \mathcal{G}_P\}$$

Case	Pseudogroup	Defining System	Defining Forms	Invariant 1-Forms	Cartan Structure	Comments
(a)	$X = f(x)$ $Y = y + a$	$X_y = 0$ $Y_x = 0$ $Y_y = 1$	$dX - u\,dx$ $dY - dy$	$\omega^1 = u\,dx$ $\omega^2 = dy$	$d\omega^1 = \pi^1 \wedge \omega^1$ $d\omega^2 = 0$	
(a$_1$)	$X = x$ $Y = y + f(x)$	$X = x$ $Y_y = 1$	$X - x$ $dX - dx$ $dY - dy - u\,dx$	$\omega^1 = dx$ $\omega^2 = dy + u\,dx$	$d\omega^1 = 0$ $d\omega^2 = \pi^1 \wedge \omega^1$	Intransitive
(a$_1'$)	$X = x + a$ $Y = y + f(x)$	$X_x = 1$ $X_y = 0$ $Y_y = 1$	$dX - dx$ $dY - dy - u\,dx$	$\omega^1 = dx$ $\omega^2 = dy + u\,dx$	$d\omega^1 = 0$ $d\omega^2 = \pi^1 \wedge \omega^1$	Isomorphic prolongation of $X = f(x)$.
(a$_2$)	$X = f(x)$ $Y = yf'(x)$	$X_x = Y/y$ $X_y = 0$ $Y_y = Y/y$	$dX - \frac{Y}{y}dx$ $dY - Yu\,dx - \frac{Y}{y}dy$	$\omega^1 = \frac{1}{y}dx$ $\omega^2 = \frac{1}{y}dy + u\,dx$	$d\omega^1 = \omega^1 \wedge \omega^2$ $d\omega^2 = \pi^1 \wedge \omega^1$	
(b$_1$)	$X = f(x)$ $Y = y + g(x)$	$X_y = 0$ $Y_y = 1$	$dX - u\,dx$ $dY - v\,dx - dy$	$\omega^1 = u\,dx$ $\omega^2 = dy + v\,dx$	$d\omega^1 = \pi^1 \wedge \omega^1$ $d\omega^2 = \pi^2 \wedge \omega^1$	
(b$_2$)	$X = f(x)$ $Y = g(y)$	$X_y = 0$ $Y_x = 0$	$dX - u\,dx$ $dY - v\,dy$	$\omega^1 = u\,dx$ $\omega^2 = v\,dy$	$d\omega^1 = \pi^1 \wedge \omega^1$ $d\omega^2 = \pi^2 \wedge \omega^2$	
(b$_3$)	$X = f(x)$ $Y = y[f'(x)]^{1/m} + g(x)$	$X_y = 0$ $Y_y = X_x^{1/m}$	$dX - u^m\,dx$ $dY - v\,dx - u\,dy$	$\omega^1 = u^m\,dx$ $\omega^2 = v\,dx + u\,dy$	$d\omega^1 = m\pi^1 \wedge \omega^1$ $d\omega^2 = \pi^2 \wedge \omega^1 + \pi^1 \wedge \omega^2$	

Table 1: Pseudogroups on $\mathbb{R}^2$ with first order defining system.

Case	Pseudogroup	Defining System	Defining Forms	Invariant 1-Forms	Cartan Structure	Comments
(c)	$X = f(x,y)$ $Y = y$	$Y = y$	$Y - y$ $dX - u\,dx - v\,dy$ $dY - dy$	$\omega^1 = u\,dx + v\,dy$ $\omega^2 = dy$	$d\omega^1 = \pi^1 \wedge \omega^1 + \pi^2 \wedge \omega^2$ $d\omega^2 = 0$	Intransitive
(c')	$X = f(x,y)$ $Y = y + a$	$Y_x = 0$ $Y_y = 1$	$dX - u\,dx - v\,dy$ $dY - dy$	$\omega^1 = u\,dx + v\,dy$ $\omega^2 = dy$	$d\omega^1 = \pi^1 \wedge \omega^1 + \pi^2 \wedge \omega^2$ $d\omega^2 = 0$	
(d$_1$)	$X = f(x)$ $Y = g(x,y)$	$X_y = 0$	$dX - u\,dx$ $dY - v\,dx - w\,dy$	$\omega^1 = u\,dx$ $\omega^2 = v\,dx + w\,dy$	$d\omega^1 = \pi^1 \wedge \omega^1$ $d\omega^2 = \pi^2 \wedge \omega^1 + \pi^3 \wedge \omega^2$	
(d$_2$)	$X = f(x,y)$ $Y = g(x,y)$, $f_x g_y - f_y g_x = 1$	$X_x Y_y - X_y Y_x = 1$	$dX - u\,dx - v\,dy$ $dY - w\,dx - z\,dy$	$\omega^1 = u\,dx + v\,dy$ $\omega^2 = w\,dx + z\,dy$	$d\omega^1 = \pi^1 \wedge \omega^1 + \pi^2 \wedge \omega^2$ $d\omega^2 = \pi^3 \wedge \omega^1 - \pi^1 \wedge \omega^2$	$uz - vw = 1$
(e)	$X = f(x,y)$ $Y = g(x,y)$	(null)	$dX - u\,dx - v\,dy$ $dY - w\,dx - z\,dy$	$\omega^1 = u\,dx + v\,dy$ $\omega^2 = w\,dx + z\,dy$	$d\omega^1 = \pi^1 \wedge \omega^1 + \pi^2 \wedge \omega^2$ $d\omega^2 = \pi^3 \wedge \omega^1 + \pi^4 \wedge \omega^2$	$uz - vw \neq 0$

Table 1: (cont.) Pseudogroups on $\mathbb{R}^2$.

is a linear Lie group – a finite (say p-dimensional) representation of the (infinite) isotropy pseudogroup. Cartan calls this the "linear stability group"[2]. The associated linear Lie algebra g_P is shown by Cartan to be the one explicitly given by the $a_{i\rho}^k$ (17) (see §4.2).

3.3.2 Transitivity

Definition 3.5 (Transitivity). Let $\mathcal{G}$ be a pseudogroup on manifold M. Then $\mathcal{G}$ is *transitive* if for any point $P \in M$ there is a neighbourhood $U \ni P$ such that if $Q \in U$ there is a diffeomorphism $\tau \in \mathcal{G}$ such that $\tau(P) = Q$.

Transitive Lie pseudogroups are those possessing no scalar invariants, that is, whose involutive defining system includes no 0-th order equations. Transitivity greatly simplifies analysis and classification, and considerably more is known about transitive Lie pseudogroups than about intransitive. Cartan [5] showed that for a transitive pseudogroup, both the c_{ij}^k and $a_{i\rho}^k$ can be chosen to be constants.

Of course, transitivity of a pseudogroup $\mathcal{G}$ is not a structural property, depending in detail on the nature of the space where $\mathcal{G}$ acts. For instance, the group action $x \mapsto x$, $y \mapsto y + a$ on $\mathbb{R}^2$ is intransitive, but is the isomorphic prolongation of a transitive group action $y \mapsto y + a$ on $\mathbb{R}$. In the finite parameter case, intransitivity can always be removed by a process of prolongation and projection. However, this is not so for infinite Lie pseudogroups. For instance, the pseudogroup

$$X = x, \qquad Y = y + f(x)$$

is intransitive, and all the prolongation and changes of variable in the world will never change this fact. The x is invariant, but can't be dropped without losing some (most) of the structure of the pseudogroup.

3.3.3 Nonconstant Structure

Here is a most interesting intransitive example (due to Cartan) which exhibits another complication.

Example 3.9. Consider the infinite Lie pseudogroup

$$X = x, \quad Y = f(y), \quad Z = z\bigl(f'(y)\bigr)^x + \phi(x, y).$$

Its defining system is

$$X = x, \quad Y_x = 0, \quad Y_z = 0, \quad Z_z = (Y_y)^x,$$

from which we obtain 1-forms

$$dX - dx, \quad dY - u\,dy, \quad dZ - v\,dx - w\,dy - u^x\,dz.$$

Hence we immediately have three invariant 1-forms:

$$\omega^1 = dx, \quad \omega^2 = u\,dy, \quad \omega^3 = v\,dx + w\,dy + u^x\,dz.$$

---

[2]In [5], Cartan confusingly called it the "adjoint group".

Taking exterior derivatives, we obtain Cartan structure equations

$$d\begin{pmatrix} \omega^1 \\ \omega^2 \\ \omega^3 \end{pmatrix} = \begin{pmatrix} 0 & 0 & 0 \\ 0 & \pi^1 & 0 \\ \pi^2 & \pi^3 & x\,\pi^1 \end{pmatrix} \wedge \begin{pmatrix} \omega^1 \\ \omega^2 \\ \omega^3 \end{pmatrix}$$

where the forms π^ρ are:

$$\begin{aligned}
\pi^1 &= du/u \\
\pi^2 &= dv - \frac{xv}{u}\,du - u^2 \ln u\,dz \\
\pi^3 &= \frac{1}{u}(dw - \frac{xw}{u}\,du).
\end{aligned}$$

The fascinating feature of this pseudogroup is that one of the structure coefficients is nonconstant, and depends on the invariant x. This dependence cannot be absorbed by redefining π^ρ.

Nonconstancy of structure is one reason why the theory of intransitive infinite Lie pseudogroups is so much more difficult than the transitive case. Resolving which pseudogroups have "truly" nonconstant structure is a nontrivial task:

Example 3.10. Consider the pseudogroup $\mathcal{G}_1$ on $\mathbb{R}^3$ given by

$$\begin{aligned}
X &= x \\
Y &= y + f(x)z + f'(x) \\
Z &= z.
\end{aligned} \tag{16}$$

This intransitive pseudogroup has two scalar invariants x, z. The defining system is

$$\begin{array}{ll}
X = x & Y_y = 1 \\
Z = z & Y_{xz} = -zY_z + Y - y \\
& Y_{zz} = 0.
\end{array}$$

After some calculation, one finds invariant 1-forms

$$\begin{aligned}
\omega^1 &= dx \\
\omega^2 &= dy + u\,dx + v\,dz \\
\omega^3 &= dz \\
\omega^4 &= du - w\,dx + (zv + y)\,dz \\
\omega^5 &= dv + (zv + y)\,dx
\end{aligned}$$

(where $u \equiv Y_x$, $v \equiv Y_z$, $w \equiv Y_{xx}$). The structure equations are

$$\begin{aligned}
d\omega^1 &= 0 \\
d\omega^2 &= \omega^4 \wedge \omega^1 + \omega^5 \wedge \omega^3 \\
d\omega^3 &= 0 \\
d\omega^4 &= \pi^1 \wedge \omega^1 + z\omega^5 \wedge \omega^3 - \omega^3 \wedge \omega^2 \\
d\omega^5 &= z\omega^5 \wedge \omega^1 + \omega^2 \wedge \omega^1.
\end{aligned}$$

Note that this pseudogroup has nonconstant structure coefficients.

However, the pseudogroup

$$X = x, \qquad Y = y + f(x)z + f'(x), \qquad Z = z, \qquad W = w + f(x)$$

is an isomorphic prolongation of $\mathcal{G}_1$. Moreover, it is an isomorphic prolongation of $\mathcal{G}_2$ given by

$$X = x, \qquad W = w + f(x).$$

But then $\mathcal{G}_1$, $\mathcal{G}_2$ are isomorphic. Thus the invariant z does not appear in every pseudogroup isomorphic to $\mathcal{G}_1$; indeed $\mathcal{G}_2$ has constant structure given by (15).

4 How it Works with Vector Fields

Calculating symmetries of differential equations or other geometric objects would scarcely be possible if we had to work with local transformations: the calculations are too difficult. Symmetry analysis [3, 28, 29] is feasible because it works infinitesimally with *vector fields*, which renders the problem linear and therefore tractable. As remarked in [28, p.43], " ... almost the entire range of applications of Lie groups to differential equations ultimately rests on this one construction". Lie [22] was in no doubt that similar considerations held for infinite Lie groups:

> Through the introduction and fundamental usage of the infinitesimal transformation the theory of infinite transformation groups acquires a surprising simplicity. Here, as in the theory of finite groups, the infinitesimal transformations form the proper foundation of the theory.

However, Lie and his circle never developed an adequate structure theory for infinite pseudogroups. When Cartan and Vessiot finally did so, they worked with the defining equations of the pseudogroup *not* its vector fields. Cartan [8] remarks:

> The generalization to infinite groups of the structure theory for finite groups due to Lie, and based on consideration of infinitesimal transformations, has proved to be very difficult, if not impossible ...

Despite the difference in starting point, Cartan's structure theory provides structure constants which directly generalize Lie's c_{ij}^k. If the c_{ij}^k are available at the infinitesimal level in the finite case why is Cartan forced to work at the pseudogroup level in the infinite case? In fact Kuranishi [20, 21] and Singer and Sternberg [37] showed that Cartan's structure constants were infinitesimal quantities. This alternative way of looking at the structure constants is our subject for the rest of these notes. Our aim is to get c_{ij}^k and $a_{i\rho}^k$ from the infinitesimal defining system, which we continue to assume is of *first order*.

4.1 Manufacturing Transitive Pseudogroups

There are close parallels between classical matrix Lie algebras and the infinite "Lie algebras" associated with infinite Lie pseudogroups. In particular many (not all!) of the series $gl(n)$, $sl(n)$, $sp(2n)$, $csp(2n)$, $so(n)$, $co(n)$, of matrix groups have analogues in the infinite case. Passage from the finite to the infinite case can be done as follows [37].

Suppose one is given a matrix Lie algebra whose matrices $A = [a^i_j]$ are defined as the solution space of a set of linear homogeneous algebraic equations for the a^i_j. Then one can pass to the defining system of a Lie algebra of vector fields simply by replacing a^i_j by $\frac{\partial \xi^i}{\partial x^j}$ in these algebraic equations. The resulting pseudogroup $\mathcal{G}$ is transitive, since the translations $\frac{\partial}{\partial x^i}$ are always solutions of the defining system. However, $\mathcal{G}$ may be finite or infinite, and the resulting defining equations may not be involutive.

Example 4.1. Consider the Lie algebra $sl(n)$, defined by $\sum_{i=1}^n a^i_i = 0$. Pass to the defining system $\sum_{i=1}^n \frac{\partial \xi^i}{\partial x^i} = 0$ which is the involutive infinitesimal defining system of an infinite Lie pseudogroup, expressing "incompressibility" (see for instance [2]).

Example 4.2. The constant Lie algebra!conformal $co(2)$ consisting of matrices $\mathbf{A} = \left(\begin{smallmatrix} \alpha & -\beta \\ \beta & \alpha \end{smallmatrix} \right)$ is defined by $a^2_1 + a^2_2 = 0$, $a^1_1 - a^2_2 = 0$. Pass to the defining system

$$\xi_x = \eta_y, \qquad \xi_y = -\eta_x$$

which are the Cauchy–Riemann equations, are first order involutive, and give an infinite Lie pseudogroup of local conformal transformations.

Example 4.3. Consider the orthogonal algebra $so(2)$, defined by $a^1_1 = a^2_2 = 0$, $a^1_2 + a^2_1 = 0$. The corresponding infinitesimal defining system

$$\xi_x = 0, \qquad \xi_y = -\eta_x, \qquad \eta_y = 0$$

is not first order involutive. Indeed, prolonging once, one deduces the equation $\eta_{xx} = 0$, and hence the pseudogroup is *finite*. Similar comments hold for $so(n)$.

The Lie algebra systems constructed in this way have $c^k_{ij} = 0$ WLOG, and are called *flat* by Singer and Sternberg [37]. There is no straightforward way to characterize whether a matrix Lie algebra gives rise to infinite or finite Lie pseudogroup in this way.

4.2   Linear Isotropy Algebra

The Lie algebra corresponding to the isotropy group $\mathcal{G}_P$ (§3.3.1) at the point P is easily characterized. Namely if $\mathcal{L}$ is the Lie algebra associated with pseudogroup $\mathcal{G}$, let

$$\mathcal{L}_P = \{\mathbf{X} \in \mathcal{L} : \mathbf{X}|_P = 0\}.$$

Call this the Lie isotropy algebra at P.

We now perform the same linearization for $\mathcal{L}_P$ as done for the isotropy group $\mathcal{G}_P$ in §3.3.1. Let $V \in T_P M$ be an arbitrary tangent vector at P. A vector field $\mathbf{X} \in \mathcal{L}_P$ generates a one-parameter group of local diffeomorphisms τ_ε, which in turn give rise to the (Jacobian) matrices B_ε in the linear isotropy group G_P. Then taking the limit

$$\lim_{\varepsilon \to 0} \frac{1}{\varepsilon}(\tau_{\varepsilon *} V - V) = \lim_{\varepsilon \to 0} \frac{1}{\varepsilon}(B_\varepsilon - I)V,$$

the matrices $A = \lim_{\varepsilon \to 0} \frac{1}{\varepsilon}(B_\varepsilon - I)$ are the matrix Lie algebra g_P associated with the matrix group G_P. Again this is of finite dimension p.

Now, Cartan showed that the geometrical meaning of his structure coefficients $a_{i\rho}^k$ was that the combinations

$$\alpha^\rho a_{i\rho}^k = \alpha^\rho A_\rho \tag{17}$$

constitute the linear isotropy algebra (with basis A_ρ). Note therefore that the number of π's in the structure equations is equal to the dimension of the linear isotropy algebra.

It is curious that the isotropy subgroup $\mathcal{G}_P$ plays such a central role in the Cartan theory, yet is itself not a Lie pseudogroup. However, note that we really only need the representation of $\mathcal{G}_P$ to leading order. We should also note that Cartan never used $T_P M$ and vector fields, instead dealing with cotangent space $T_P^* M$ and differential forms. There is no great significance to this; at worst, one ends up with the contragredient representation of the matrix Lie algebra.

4.3 How-to Guide to Structure Constants c_{ij}^k and $a_{i\rho}^k$

As well as the geometrical interpretation, we need actual computational procedures for finding the constants c_{ij}^k and $a_{i\rho}^k$.

4.3.1 Linear Isotropy Algebra

Let's rework that construction of the linear isotropy algebra explicitly via Taylor series. Our font conventions: curly $\mathcal{G}$, $\mathcal{G}_P$ for a group, curly $\mathcal{L}$, $\mathcal{L}_P$ for a Lie algebra, bold $\mathbf{X}$ for an operator in a Lie algebra, ordinary G, GL_n for a matrix Lie group, ordinary g, gl_n for the corresponding matrix Lie algebra.

Let the point P above have coordinates $x_0 = (x_0^1, \dots, x_0^n)$. Let $\tau \in \mathcal{G}_{x_0}$ be in the isotropy subgroup at x_0. Then Taylor expansion gives:

$$\tau^i(x) = x_0^i + b_j^i(x^j - x_0^j) + O(|x - x_0|)^2$$

where $b_j^i = \frac{\partial \tau^i}{\partial x^j}\Big|_{x=x_0}$. As τ ranges over the isotropy subgroup $\mathcal{G}_P$, the matrix $B = [b_j^i]$ ranges over a matrix Lie group, and this is the linear isotropy group G_P.

Now let $\tau = \tau_\varepsilon$ be generated by a vector field $\mathbf{X}$ satisfying $\mathbf{X}_{x_0} = 0$. We have two Taylor expansions to do, one in ε about $\varepsilon = 0$, and one in x about $x = x_0$. First about $x = x_0 \dots$

$$\tau_\varepsilon^i(x) = x_0^i + b_{\varepsilon j}^{\ i}(x^j - x_0^j) + O(x - x_0)^2,$$

where $b_{\varepsilon j}^{\ i} = \frac{\partial \tau_\varepsilon^i}{\partial x^j}\Big|_{x=x_0}$. Now expand about $\varepsilon = 0$: we let $\mathbf{X} = \xi^i \frac{\partial}{\partial x^i}$, so

$$\tau_\varepsilon^i(x) = x^i + \varepsilon \xi^i(x) + O(\varepsilon^2)$$

and the components of the matrices $B_\varepsilon \in \mathcal{G}_P$ are

$$b_{\varepsilon j}^{\ i} = \frac{\partial \tau_\varepsilon^i}{\partial x^j}\Big|_{x=x_0} = \delta_j^i + \varepsilon \frac{\partial \xi^i}{\partial x^j}\Big|_{x=x_0} + O(\varepsilon^2).$$

So the corresponding matrices A in the linear isotropy algebra g_P are

$$a^i_j = \lim_{\varepsilon \to 0} \frac{1}{\varepsilon}(b_{\varepsilon}{}^i_j - \delta^i_j) = \frac{\partial \xi^i}{\partial x^j}\bigg|_{x=x_0}.$$

In even more elementary terms, any operator in the isotropy Lie algebra at x_0 is of the form

$$\mathbf{X} = a^i_j(x^j - x^j_0)\partial_{x^i} + \text{h.o.t.s.}$$

Taking a second one

$$\mathbf{Y} = b^i_j(x^j - x^j_0)\partial_{x^i} + \text{h.o.t.s},$$

the commutator is

$$[\mathbf{X}, \mathbf{Y}] = (b^i_k a^k_j - a^i_k b^k_j)(x^j - x^j_0)\partial_{x^i} + O(x - x_0)^2.$$

In particular, the leading order terms of $[\mathbf{X}, \mathbf{Y}]$ depend only on the leading order terms of $\mathbf{X}$, $\mathbf{Y}$. The commutator bracket on the isotropy algebra induces a commutator bracket on the first order terms. This "leading order" bracket in turn induces one on the matrices A, B: it's just the matrix commutator $BA - AB$. This matrix Lie algebra is the linear isotropy algebra.

Example 4.4. Consider the infinite group (12)

$$X = x + a, \qquad Y = y + f(x),$$

which has defining equations of first order. The corresponding vector fields are

$$c_1 \frac{\partial}{\partial x} + f(x) \frac{\partial}{\partial y}.$$

At a point (x_0, y_0), the isotropy Lie algebra consists of vector fields vanishing there, that is $c_1 = 0$, $f(x_0) = 0$, so if $\mathbf{X} \in \mathcal{L}_P$,

$$\mathbf{X} = \xi \partial_x + \eta \partial_y = f(x)\frac{\partial}{\partial y}, \qquad \text{where } f(x_0) = 0.$$

The linear isotropy algebra is one-dimensional:

$$A = \begin{pmatrix} \frac{\partial \xi}{\partial x} & \frac{\partial \xi}{\partial y} \\ \frac{\partial \eta}{\partial x} & \frac{\partial \eta}{\partial y} \end{pmatrix}\bigg|_{(x_0, y_0)} = \begin{pmatrix} 0 & 0 \\ a & 0 \end{pmatrix}$$

where the value $a = f'(x_0)$ can be assigned arbitrarily.

4.3.2 Finite Lie Algebra

Before showing how c^k_{ij} and $a^k_{i\rho}$ can be calculated from the Lie algebra $\mathcal{L}$ in the infinite case, we rehearse with a finite pseudogroup. The first order involutive infinitesimal defining equations in this case look like

$$\frac{\partial \xi^i}{\partial x^l} = \beta^i_{lj}(x)\xi^j, \qquad i = 1, \dots, n, \quad l = 1, \dots, n.$$

There are n linearly independent solutions in the neighbourhood of a point x_0, namely

$$\mathbf{X}_i = \partial_{x^i} + \beta_{li}^k(x_0)(x^l - x_0^l)\partial_{x^k} + O(x - x_0)^2, \qquad i = 1, \dots, n.$$

The commutator of two solutions is

$$[\mathbf{X}_i, \mathbf{X}_j] = \underbrace{\left(b_{ji}^k(x_0) - b_{ij}^k(x_0)\right)}_{c_{ij}^k}\partial_{x^k} + O(x - x_0)$$

$$= c_{ij}^k \mathbf{X}_k$$

where the last equality follows closure of the Lie algebra under commutator bracket.

4.3.3 Infinite Case: c_{ij}^k

Now consider the case of the infinitesimal defining equations for an infinite Lie algebra. The first order involutive infinitesimal defining system now has some first order derivatives $\frac{\partial \xi^i}{\partial x^l}$ missing from the left-hand side. Pose the initial data $\xi^k(x_0) = \delta_i^k$, and, for convenience, set any first order parametric derivatives to 0 at x_0. We are assured of the existence of n solutions

$$\mathbf{X}_i = \partial_{x^i} + b_{il}^k(x_0)(x^l - x_0^l)\partial_{x^k} + O(x - x_0)^2$$

where $b_{il}^k(x_0)$ have the same meaning as above (though not uniquely defined). Taking commutators, we get

$$[\mathbf{X}_i, \mathbf{X}_j] = \underbrace{\left(b_{ji}^k(x_0) - b_{ij}^k(x_0)\right)}_{c_{ij}^k}\partial_{x^k} + O(x - x_0)^2$$

$$\equiv c_{ij}^k \mathbf{X}_k \pmod{\mathcal{L}_{x_0}}$$

where the congruence acknowledges the fact that solutions matching at zeroth order may not match at first and higher orders.

According to results derived in [37], the c_{ij}^k constructed in this way are those of Cartan.

4.3.4 Infinite Case: $a_{i\rho}^k$

Because the infinitesimal defining system of an infinite Lie pseudogroup has some (say p) parametric first derivatives, it follows that there are solutions whose Taylor series start at first order. Let $\mathbf{Z}_\rho$ be a vector field resulting from solving the defining system defining system!infinitesimal with initial values $\xi^i(x_0) = 0$ and all first order parametric derivatives 0, except the ρ-th, set equal to 1. Then

$$\mathbf{Z}_\rho = a_{i\rho}^k(x^i - x_0^i)\partial_{x^k} + O(x - x_0)^2$$

for some well-defined numbers $a_{i\rho}^k$. Taking commutators with $\mathbf{X}_i$ neatly picks off the desired coefficients:

$$[\mathbf{X}_i, \mathbf{Z}_\rho] = a_{i\rho}^k \partial_{x^k} + O(x - x_0)^2$$

$$\equiv a_{i\rho}^k \mathbf{X}_k, \pmod{\mathcal{L}_{x_0}^0}.$$

We observe that $a_{i\rho}^k$ here give rise to the linear isotropy algebra, and can be identified with those of Cartan.

The upshot of the above calculations is that to calculate the structure coefficients, one needs only the Taylor expansions to first order of solutions of the infinitesimal defining system with nonvanishing zeroth or first order initial data.

4.3.5   Finding $a_{i\rho}^k$ from Infinitesimal Defining System

It is now straightforward to get the $a_{i\rho}^k$ from the infinitesimal defining system. The crucial point is that they depend only on Taylor coefficients of the vector fields in $\mathcal{L}_x$ *to first order*. But it's straightforward to construct a Taylor approximation of a solution to the defining system (see [31]).

The net result is that $a_{i\rho}^k$ can be found from the infinitesimal defining system by the following steps:

1. Form the matrix $\left[\frac{\partial \xi^i}{\partial x^j}\right]$.

2. Simplify it modulo the infinitesimal defining system.

3. Fix an initial data point x_0 and evaluate the matrix at x_0.

4. Set all 0-th order terms (i.e. ξ^i) to 0.

5. Replace the remaining first order partials $\left.\frac{\partial \xi^i}{\partial x^j}\right|_{x_0}$ by arbitrary values α^ρ.

6. Read off the matrix Lie algebra $\alpha^\rho a_{i\rho}^k$.

Recall, however, our blanket assumption: for this to work, the defining system must be of first order, and must have no invariants (zeroth order equations).

Example 4.5. Consider the infinitesimal defining system

$$\xi_x = \frac{1}{y}\eta, \qquad \xi_y = 0, \qquad \eta_y = \frac{1}{y}\eta$$

for an infinite Lie algebra. We form the matrix and simplify mod the defining system:

$$\begin{pmatrix} \xi_x & \xi_y \\ \eta_x & \eta_y \end{pmatrix} \quad \rightarrow \quad \begin{pmatrix} \frac{1}{y}\eta & 0 \\ \eta_x & \frac{1}{y}\eta \end{pmatrix}.$$

Now evaluate at an initial data point (x_0, y_0) (with $y_0 \neq 0$), setting the zeroth order values $\xi(x_0, y_0) = \eta(x_0, y_0) = 0$ (so that we have vector fields in the isotropy algebra) and prescribing the initial value $\eta_x(x_0, y_0) = \alpha$:

$$\begin{pmatrix} \frac{1}{y}\eta & 0 \\ \eta_x & \frac{1}{y}\eta \end{pmatrix} \quad \rightarrow \quad \begin{pmatrix} 0 & 0 \\ \alpha & 0 \end{pmatrix}.$$

These last matrices constitute the linear isotropy algebra, and will therefore appear in the Cartan structure equations for the pseudogroup.

Case	Pseudogroup	Vector Fields $\mathbf{X} = \xi\frac{\partial}{\partial x} + \eta\frac{\partial}{\partial y}$	Infinitesimal Defining System	Linear Isotropy Algebra	Comments
(a)	$X = f(x)$ $Y = y + a$	$\xi = \phi(x)$ $\eta = a$	$\xi_y = 0$ $\eta_x = 0$ $\eta_y = 0$	$\begin{pmatrix} a & 0 \\ 0 & 0 \end{pmatrix}$	
(a$_1$)	$X = x$ $Y = y + f(x)$	$\xi = 0$ $\eta = \phi(x)$	$\xi = 0$ $\eta_y = 0$	$\begin{pmatrix} 0 & 0 \\ a & 0 \end{pmatrix}$	Intransitive
(a$_1'$)	$X = x + a$ $Y = y + f(x)$	$\xi = a$ $\eta = \phi(x)$	$\xi_x = 0$ $\xi_y = 0$ $\eta_y = 0$	$\begin{pmatrix} 0 & 0 \\ a & 0 \end{pmatrix}$	
(a$_2$)	$X = f(x)$ $Y = yf'(x)$	$\xi = \phi(x)$ $\eta = y\phi'(x)$	$\xi_x = \frac{1}{y}\eta$ $\xi_y = 0$ $\eta_y = \frac{1}{y}\eta$	$\begin{pmatrix} 0 & 0 \\ a & 0 \end{pmatrix}$	$[\mathbf{X}_1, \mathbf{X}_2] \equiv \mathbf{X}_1$
(b$_1$)	$X = f(x)$ $Y = y + g(x)$	$\xi = \phi(x)$ $\eta = \psi(x)$	$\xi_y = 0$ $\eta_y = 0$	$\begin{pmatrix} a & 0 \\ b & 0 \end{pmatrix}$	
(b$_2$)	$X = f(x)$ $Y = g(y)$	$\xi = \phi(x)$ $\eta = \psi(y)$	$\xi_y = 0$ $\eta_x = 0$	$\begin{pmatrix} a & 0 \\ 0 & b \end{pmatrix}$	
(b$_3$)	$X = f(x)$ $Y = y[f'(x)]^{1/m}$ $+ g(x)$	$\xi = \phi(x)$ $\eta = \frac{1}{m}y\phi'(x)$ $+ \psi(x)$	$\xi_y = 0$ $\eta_y = \frac{1}{m}\xi_x$	$\begin{pmatrix} ma & 0 \\ b & a \end{pmatrix}$	
(c)	$X = f(x,y)$ $Y = y$	$\xi = \phi(x,y)$ $\eta = 0$	$\eta = 0$	$\begin{pmatrix} a & b \\ 0 & 0 \end{pmatrix}$	Intransitive
(c$'$)	$X = f(x,y)$ $Y = y + a$	$\xi = \phi(x,y)$ $\eta = a$	$\eta_x = 0$ $\eta_y = 0$	$\begin{pmatrix} a & b \\ 0 & 0 \end{pmatrix}$	
(d$_1$)	$X = f(x)$ $Y = g(x,y)$	$\xi = \phi(x)$ $\eta = \psi(x,y)$	$\xi_y = 0$	$\begin{pmatrix} a & 0 \\ b & c \end{pmatrix}$	
(d$_2$)	$X = f(x,y)$ $Y = g(x,y)$, $f_x g_y - f_y g_x = 1$	$\xi = \phi(x,y)$ $\eta = \psi(x,y)$, $\phi_x + \psi_y = 0$	$\xi_x + \eta_y = 0$	$\begin{pmatrix} a & b \\ c & -a \end{pmatrix}$	Algebra is sl_2
(e)	$X = f(x,y)$ $Y = g(x,y)$	$\xi = \phi(x,y)$ $\eta = \psi(x,y)$	(null)	$\begin{pmatrix} a & b \\ c & d \end{pmatrix}$	Algebra is gl_2

Table 2: Vector fields and infinitesimal defining systems associated with each of the pseudogroups acting on $\mathbb{R}^2$. In all cases except (a$_2$), the commutation relations are $[\mathbf{X}_1, \mathbf{X}_2] \equiv 0$ modulo the isotropy algebra L^0.

4.3.6   Examples: Pseudogroups on $\mathbb{R}^2$

In Table 2, we give a catalogue (following Cartan) of the pseudogroups on $\mathbb{R}^2$ with first order defining system. The above Example 4.5 is case (a_2).

To see the matrix Lie algebras more clearly, write Cartan's structure equations in matrix form. For example, for (d_2) above, write

$$d\begin{pmatrix} \omega^1 \\ \omega^2 \end{pmatrix} = \begin{pmatrix} \pi^1 & \pi^2 \\ \pi^3 & -\pi^1 \end{pmatrix} \wedge \begin{pmatrix} \omega^1 \\ \omega^2 \end{pmatrix}.$$

The π go in a matrix which is related in an obvious way to the linear isotropy algebra.

4.4   Symmetries of Some d.e.s

In this section, we demonstrate the feasibility of the methods described above for calculating Cartan structure. We begin with a "toy" symmetry calculation, which we carry out by hand:

Example 4.6 (1st order o.d.e.). We seek symmetries of the scalar o.d.e.

$$y' = f(x, y).$$

Now it is well known that in a neighbourhood of a nonsingular point this equation can be mapped to $y' = 0$, whose symmetry pseudogroup is infinite:

$$X = f(x, y), \qquad Y = g(y).$$

In fact this is pseudogroup (d_1) of §4.3.6. Its structure matrix is $\begin{pmatrix} a & 0 \\ b & c \end{pmatrix}$. Thus we know before we start what the answer *should* be: we must get a pseudogroup similar to this one.

Now to it. Seeking symmetry vector fields $\xi\frac{\partial}{\partial x} + \eta\frac{\partial}{\partial y}$, we obtain the first order defining system

$$\eta_x = -f\eta_y + f\xi_x + f^2\xi_y + f_x\xi + f_y\eta. \tag{18}$$

We see three parametric first order derivatives, so the linear isotropy Lie algebra will be of dimension three. The linear isotropy algebra at a point (x_0, y_0) is found by simplifying the infinitesimal Jacobian modulo (18):

$$\begin{pmatrix} \xi_x & \xi_y \\ \eta_x & \eta_y \end{pmatrix} \to \begin{pmatrix} \xi_x & \xi_y \\ -f\eta_y + f\xi_x + f^2\xi_y & \eta_y \end{pmatrix}$$

and evaluating at (x_0, y_0) to get

$$\begin{pmatrix} a & b \\ -f_0c + f_0a + f_0^2b & c \end{pmatrix}$$

where f_0 means $f(x_0, y_0)$. These are the $a_{i\rho}^k$.

To calculate c_{ij}^k, we take the basis $\{\partial_x, \partial_y\}$ to tangent space at (x_0, y_0) and construct solutions of (18) by Taylor expansion:

$$\mathbf{X}_1 = \partial_x + f_{x0}(x - x_0)\partial_y + O(x - x_0)^2$$
$$\mathbf{X}_2 = \partial_y + f_{y0}(x - x_0)\partial_y + O(x - x_0)^2$$

Taking commutators and modding out higher order terms, we have

$$[\mathbf{X}_1, \mathbf{X}_2] = f_{y0}\partial_y + O(x - x_0) \equiv f_{y0}\mathbf{X}_2 \quad (\mathrm{mod}\ \mathcal{L}_{(x_0, y_0)})$$

which gives $c_{12}^2 = f_{y0}$, and all other c_{ij}^k vanishing.

The $a_{i\rho}^k$ and c_{ij}^k just found give Cartan structure

$$d\omega^1 = \pi^1 \wedge \omega^1 + \pi^2 \wedge \omega^2$$
$$d\omega^2 = f_0\pi^1 \wedge \omega^1 + f_0^2\pi^2 \wedge \omega^1 - f_0\pi^3 \wedge \omega^1 + \pi^3 \wedge \omega^2 - f_{y0}\omega^1 \wedge \omega^2.$$

The torsion term is absorbed on setting $\bar{\pi}^3 = \pi^3 - f_{y0}\omega^1$, yielding

$$d\begin{pmatrix} \omega^1 \\ \omega^2 \end{pmatrix} = \begin{pmatrix} \pi^1 & \pi^2 \\ f_0\pi^1 + f_0^2\pi^2 - f_0\bar{\pi}^3 & \bar{\pi}^3 \end{pmatrix} \wedge \begin{pmatrix} \omega^1 \\ \omega^2 \end{pmatrix}.$$

Further simplification is possible if we choose an appropriate initial point (e.g. choose a point (x_0, y_0) where $f(x_0, y_0) = 0$). Alternatively, introduce the change of basis

$$\bar{\omega}^2 = \omega^2 - f_0\omega^1$$
$$\bar{\pi}^1 = \pi^1 + f_0\pi^2$$
$$\bar{\bar{\pi}}^3 = \bar{\pi}^3 - f_0\pi^2$$

after which the structure equations become

$$d\begin{pmatrix} \omega^1 \\ \bar{\omega}^2 \end{pmatrix} = \begin{pmatrix} \bar{\pi}^1 & \pi^2 \\ 0 & \bar{\bar{\pi}}^3 \end{pmatrix} \wedge \begin{pmatrix} \omega^1 \\ \bar{\omega}^2 \end{pmatrix}.$$

We note that this is identical to the structure of case (d_1) of §4.3.6. Hence the infinitesimal method gives the same result as Cartan.

Example 4.7 (Liouville equation). Liouville's equation

$$u_{xy} = e^u$$

admits an infinite Lie pseudogroup of symmetries with two arbitrary functions of one variable. If we seek symmetry vector fields of the form

$$\xi\partial_x + \tau\partial_y + \eta\partial_u$$

then we obtain the standard form of the infinitesimal defining system

$\xi_x = -\tau_y - \eta$	$\tau_{yy} = -\eta_y$	$\eta_{xy} = 0$
$\xi_y = 0$	$\tau_x = 0$	$\eta_u = 0$
$\xi_u = 0$	$\tau_u = 0.$	

After reducing to first order as described in [23, 24], the method of §4 yields Cartan structure equations

$$d\omega^1 = -\omega^1 \wedge \omega^6$$
$$d\omega^2 = -\omega^2 \wedge \omega^3 + \omega^2 \wedge \omega^6$$
$$d\omega^3 = -\omega^1 \wedge \omega^4 - \omega^2 \wedge \omega^5$$
$$d\omega^4 = \pi^1 \wedge \omega^1 + \omega^4 \wedge \omega^6$$
$$d\omega^5 = \pi^2 \wedge \omega^2 - \omega^3 \wedge \omega^5 - \omega^5 \wedge \omega^6$$
$$d\omega^6 = -\omega^1 \wedge \omega^4.$$

In [32] we derive the above structure equations using a different method.

Example 4.8 (KP equation). The Kadomtsev–Petviashvili equation

$$u_{yy} + \left(u_t + u_{xxx} + 2uu_x\right)_x = 0$$

has an infinite Lie pseudogroup of symmetries depending on three arbitrary functions of one variable. Its structure equations are

$$d\omega^1 = \omega^1 \wedge \omega^8 + 2\omega^3 \wedge \omega^9$$
$$d\omega^2 = -\omega^1 \wedge \omega^9 + \tfrac{1}{2}\omega^2 \wedge \omega^8 + 2\omega^3 \wedge \omega^4$$
$$d\omega^3 = \tfrac{3}{2}\omega^3 \wedge \omega^8$$
$$d\omega^4 = -\omega^1 \wedge \omega^5 - \omega^2 \wedge \omega^6 - \omega^3 \wedge \omega^7 - \omega^4 \wedge \omega^8$$
$$d\omega^5 = \pi^1 \wedge \omega^1 + \pi^2 \wedge \omega^3 - 2\omega^5 \wedge \omega^8 + \omega^6 \wedge \omega^9$$
$$d\omega^6 = -\pi^1 \wedge \omega^3 - \tfrac{3}{2}\omega^6 \wedge \omega^8$$
$$d\omega^7 = \pi^2 \wedge \omega^1 - \pi^1 \wedge \omega^2 + \pi^3 \wedge \omega^3 - 2\omega^4 \wedge \omega^6$$
$$\qquad - 2\omega^5 \wedge \omega^9 - \tfrac{5}{2}\omega^7 \wedge \omega^8$$
$$d\omega^8 = -4\omega^3 \wedge \omega^6$$
$$d\omega^9 = -2\omega^1 \wedge \omega^6 + 2\omega^3 \wedge \omega^5 + \tfrac{1}{2}\omega^8 \wedge \omega^9.$$

In [13] the explicit form of the infinitesimal generators of symmetries of the KP equation is given.

Example 4.9 (Steady boundary layer equations). The equations [29]

$$uu_x + vu_y + p_x = u_{yy}, \qquad p_y = 0, \qquad u_x + v_y = 0$$

for a steady state boundary layer have an infinite Lie pseudogroup of symmetries depending on one arbitrary function of one variable. Its Cartan structure equations are

$$d\omega^1 = \omega^1 \wedge \omega^3 - 2\omega^1 \wedge \omega^7$$
$$d\omega^2 = \omega^1 \wedge \omega^4 - \omega^2 \wedge \omega^7$$
$$d\omega^3 = 0$$
$$d\omega^4 = -\omega^1 \wedge \omega^4 - \omega^1 \wedge \omega^6 + \omega^3 \wedge \omega^4 + \omega^4 \wedge \omega^7$$
$$d\omega^5 = -2\omega^3 \wedge \omega^5$$
$$d\omega^6 = \pi^1 \wedge \omega^1 + \omega^3 \wedge \omega^4 + 2\omega^3 \wedge \omega^6 + 2\omega^4 \wedge \omega^7 + 3\omega^6 \wedge \omega^7$$
$$d\omega^7 = 0.$$

In [33] the commutation relations of this infinite dimensional Lie symmetry algebra are parametrized using arbitrary functions.

5 Conclusions

The approach explained here is highly simplified. In particular, the assumptions that the pseudogroup is transitive with first order involutive defining system are unduly restrictive. However, overcoming these restrictions is mainly a technical matter. Reduction of higher order defining systems to first order is described in

[23], and in more detail in [24]. Transitivity is more delicate. Ideally one would like a method which could reliably distinguish those intransitive Lie pseudogroups not isomorphic to any transitive pseudogroup. As a first step in that direction, a method is described in [23] for identifying and eliminating inessential scalar invariants. If essential invariants are present, the situation is considerably more complex.

Acknowledgments

It is a pleasure to acknowledge the influence that Niky Kamran's seminars at the West Coast Symmetry Workshop at UBC in 1992 had in stimulating our interest in the subject.

References

[1] E.A. Araïs, V.P. Šapeev, and N.N. Janenko, Realization of Cartan's method of exterior differential forms on an electronic computer, *Sov. Math. Dokl.* **15**, 203–205, 1974.

[2] V.I. Arnold, *Mathematical Methods of Classical Mechanics*, Springer-Verlag, New York, 1978.

[3] G.W. Bluman and S. Kumei, *Symmetries and Differential Equations*, Springer-Verlag, New York, 1989.

[4] R.L. Bryant, S.S. Chern, R.B. Gardner, H.L. Goldschmidt, and P.A. Griffiths, *Exterior differential systems*, Springer-Verlag, New York, 1991.

[5] É.J. Cartan, Sur la structure des groupes infinis de transformations, *Annales de l'École Normale* **22**, 219–308, 1905, [11, p.571–714] .

[6] É.J. Cartan, Les sous-groupes des groupes continus de transformations, *Annales de l'École Normale* **25**, 57–194, 1908, [11, p.719–856].

[7] É.J. Cartan, Les groupes de transformations continus, infinis, simples, *Annales de l'École Normale* **26**, 93–161, 1909, [11, p.857–925] .

[8] É.J. Cartan, La structure des groupes infinis, *Séminaire de Math.*, 1937, [11, p.1335–1384].

[9] É.J. Cartan, Les problèmes d'équivalence, *Séminaire de Math.*, 1937, [11, p.1311–1334].

[10] É.J. Cartan, *Les Systèmes Différentiels Exteriéurs et leurs Applications Géometriques*, Hermann, Paris, 1946.

[11] É.J. Cartan, *Oeuvres Complètes*, volume 2, Part II, Gauthier-Villars, Paris, 1953.

[12] C. Chevalley, *Theory of Lie Groups*, Princeton Univ. Press, Princeton, NJ, 1946.

[13] D. David, N. Kamran, D. Levi, and P. Winternitz, Symmetry reduction for the Kadomtsev-Petviashvili equation using a loop algebra, *JMP* **27**, 1225–1337, 1986.

[14] V. Guillemin, A Jordan-Hölder decomposition for certain classes of infinite-dimensional Lie algebras, *J. Diff. Geom.* **2**, 313–345, 1966.

[15] D. Hartley and R.W. Tucker, A constructive implementation of the Cartan-Kähler theory of exterior differential systems, *J. Symb. Comp.* **12**, 655–667, 1991.

[16] R. Hermann, *Cartanian geometry, nonlinear waves and solitons, Part B*, MathSci Press, Brookline, MA, 1980.

[17] M. Janet, Sur les systèmes d'équations aux dérivées partielles, *J. de Math* **3**, 65–151, 1920.

[18] M. Janet, *Leçons sur les systèmes d'équations aux derivées partielles*, Gauthier-Villars, Paris, 1929.

[19] S. Kumei and G.W. Bluman, When nonlinear differential equations are equivalent to linear differential equations, *SIAM J. Appl. Math.* **42**, 1157–1173, 1982.

[20] M. Kuranishi, On the local theory of continuous infinite pseudo-groups. I, *Nagoya Math. J.* **15**, 225–260, 1959.

[21] M. Kuranishi, On the local theory of continuous infinite pseudo-groups. II, *Nagoya Math. J.* **19**, 55–91, 1961.

[22] S. Lie, Grundlagen für die Theorie der unendlichen continuerliche transformationsgruppen I, *Leipziger Berichte*, 1891, (English translation in [16]).

[23] I.G. Lisle and G.J. Reid, Geometry and structure of Lie pseudogroups from infinitesimal defining systems, *J. Symb. Comp.* **26**, 355–379, 1998.

[24] I.G. Lisle, G.J. Reid, and A. Boulton, Algorithmic determination of structure of infinite Lie pseudogroups of symmetries of PDEs, In *Proc. ISSAC '95*, New York, 1995. ACM Press.

[25] E. Mansfield, *Differential Gröbner bases*, PhD thesis, Univ. of Sydney, Australia, 1991.

[26] E. Mansfield and E.D. Fackerell, Differential Gröbner bases, School of Mathematics, Physics, Computer Science and Electronics Preprint 92-108, Macquarie Univ, Australia, 1992.

[27] E.L. Mansfield, A simple criterion for involutivity, *J. London Math. Soc.* **54**, 323–345, 1996.

[28] P.J. Olver, *Application of Lie groups to differential equations*, Springer-Verlag, New York, 1986.

[29] L.V. Ovsiannikov, *Group analysis of differential equations*, Academic Press, New York, 1982.

[30] J.-F. Pommaret, *Systems of partial differential equations and Lie pseudo-groups*, Gordon and Breach, New York, 1978.

[31] G.J. Reid, Algorithms for reducing a system of PDEs to standard form, determining the dimension of its solution space and calculating its Taylor series solution, *Euro. J. Appl. Maths.* **2**, 293–318, 1991.

[32] G.J. Reid, A. Boulton, and I.G. Lisle, Characterising Lie systems by their infinitesimal symmetries, Preprint, Univ. of British Columbia, 1996.

[33] G.J. Reid, I.G. Lisle, A. Boulton, and A.D. Wittkopf, Algorithmic determination of commutation relations for Lie symmetry algebras of PDEs,, In *Proc. ISSAC '92*, 63–68, New York, 1992. ACM Press.

[34] G.J. Reid, A.D. Wittkopf, and A. Boulton, Reduction of systems of nonlinear partial differential equations to simplified involutive forms, IAM Report 94-14, Univ. of British Columbia, Canada, 1994.

[35] C. Riquier, *Les systémes d'équations aux dérivées partielles*, Gauthier-Villars, Paris, 1910.

[36] F. Schwarz, Reduction and completion algorithms for partial differential equations, In *Proc ISSAC '92*, 49–56, New York, 1992. ACM Press.

[37] I.M. Singer and S. Sternberg, The infinite groups of Lie and Cartan. I. The transitive groups, *J. d'Analyse Math.* **15**, 1–115, 1965.

[38] A. Tresse, Sur les invariants différentiels des groupes continus de transformations, *Acta Math.* **18**, 1–88, 1894.

[39] E. Vessiot, Sur la théorie des groupes continus, *Ann. de l'École normale*, 1903.

[40] E. Vessiot, Sur l'intégration des systèmes différentiels qui admettent des groupes continus de transformations, *Acta Math.* **28**, 307–349, 1904.

[41] T. Wolf, A package for the analytic investigation and exact solutions of differential equations, In J.H. Davenport, editor, *Proc. EUROCAL '87 (Lecture Notes in Computer Science 378)*, 479–491. Springer-Verlag, Berlin, 1989.

[42] H. Wussing, *The Genesis of the Abstract Group Concept: A contribution to the history of the origin of abstract group theory*, MIT Press, Cambridge, MA, 1984.

Cartan's Method of Equivalence

DAVID H. HARTLEY

Keywords: Cartan equivalence, differential systems

1 Introduction

Cartan's *method of equivalence* addresses the following question:

Equivalence problem. *Given local coframes $\boldsymbol{\omega} = \{\omega^i\}$ and $\bar{\boldsymbol{\omega}} = \{\bar{\omega}^i\}$ on open subsets U and $\bar{U}$ of n-dimensional manifolds M and $\bar{M}$, and a prescribed subgroup $G \subset GL(n)$, does there exist a local diffeomorphism $\Phi : U \to \bar{U}$ such that*

$$\Phi^* \bar{\boldsymbol{\omega}} = \gamma \, \boldsymbol{\omega} \tag{1}$$

for some G-valued function $\gamma = (\gamma^i{}_j)$ on U?

There are two things which make this somewhat special looking problem extremely interesting to study. First, it is possible to prescribe many geometrical and differential structures in terms of a particular set of 1-forms fixed up to a certain transformation group. Among the many areas which have been studied using this method are symplectic structures, metric tensors, differential operators, differential equations, Lagrangians, and control systems. The method of equivalence thus tackles the equivalence question for a diverse range of problems.

Second, Cartan's method is the means for tackling a larger question. Suppose we are given not two specific coframes $\boldsymbol{\omega}$ and $\bar{\boldsymbol{\omega}}$, but a whole family of coframes acted upon by the group G. The method of equivalence allows us to divide the family of coframes into equivalence classes $[\boldsymbol{\omega}]$ with respect to the equivalence relation (1). Each class is characterised by a set of invariant functions thrown up in the course of calculation, and represented by a so-called *invariant coframe*. Using such an invariant coframe can vastly simplify other computations, such as symmetry analysis (cf. the adapted moving frames in I. Lisle's contribution to this volume [7]).

In the remainder of this chapter, the equivalence problem for a given pair of coframes will be called the *specific problem*, while the more general search for equivalence classes within a family of coframes will be called the *classification problem*.

The aim of this chapter is to provide an introduction to Cartan's method accessible to non-experts. As such, attention will be focussed on the simplest problems and broadest outline of the method, ignoring some of the interesting side-branches and difficulties which can arise. In particular, only the local equivalence problem will be discussed, so that the open set $U \subset M$ will be identified with the manifold M itself. The missing details, along with historical accounts of the method and its applications, can be filled in by consulting the seminal expositions by Gardner [2] and Kamran [4], as well as the earlier and original works cited therein, or the more recent treatments by Yang [12] and Olver [9]. The present chapter owes much to these references.

In order to give as comprehensive an overview as possible within the above constraints, the language of exterior differential systems will be used throughout. An introduction to this formalism is contained in N. Kamran's contribution to this volume [5], and further details may be found in other places already cited [9, 12] or the standard reference [1].

The remainder of these notes is organised as follows. Section 2 deals with the initial formulation of a general equivalence problem in terms of a coframe and equivalence group. The specific equivalence problem is treated in section 3, while the more general classification problem is the subject of sections 4 and 5. Finally, section 6 treats two non-trivial examples, the equivalence of Lagrangians under a change of variables and the local isometry problem for pseudo-Riemannian manifolds, to illustrate the general discussion.

2 Formulation of Equivalence Problems

As already mentioned, equivalence problems arise for many different structures in geometry and analysis. Some, such as the equivalence of Riemannian or symplectic structures, have an obvious geometrical character, while others, such as differential operator equivalence under a change of variables, are less apparently geometrical. None of the examples mentioned is stated directly as a coframe equivalence in the sense quoted in the introduction, so the first step in applying Cartan's method is to work the given problem into the required form. This is a non-trivial matter, requiring some geometrical insight into the structure being studied: the "methodical" part of Cartan's method cannot commence until this has been done.

Two aspects of this reformulation need attention. First, a suitable differentiable manifold must be found, together with a set of 1-forms encoding the desired mathematical object. If the encoding 1-forms form a coframe for M as they stand, the equivalence problem is said to be *determined*. Otherwise, it is either *under-determined* or *over-determined* according to whether the encoding 1-forms are too few or too many, respectively. For an under-determined problem, additional 1-forms can be freely chosen to complete a local coframe. For an over-determined problem, there is a systematic procedure for reducing the problem to a determined one, but it will not be described here.

The other aspect of formulating equivalence problems is the choice of equivalence group G. This choice may have its origins in the original statement of the equivalence problem, but will typically be affected by the way the 1-form representation is carried out. For under-determined problems, choosing any set of additional 1-forms forming a coframe leaves the original structure unaffected, so any transformation of this subset must be admitted.

To illustrate the above ideas, two simple examples will be treated. These equivalence problems will be completed in section 6.

Example: Riemannian geometry. Two n-dimensional Riemannian manifolds $(M, \mathrm{d}s^2)$ and $(\bar{M}, \mathrm{d}\bar{s}^2)$ are isometric if there exists a diffeomorphism $\Phi: M \to \bar{M}$ such that $\Phi^* \mathrm{d}\bar{s}^2 = \mathrm{d}s^2$. In order to cast the isometry question in Cartan's framework, it is necessary to represent the metric tensor ds^2 by a set of 1-forms. This is achieved naturally by fixing a local orthonormal coframe ω on an open set $U \subset M$ so that

$$ds^2 = \sum_{i=1}^{n} \omega^i \otimes \omega^i \tag{2}$$

on U. This representation is, of course, not unique: rotating the coframe independently at each point of U will produce a new coframe for the same metric. This is extra freedom in the equivalence problem not present in the original isometry problem. Restricting to local subsets, the local isometry condition thus becomes the equivalence relation (1)

$$\Phi^* \bar{\omega} = \gamma \omega, \tag{3}$$

where ω is an orthonormal coframe for ds^2 on $U \subset M$, $\bar{\omega}$ is an orthonormal coframe for $d\bar{s}^2$ on $\bar{U} \subset \bar{M}$, and γ takes values in $SO(n) \subset GL(n)$. This yields a determined equivalence problem on M with equivalence group $G = SO(n)$. Application of Cartan's method leads to a classification of Riemannian metrics together with adapted coframes representing the equivalence classes.

Example: Lagrangian mechanics. The state of a simple mechanical system with a single degree of freedom is governed by a Lagrangian L which is a function of time t, the generalised coordinate q and generalised velocity v. It is interesting to know whether two different Lagrangians really describe the same mechanical system in different variables. More generally, it is interesting to classify Lagrangians into equivalence classes under a change of variables.

The natural geometrical arena for this problem is the jet bundle $M = J^1(\mathbb{R}, \mathbb{R})$ of first jets of maps $\mathbb{R} \to \mathbb{R}$ (see P. Vassiliou's chapter in this volume). Let (t, q, v) be natural coordinates on M, in terms of which the contact form is

$$\theta = dq - v\,dt. \tag{4}$$

Any differentiable map $\xi: I \subset \mathbb{R} \to \mathbb{R}$, $t \mapsto q = \xi(t)$ prolongs to give $j^1\xi: I \to M$, $t \mapsto (t = t, q = \xi(t), v = \dot{\xi}(t))$, which satisfies $j^1\xi^*\theta = 0$. The Lagrangian is then a function on M, and the mechanical action associated with a trajectory ξ is

$$\int_{j^1\xi(I)} L\,dt. \tag{5}$$

In this example, we will take a "change of variables" to mean a diffeomorphism $\Phi: M \to \bar{M}$, and consider two Lagrangians L and $\bar{L}$ equivalent if they generate the same action on all trajectories:

$$\int_{\Phi(j^1\xi(I))} \bar{L}\,d\bar{t} = \int_{j^1\xi(I)} L\,dt \qquad \forall\,(\xi, I). \tag{6}$$

Our task is to convert this condition into a statement about 1-forms on M.

Since the trajectory is arbitrary, condition (6) is equivalent to

$$j^1\xi^*\Phi^*(\bar{L}\,d\bar{t}) = j^1\xi^*(L\,dt) \qquad \forall\,(\xi, I), \tag{7}$$

which, observing that the contact structure is uniquely specified (up to rescaling) by the requirement $j^1\xi^*\theta = 0$ for all ξ, allows us to recast the equivalence condition without ξ appearing:

$$\Phi^*(\bar{L}\,d\bar{t}) = L\,dt + \beta\theta \qquad (8)$$

for some function β on M. Demanding transitivity of the equivalence relation expressed by (8) yields a further stipulation that the contact structure itself be preserved up to scale:

$$\Phi^*\theta = \alpha\theta \qquad (9)$$

for some nowhere-vanishing function α on M.

These considerations have singled out two 1-forms, θ and $L\,dt$, which encode the mechanical system as well as fixing the corresponding parts of the equivalence group. For simplicity, we will assume that L is nowhere vanishing, so that we have a two-dimensional distribution of 1-forms. However, $\dim M = 3$, making this an under-determined problem, and so the coframe must be completed with a further 1-form. A convenient choice in the given coordinates is dv, but any other 1-form in which the coefficient of dv is non-vanishing would be equally valid. This fixes the freedom under the diffeomorphism Φ – the complementary 1-form $d\bar{v}$ must pull back to some other complementary 1-form:

$$\Phi^*\,d\bar{v} = \rho\theta + \sigma L\,dt + \tau\,dv, \qquad (10)$$

where ρ, σ and τ are functions on M with τ nowhere-vanishing.

With relations (8), (9) and (10), we have recast the equivalence of Lagrangians under changes of variables as the Cartan equivalence problem

$$\Phi^*\bar{\omega} = \gamma\,\omega \qquad (11)$$

where

$$\omega = \begin{pmatrix} \omega^1 \\ \omega^2 \\ \omega^3 \end{pmatrix} = \begin{pmatrix} dq - v\,dt \\ L\,dt \\ dv \end{pmatrix} \qquad L \neq 0 \qquad (12)$$

is a coframe for M, and

$$\gamma = \begin{pmatrix} \alpha & 0 & 0 \\ \beta & 1 & 0 \\ \rho & \sigma & \tau \end{pmatrix} \qquad \alpha \neq 0 \quad \tau \neq 0 \qquad (13)$$

is a function on M taking values in a 5-parameter Lie group.

3 Equivalence of Given Coframes

In this section, we will examine the *specific equivalence problem*: the equivalence question for two specified coframes. We will assume that the coframes ω on M and $\bar{\omega}$ on $\bar{M}$ and the equivalence group G are all given, so that the problem is to decide the existence of a diffeomorphism $\Phi\colon M \to \bar{M}$ and a map $\gamma\colon M \to G$ satisfying the equivalence relation (1). Although the more general *classification problem* for families of coframes is intrinsically more interesting, studying the simpler problem brings out the connection between equivalence problems and exterior differential systems more clearly. The classification problem will be described in the next section.

In local coordinates, the equivalence relation (1) becomes a system of partial differential equations for the components of Φ and γ. The equivalence question is then the *existence* question for solutions of these partial differential equations. The most natural step is to view the equivalence relation as an exterior differential system and apply the results on existence of solutions, as described in N. Kamran's contribution to this volume [5], such as the Frobenius and Cartan–Kähler theorems.

For many examples, additional techniques such as prolongation and reduction are required in order to bring the problem to a form where a decisive answer is possible. These are described later in this section. As mentioned in the introduction, we will restrict M and $\bar{M}$ if necessary so that the coframes are global.

3.1   Exterior Systems Formulation

Given the quantities appearing in the equivalence relation (1), the smallest space on which an exterior system could be formulated is $N = \bar{M} \times M \times G$ with (local) coordinates $(\bar{x}, x, g)$. The coframe 1-forms $\boldsymbol{\omega}$ and $\bar{\boldsymbol{\omega}}$ pull back via the canonical projections π_M and $\pi_{\bar{M}}$ to give independent 1-forms on N which we can continue to denote by the same symbols without risking confusion.

Consider the map $f\colon M \to N$, $x \mapsto (\Phi(x), x, \gamma(x))$. If Φ and γ satisfy the equivalence relation, then $f^*(\bar{\boldsymbol{\omega}} - g\,\boldsymbol{\omega}) = 0$. This suggests that the solutions of the equivalence relation are in one-to-one correspondence with the n-dimensional integral manifolds of the exterior system S with independence condition Ω given by

$$S = \{\bar{\boldsymbol{\omega}} - g\,\boldsymbol{\omega}\} \qquad \Omega = \{\boldsymbol{\omega}\}. \tag{14}$$

To confirm this, consider an n-dimensional integral manifold $f\colon \Sigma \to N$, $y \mapsto (\bar{x} = \phi(y), x = \psi(y), g = \chi(y))$. The independence condition implies that the 1-forms $f^*\boldsymbol{\omega}$ are independent at every point of Σ. Since the 1-forms $\boldsymbol{\omega}$ are independent on N, we find that ψ is locally invertible. Setting $\Phi = \phi \circ \psi^{-1}$ and $\gamma = \chi \circ \psi^{-1}$, we recover the original equivalence relation since $0 = \psi^{-1\,*}f^*(\bar{\boldsymbol{\omega}} - g\,\boldsymbol{\omega}) = \Phi^*\bar{\boldsymbol{\omega}} - \gamma\,\boldsymbol{\omega}$.

It is convenient to introduce the 1-forms $\boldsymbol{\theta} = \{\theta^i\}$ on $M \times G$ defined by

$$\boldsymbol{\theta} = g\,\boldsymbol{\omega}, \qquad \text{or} \qquad \theta^i = g^i{}_j\omega^j. \tag{15}$$

There is more than just shorthand behind this definition: in a global setting, $\boldsymbol{\theta}$ is the vector-valued soldering form on the principal G-bundle obtained by reducing the linear frame bundle FM. Since the independence condition is essentially algebraic, and the matrices $(g^i{}_j)$ are invertible, we can modify the exterior system above and write

$$S = \{\boldsymbol{\theta} - \bar{\boldsymbol{\omega}}\} \qquad \Omega = \{\boldsymbol{\theta}\}. \tag{16}$$

The analysis of this exterior system proceeds in the usual fashion. In general, we look for n-dimensional integral elements $E \subset T_{(\bar{x},x,g)}N$ of (S, Ω). These must satisfy $S|_E = 0$ and $\hat{\Omega}|_E \neq 0$, where $\hat{\Omega}$ is the n-form exterior product of the 1-forms in Ω and, for any p-form ϕ, $\phi|_E = \{\phi(u_1, \ldots, u_p) \mid u_1, \ldots, u_p \in$

E}. In addition, we must impose $dS|_E = 0$, if E is to be an integral element tangent to an integral manifold. (For an integral manifold, $f^* dS = 0$ follows automatically from $f^* S = 0$, but for an arbitrary integral element, $dS|_E = 0$ is not a consequence of $S|_E = 0$.) So we must study the exterior derivatives of the forms in S.

Since each ω^i is the lift of a 1-form from M, it follows that

$$d\omega^i = \tfrac{1}{2} C^i{}_{jk}\, \omega^j \wedge \omega^k \tag{17}$$

where $C^i{}_{jk}$ are functions lifted from M. A parallel relation holds for $d\bar{\omega}^i$. This gives the *structure equations*

$$
\begin{aligned}
d(\theta^i - \bar{\omega}^i) &= dg^i{}_j \wedge \omega^j + g^i{}_j\, d\omega^j - d\bar{\omega}^i \\
&\simeq \pi^i{}_j \wedge \theta^j + \tfrac{1}{2}(T^i{}_{jk} - \bar{C}^i{}_{jk})\theta^j \wedge \theta^k \qquad (\mathrm{mod}\ S),
\end{aligned}
\tag{18}
$$

where $(\pi^i{}_j) = \boldsymbol{\pi}_{MC}$ is the matrix of Maurer–Cartan 1-forms on G defined by

$$\boldsymbol{\pi}_{MC} = dg\, g^{-1}, \tag{19}$$

and

$$T^i{}_{jk} = g^i{}_l C^l{}_{mn} (g^{-1})^m{}_j (g^{-1})^n{}_k. \tag{20}$$

If G is a proper subgroup of $GL(n)$, the n^2 1-forms $\boldsymbol{\pi}_{MC}$ are not all independent. Adopting a maximal set $\boldsymbol{\pi} = \{\pi^\rho\}$ of right-invariant Maurer–Cartan 1-forms on G, we can write

$$\pi^i{}_j = A^i{}_{j\rho}\pi^\rho, \tag{21}$$

where the coefficients $A^i{}_{j\rho}$ are constants (the matrices $A_\rho = (A^i{}_{j\rho})$ form a basis for the Lie algebra $\mathfrak{g}$ of G). The structure equations become

$$d(\theta^i - \bar{\omega}^i) \simeq A^i{}_{j\rho}\pi^\rho \wedge \theta^j + \tfrac{1}{2}(T^i{}_{jk} - \bar{C}^i{}_{jk})\theta^j \wedge \theta^k \qquad (\mathrm{mod}\ S). \tag{22}$$

Recalling that the independence condition is $\Omega = \{\boldsymbol{\theta}\}$, it follows from the structure equations (22) that (S, Ω) is a quasilinear exterior differential system. As explained in N. Kamran's chapter [5], we can use Cartan's test to help determine if the system is in involution. If it is, then the Cartan–Kähler theorem guarantees the existence of local integral manifolds. In other words, the original coframes ω and $\bar{\omega}$ are equivalent.

If the system (S, Ω) is not immediately in involution (and an equivalence problem rarely is), we do not know whether or not integral manifolds exist. The Cartan–Kähler theorem provides a one-sided test only. However, all is not lost. By the techniques of reduction and prolongation (described below), we can bring the system into involution or inconsistency, deciding the equivalence question.

3.2 Reduction

Reduction and prolongation are standard techniques applying to any exterior differential system. The following description is biassed towards equivalence problems, but the general methods are very similar (see *eg* [1]).

An n-dimensional integral element E of the Pfaffian system (S, Ω) satisfies the n linear conditions $S|_E = 0$. However, the underlying manifold N has dimension $2n + \dim G$, so E must annul some additional $\dim G$ 1-forms. Since the independence condition requires $\hat{\Omega}|_E \neq 0$, these additional 1-forms must be spanned by

$$\pi - v\,\theta, \tag{23}$$

for some matrix of functions $v = (v_i^\rho)$ dependent on the particular integral element E. We can now see the algebraic consequences of the conditions $dS|_E = 0$, since these conditions are readily obtained by calculating the structure equations modulo the new 1-forms. By virtue of this calculation, the components of v must satisfy

$$2A^i{}_{[j\rho}v^\rho_{k]} = T^i{}_{jk} - \bar{C}^i{}_{jk}, \tag{24}$$

where the square brackets denote weighted antisymmetrisation of the indices. These are inhomogeneous linear equations for v which can be written more succinctly

$$\mathcal{A}v = T - \bar{C} \tag{25}$$

in terms of a linear operator $\mathcal{A}$ and functions $T = (T^i{}_{jk})$ and $\bar{C} = (\bar{C}^i{}_{jk})$. The components of $\mathcal{A}$ are

$$\mathcal{A}^i{}_{jk\rho}{}^l = 2A^i{}_{[j\rho}\delta^l_{k]}. \tag{26}$$

The nullspace $\ker \mathcal{A}$ is sometimes denoted $\mathfrak{g}^1$ and called the *first prolongation* of the Lie algebra $\mathfrak{g}$.

If $\lambda = (\lambda^{jk}{}_i)$ is a non-zero vector in the left nullspace of $\mathcal{A}$ (i.e. the kernel of the transpose $\mathcal{A}^t$), then

$$0 = \lambda \cdot \mathcal{A}v = \lambda \cdot (T - \bar{C}). \tag{27}$$

The right-hand side is independent of v, being comprised of functions on N alone, so there are no integral elements (S,Ω) away from points on N satisfying equation (27). Collecting all such conditions together

$$\tau = \{\lambda \cdot (T - \bar{C}) \mid \lambda \cdot \mathcal{A} = 0\}, \tag{28}$$

gives what is called the *torsion* of the exterior system (S,Ω). A slightly different definition of torsion (as above, but with $\bar{C}$ missing) will be used in the classification problem in section 4. The zero set $\tau = 0$ defines a variety in N to which we must restrict in order to find integral elements. In practice, we usually uncover the variety τ by Gaussian elimination on the equations (25) rather than looking directly for left null vectors λ. In fact, it is often possible to pick out the functions τ by eye from the structure equations. We follow each branch of the variety separately, in order that we continue to work on analytic manifolds.

There are now two possibilities. If the branch under consideration does not map onto M under the projection $\pi_M: N \to M$, then no integral elements can be found without restricting the space M and compromising the independence condition Ω. This branch must be discarded. A similar argument applies if the branch does not map onto $\bar{M}$ under $\pi_{\bar{M}}$. If this happens on all branches of the variety, there are then no admissible integral elements, no integral manifolds, and the system is said to be *inconsistent*. The original coframes are not equivalent. Referring back to the origin (20) of the functions T, this will happen if there is no element $g \in G$ of the equivalence group transforming the relevant combinations of $C^i{}_{jk}$ into the same combinations of $\bar{C}^i{}_{jk}$.

On the other hand, if a branch N_1 of $\tau = 0$ does surject onto $\bar{M} \times M$, then the local immersion $i: N_1 \to N$ specifies some local coordinates on G as functions on $\bar{M} \times M$ and the rest of G (i.e. on $N_1 \simeq \bar{M} \times M \times G_1$ where $G_1 \subset G$). To find integral elements, we must restrict to this immersion, which we achieve by pulling back the system (S, Ω) to give the *reduced* or *projected* system (S_1, Ω_1) on N_1.

It is conventional to re-cycle the symbols in an equivalence method calculation. Thus, once the reduction is performed, the same notation $N = \bar{M} \times M \times G$, $S = \{\boldsymbol{\theta} - \bar{\boldsymbol{\omega}}\}$, $\Omega = \{\boldsymbol{\theta}\}$ etc. will be used for the *reduced* equivalence problem.

This process of reducing the space (and particularly the group G) is typically repeated several times in equivalence problems. Eventually, if the system does not become inconsistent, we arrive at a point where the right-hand side of (27) vanishes for all left null vectors λ. At this stage, there are no further consistency conditions to be satisfied, and integral elements of (S, Ω) exist over all points of N. Unfortunately, this may not be the end of the story, but before going further, an example is in order.

Example: Reduction. Reduction or projection of an exterior differential system occurs in contexts other than the equivalence problem. Roughly speaking, the need for reduction corresponds to the presence of first-order integrability conditions. As a trivial example, the partial differential equations

$$
\begin{aligned}
u_z &= y u_x \\
u_y &= 0
\end{aligned}
\tag{29}
$$

can be encoded in the exterior differential system

$$
S = \{\mathrm{d}u - p\,\mathrm{d}x - yp\,\mathrm{d}z\} \qquad \Omega = \{\mathrm{d}x, \mathrm{d}y, \mathrm{d}z\},
\tag{30}
$$

on a five-dimensional submanifold M of the first jet bundle $J^1(\mathbb{R}^3, \mathbb{R})$. Any integral element of (S, Ω) must annihilate the additional 1-form

$$
\mathrm{d}p - v_x^p\,\mathrm{d}x - v_y^p\,\mathrm{d}y - v_z^p\,\mathrm{d}z.
\tag{31}
$$

The structure equations for the system (30) are

$$
\mathrm{d}S \simeq \{-\mathrm{d}p \wedge \mathrm{d}x - y\,\mathrm{d}p \wedge \mathrm{d}z - p\,\mathrm{d}y \wedge \mathrm{d}z\} \pmod{S},
\tag{32}
$$

from which we see that the unknown coefficients v must satisfy the inhomogeneous linear system

$$
\begin{pmatrix} y & 0 & -1 \\ 0 & 1 & 0 \\ 0 & y & 0 \end{pmatrix}
\begin{pmatrix} v_x^p \\ v_y^p \\ v_z^p \end{pmatrix}
= \begin{pmatrix} 0 \\ 0 \\ p \end{pmatrix}
\tag{33}
$$

corresponding to (25). Clearly $\lambda = (0, -y, 1)$ spans the left nullspace of the coefficient matrix, giving $\tau = \{p\}$. Integral elements can only be found over points in M where $p = 0$, corresponding to the integrability condition $u_x = 0$ of the original partial differential equations.

3.3 Prolongation

Reduction of an exterior system may not suffice to bring it into involution, even when solutions are known to exist. A system which fails to be involutive must be analysed further to decide the existence of solutions. The technique used is known as *prolongation* and consists of finding a new Pfaffian system satisfying the (higher-degree) structure equations of the original exterior system. Since the structure equations arise through exterior differentiation, prolonging an exterior system is closely related to differentiating a system of differential equations to obtain a higher-order system.

Assuming that all reductions possible have been performed, the integral elements of S are characterised by a consistent system of inhomogeneous linear equations (25) for v. This implies that the set of integral elements $V_\Omega(S)$ forms an affine bundle over the base space N. The linear equations (25) fix rank $\mathcal{A}$ of the components v_i^ρ of v, but the remainder (if any) can be regarded as new coordinates parameterising $V_\Omega(S)$. Setting $M^1 = V_\Omega(S)$, and writing the general parametric solution of equations (25) as $v = w(t)$, where t denotes the set of free parameters, we arrive at the *first prolonged* exterior differential system

$$S^1 = S \cup \{\pi - w(t)\,\theta\} \qquad \Omega^1 = \{\theta\} \qquad (34)$$

on M^1.

Example: Prolongation. As a trivial example of the prolongation process (outside of the equivalence problem framework), consider the contact system

$$S = \{du - p\,dx - q\,dy\} \qquad \Omega = \{dx, dy\} \qquad (35)$$

on the first jet bundle $M = J^1(\mathbb{R}^2, \mathbb{R})$ with coordinates (x, y, u, p, q). Any two-dimensional integral element E must annul an additional two 1-forms (dim $M = 5$) which, taking into account the independence condition, must be of the form

$$\begin{cases} dp - r\,dx - s\,dy \\ dq - s'\,dx - t\,dy. \end{cases} \qquad (36)$$

In the above notation,

$$v = \begin{pmatrix} r & s \\ s' & t \end{pmatrix}. \qquad (37)$$

The structure equations are given by

$$dS = \{-dp \wedge dx - dq \wedge dt\}. \qquad (38)$$

Modulo the 1-form (36) we have a single linear equation for v:

$$s' - s = 0. \qquad (39)$$

This fixes s' in terms of s, so the integral elements are parameterised by

$$w(r, s, t) = \begin{pmatrix} r & s \\ s & t \end{pmatrix}. \qquad (40)$$

The prolonged system is thus

$$S^1 = \begin{cases} du - p\,dx - q\,dy \\ dp - r\,dx - s\,dy \\ dq - s\,dx - t\,dy \end{cases} \qquad (41)$$

on $M^1 = M \times \mathbb{R}^3$. This is nothing other than the contact system on the second jet bundle $J^2(\mathbb{R}^2, \mathbb{R})$.

There is a one-to-one relationship between the integral manifolds of (S, Ω) and those of (S^1, Ω^1), so we can use the prolonged system to pursue the equivalence problem further. Again, we generate the structure equations dS^1, and look at the system of linear equations for integral elements. Non-trivial consistency conditions in this linear system can imply a violation of the independence condition, in which case there are no solutions, or require reduction to a system S^1_1. If the consistency conditions are void, then Cartan's test may indicate the system is in involution, in which case we are finished, or we may need to prolong again to S^2.

Should S^1 need further reduction or prolongation, we simply repeat the procedure on the resulting exterior system. Assuming we have a sufficiently regular problem, the Cartan–Kuranishi theorem (see [1]) assures us that in a finite number of steps, we will arrive at a system which is either inconsistent with the independence condition or involutive. In this way, we achieve a decisive test of equivalence.

We have now seen how to decide the specific equivalence problem for two given coframes. Reformulating the equivalence relation as an exterior differential system on $N = \bar{M} \times M \times G$ allows us to use the standard Cartan–Kähler and Frobenius tests for existence of solutions, combined with the techniques of prolongation and reduction to give a decisive test of equivalence.

This aspect of the equivalence problem is constructive, at least for those coefficient fields with constructive algorithms for variety decomposition, such as rational functions. Algorithms for reduction and prolongation to involution have been implemented in computer algebra systems, both for exterior differential systems [3] and the dual "formal theory" of partial differential equations [11].

4 Classification of Coframes

Having seen how the general methods of exterior differential systems can be applied to solve the equivalence problem for specific given coframes, we are equipped to examine the more general task of classifying a family of coframes under equivalence. The tools used here are broadly the same as those already discussed: the Frobenius, Cartan–Kähler and Cartan–Kuranishi theorems are all essential elements (although they may not all be required for a given problem). The basic procedures of reduction, Cartan's test for involution, and prolongation still direct the steps taken, but some of the details are adapted to the more general setting and the goal is different. Of course, any coframe ω is equivalent to itself under the identity transformation, so there is always a solution to the equivalence relation: the process outlined in section 3 serves here to determine the full extent of the equivalence class $[\omega]$. There is also the matter of extracting the final set of invariants characterising the equivalence classes, and determining the remaining freedom within each class. These topics will be treated in section 5.

In this section, we will be concerned with exploring the consequences of the equivalence relation with the given group and set of coframes. In practical applications, as will be seen from the examples, attention is restricted to just half of the problem: $\boldsymbol{\theta}$ on $P = M \times G$, say, but in describing the reasoning underlying the method, it is useful to stay on a larger space.

For clarity, the problem will be re-stated in this new setting. Suppose we are given a family $\{(\boldsymbol{\omega}, U)\}$ of coframes $\boldsymbol{\omega}$ on open sets $U \subset M$, and a linear Lie group $G \subset GL(n)$ where $n = \dim M$. The equivalence problem is then to divide the coframes into equivalence classes under the equivalence relation $(\boldsymbol{\omega}, U) \equiv (\bar{\boldsymbol{\omega}}, \bar{U})$ if and only if there is a local diffeomorphism $\Phi: U \to \bar{U}$ and a function $\gamma: U \to G$ such that

$$\Phi^* \bar{\boldsymbol{\omega}} = \gamma \, \boldsymbol{\omega}. \tag{42}$$

As before, we will restrict our attention entirely to local issues, and so take $M = U$ and $\bar{M} = \bar{U}$.

4.1 G-structures and Exterior Systems Formulation

In examining the specific equivalence problem in section 3, it was helpful to introduce the product space $P = M \times G$ with projection $\pi: P \to M$. To make the symmetry of the equivalence relation (42) more transparent, it is helpful to introduce the product space $\bar{P} = \bar{M} \times \bar{G}$, where $\bar{G} = G$, with corresponding projection $\bar{\pi}$. In parallel with the equivariant 1-forms $\boldsymbol{\theta} = g \, \boldsymbol{\omega}$ on P, we define the 1-forms $\bar{\boldsymbol{\theta}} = \bar{g} \, \bar{\boldsymbol{\omega}}$. The pair $(P, \boldsymbol{\theta})$ is called a G-structure. Using these spaces, we can restate the equivalence relation in a manifestly symmetrical form.

Given a pair of coframes $\boldsymbol{\omega}$ and $\bar{\boldsymbol{\omega}}$, equivalent under the maps Φ and γ, the 1-forms $\boldsymbol{\theta}$ and $\bar{\boldsymbol{\theta}}$ are directly related under pullback by the map $\tilde{\Phi}: P \to \bar{P}$ given by $\tilde{\Phi}(x, g) = (\Phi(x), g \, \gamma(x)^{-1})$. This follows easily from $\tilde{\Phi}^* \bar{\boldsymbol{\theta}} = \tilde{\Phi}^*(\bar{g} \, \bar{\boldsymbol{\omega}}) = g \, \gamma(x)^{-1} \, \Phi^* \bar{\boldsymbol{\omega}} = g \, \boldsymbol{\omega} = \boldsymbol{\theta}$. This suggests that we can convert the equivalence relation (42) into the following form: $\boldsymbol{\theta} \equiv \bar{\boldsymbol{\theta}}$ if and only if there is a local fibre-preserving diffeomorphism $\tilde{\Phi}: P \to \bar{P}$ such that

$$\tilde{\Phi}^* \bar{\boldsymbol{\theta}} = \boldsymbol{\theta}. \tag{43}$$

To confirm this, consider a fibre-preserving map $\tilde{\Phi}: P \to \bar{P}$ satisfying (43). The fibre-preserving condition allows us to determine $\Phi: M \to \bar{M}$ and $\eta: P \to \bar{G}$ such that $\tilde{\Phi}(x, g) = (\Phi(x), \eta(x, g))$. Then the equivalence (43) of $\boldsymbol{\theta}$ and $\bar{\boldsymbol{\theta}}$ implies that $g \boldsymbol{\omega} = \tilde{\Phi}^*(\bar{g} \bar{\boldsymbol{\omega}}) = \eta \Phi^* \bar{\boldsymbol{\omega}}$. From this we get the equivalence (42) of the base forms $\boldsymbol{\omega}$ and $\bar{\boldsymbol{\omega}}$ using $\gamma(x) = \eta(x, g)^{-1} g$, which is necessarily independent of g.

Using a variation on the arguments employed in section 3, we further alter condition (43) to the existence of solutions (integral manifolds) of the exterior differential system

$$S = \{\bar{\boldsymbol{\theta}} - \boldsymbol{\theta}\} \qquad \Omega = \{\boldsymbol{\theta}\} \tag{44}$$

on $P \times \bar{P}$. As before, we use the same symbols for 1-forms lifted to the product space.

In this form of the equivalence relation, $\boldsymbol{\theta}$ and $\bar{\boldsymbol{\theta}}$ appear on equal terms (even $\boldsymbol{\theta}$ in the independence condition could be swapped for $\boldsymbol{\theta} + \bar{\boldsymbol{\theta}}$ without changing the results). In the rest of this treatment, all operations on the P and $\bar{P}$ factors will be performed in parallel, so this balance will be maintained.

To refine the equivalence relation further, we once again examine the conditions governing integral elements of (S, Ω). The structure equations on P are much as before:

$$d\theta^i \simeq A^i_{j\rho} \pi^\rho \wedge \theta^j + \tfrac{1}{2} T^i_{jk} \theta^j \wedge \theta^k, \tag{45}$$

where $A^i{}_{j\rho}$ are constants and

$$T^i{}_{jk} = g^i{}_l C^l{}_{mn} (g^{-1})^m{}_j (g^{-1})^n{}_k \tag{46}$$

as before with $C^i{}_{jk}$ defined by

$$d\omega^i = \tfrac{1}{2} C^i{}_{jk} \omega^j \wedge \omega^k \tag{47}$$

on M.

The parallel relation on $\bar{P}$ is

$$d\bar{\theta}^i \simeq A^i{}_{j\rho} \bar{\pi}^\rho \wedge \bar{\theta}^j + \tfrac{1}{2} \bar{T}^i{}_{jk} \bar{\theta}^j \wedge \bar{\theta}^k, \tag{48}$$

in which it is convenient to use the same Lie algebra basis $\{A_\rho\}$.

Integral elements E of (S, Ω) are characterised by the additional annihilating 1-forms

$$\begin{cases} \boldsymbol{\pi} - \boldsymbol{v}\,\boldsymbol{\theta} \\ \bar{\boldsymbol{\pi}} - \bar{\boldsymbol{v}}\,\bar{\boldsymbol{\theta}}. \end{cases} \tag{49}$$

By virtue of the structure equations (45) and (48), the coefficients $\boldsymbol{v}$ and $\bar{\boldsymbol{v}}$ must satisfy a set of linear equations which may be written

$$\mathcal{A}(\boldsymbol{v} - \bar{\boldsymbol{v}}) = \boldsymbol{T} - \bar{\boldsymbol{T}} \tag{50}$$

in terms of the linear operator $\mathcal{A}$ with components $\mathcal{A}^i{}_{jk\rho}{}^l = 2A^i{}_{[j\rho} \delta^l_{k]}$ and the functions $\boldsymbol{T} = (T^i{}_{jk})$ and $\bar{\boldsymbol{T}} = (\bar{T}^i{}_{jk})$.

If $\lambda = (\lambda^{jk}{}_i)$ is a non-zero left null vector of $\mathcal{A}$, then

$$0 = \lambda \cdot \mathcal{A}(\boldsymbol{v} - \bar{\boldsymbol{v}}) = \lambda \cdot (\boldsymbol{T} - \bar{\boldsymbol{T}}), \tag{51}$$

which again provides a set of consistency conditions for the existence of integral elements. Integral elements can exist only at points on $P \times \bar{P}$ where the right-hand side $\lambda \cdot (\boldsymbol{T} - \bar{\boldsymbol{T}})$ vanishes. The set of functions

$$\tau = \{\lambda \cdot \boldsymbol{T} \mid \lambda \cdot \mathcal{A} = 0\} \tag{52}$$

is called the *intrinsic* or *essential torsion* of $\boldsymbol{\theta}$.

This definition of intrinsic torsion, while convenient for understanding the method, does not give the most practical computational tool. What is usually done is to solve the equations $\mathcal{A}\boldsymbol{v} = \boldsymbol{T}$ as far as possible, giving

$$\boldsymbol{v} = \boldsymbol{w}(t), \tag{53}$$

where the variables t parameterise $\ker \mathcal{A}$. The 1-forms $\boldsymbol{\pi}$ are then replaced by

$$\boldsymbol{\pi}' = \boldsymbol{\pi} - \boldsymbol{w}(t)\boldsymbol{\theta}. \tag{54}$$

This change of basis is called *absorption* of the non-intrinsic torsion. Writing the structure equations in terms of the new 1-forms $\boldsymbol{\pi}'$, we have

$$d\theta^i \simeq A^i{}_{j\rho} \pi'^\rho \wedge \theta^j + \tfrac{1}{2} T'^i{}_{jk} \theta^j \wedge \theta^k, \tag{55}$$

in which any remaining functions $\boldsymbol{T}'$ are intrinsic torsion.

In some cases, the group parameterisation is unknown or unwieldy in computations. However, without a group parameterisation, it is not possible to specify the coframe elements π^ρ explicitly. None the less, these coframe elements are often conveniently expressed implicitly by the linear relations satisfied by the (linearly dependent) components of π_{MC}. Gardner's *implicit approach* [2] allows us to work with this linearly dependent set, often simplifying computations even when a group parameterisation is available. The process of computing a replacement for π_{MC} to isolate the intrinsic torsion, known as *Lie algebra-compatible absorption* is followed in one of the examples in section 6.2.

Unlike the specific equivalence problem, the exterior system for the classification problem must always have a solution, since any coframe is equivalent to itself. Thus the consistency conditions (51) cannot violate the independence condition. None the less, it is possible that they imply a relation between the coordinates on the base $M \times \bar{M}$. That is, there is some function I in the module generated by τ which is a function on M alone, and a corresponding function $\bar{I}$ on $\bar{M}$. In terms of the equivalence map, the consistency conditions demand

$$\Phi^* \bar{I} = I. \tag{56}$$

Any such functions I are thus invariants under equivalence transformations, characterising equivalence classes.

4.2   Normalisation

Leaving aside such invariants, the remaining conditions in (51) are functions on $P \times \bar{P}$ whose zero sets specify surfaces lying transverse to the fibres. At this stage, a brute-force application of the general reduction scheme outlined in section 3.2 might suggest solving these conditions for some parameters from G as functions on $M \times \bar{P}$ and the rest of G. However, in the ultimate integral manifold for equivalent coframes, both g and $\bar{g}$ will be given as functions on M. Furthermore, in terms of the lifted equivalence map $\tilde{\Phi}$, we will have

$$\tilde{\Phi}^*(\lambda \cdot \bar{\boldsymbol{T}}) = \lambda \cdot \boldsymbol{T}, \tag{57}$$

so a more elegant (and more instructive) way to proceed is to fix parameters from G and $\bar{G}$ in parallel. This has the effect of reducing the equivalence group to a subgroup $G_1 \subset G$, and is called *normalisation of torsion*.

Suppose we fix a point $x_0 \in M$ and set $\tau_0 = \tau(x_0, e)$. If there is a local G-valued function γ on a neighbourhood V of each point in M such that $\tau(x, \gamma(x)) = \tau_0$ for all $x \in V$, then τ is said to define a *normalisation of constant type*. This is the best of all possibilities, since we will see that it allows the equivalence problem as a whole to be reduced to a similar one with a smaller group. For simplicity, we will consider no other type of normalisation, and assume that the function γ is globally defined on M (there are, however, important problems such as the equivalence of Lagrangian field theories which lead to normalisations of non-constant type [9]).

The function γ is defined only up to multiplication by the stability group $G_1 = \{g \in G \mid \tau(x_0, g) = \tau_0\}$. Writing $P_1 = M \times G_1$, there is a function $\rho: P_1 \to P$, $(x, h) \mapsto (x, \gamma(x)h)$ such that $\rho^*\tau = \tau_0$.

Repeating the same considerations on $\bar{P}$ (and choosing the same τ_0), we obtain another map $\bar{\rho}: \bar{P}_1 \to \bar{P}$ with $\bar{\rho}^*\bar{\tau} = \tau_0$. These two maps then solve the consistency conditions (51) for integral elements of (S, Ω) on $P \times \bar{P}$ since $\rho^*\tau - \bar{\rho}^*\bar{\tau} = 0$.

The power of this approach to resolving the consistency conditions (51) lies in the fact that the Cartan equivalence problems

$$S = \{\bar{\boldsymbol{\theta}} - \boldsymbol{\theta}\} \qquad \Omega = \{\boldsymbol{\theta}\} \tag{58}$$

on $P \times \bar{P}$ and

$$S_1 = \{\bar{\rho}^*\bar{\boldsymbol{\theta}} - \rho^*\boldsymbol{\theta}\} \qquad \Omega_1 = \{\rho^*\boldsymbol{\theta}\} \tag{59}$$

on $P_1 \times \bar{P}_1$ share the same solution set. As a result, we abandon the original equivalence group, and see what consequences the equivalence relation has for the reduced equivalence problem.

In order to gain a better understanding of the normalisation type, we note that since the matrices $A_\rho = (A^i{}_{j\rho})$ form a basis for $\mathfrak{g}$, they transform under the adjoint representation of G. It is then not difficult to see that there is a natural action of G on the left nullspace of $\mathcal{A}$ and hence on the codomain of τ. Each fibre of $P = M \times G$ corresponds under the map τ to an orbit of G under this natural action. For a normalisation of constant type, the image $\tau(P)$ is a single orbit. As long as the action of G on $\tau(P)$ is non-trivial, the reduced equivalence group G_1 is a proper subgroup of G.

To choose τ_0, it is helpful to recognise the representation corresponding to the action of G on the intrinsic torsion, and then fix a convenient point on the orbit. Without knowing the explicit group parameterisation, this can be very difficult, but Gardner's implicit approach [2] provides an answer. From the identities $0 = \mathrm{d}^2\boldsymbol{\theta}$, it is possible to recognise the action of the Lie algebra $\mathfrak{g}$ (as opposed to the Lie group G) on the torsion and thus deduce a convenient normal form τ_0, at least for the connected part of G. This technique can often help to minimise computation even when an explicit parameterisation is known.

In practice, it is often even possible to see by inspection what values (usually 0 or ± 1) the components of τ_0 should take. The map ρ is constructed by solving the equations $\tau(x, g) = \tau_0$ for some of the group parameters $g^i{}_j$. The remaining, free group parameters then parametrise the stability subgroup G_1.

Having performed the group reduction to $G_1 \subset G$, we can repeat the process thus far, computing structure equations and intrinsic torsion for the reduced equivalence problem. We may uncover further invariants I_1 (functions on M alone) within the intrinsic torsion at this level. After perhaps several repetitions, the chain $G \supset G_1 \supset \cdots \supset G_k$ of group reductions must stabilise so that $G_{k+1} = G_k$.

Re-cycling the symbols as usual, we may thus assume that the equivalence problem has been brought into a form where the action of G on the intrinsic torsion in trivial. The remaining intrinsic torsion $\boldsymbol{T}$ consists entirely of invariants I resolving the consistency conditions (51).

At this point, it is worth distinguishing between so-called *e-structures* where $G = \{e\}$ is itself trivial, and other G-structures with $G \neq \{e\}$. If $G = \{e\}$, then the 1-forms $\boldsymbol{\theta}$ form a coframe for $P \simeq M$. Local solutions always exist, by virtue of the Frobenius theorem. This case is dealt with in section 5. For $G \neq \{e\}$, we cannot use the Frobenius theorem but, having eliminated the consistency conditions (51), we are in a position to use Cartan's test for involution on the system (S, Ω). In the language of quasi-linear Pfaffian systems [1], $\boldsymbol{\pi}_{MC} = (\pi^i{}_j)$ is the *tableau* matrix, from which the reduced Cartan characters can easily be calculated, as explained in the exterior systems chapter [5]. In fact, it makes no difference for this purpose if we use the original $\boldsymbol{\pi}_{MC}$ here or the replacement with the non-intrinsic torsion absorbed. Let $s_1, s_2, \ldots, s_n$ be the reduced Cartan characters. Then Cartan's test is $\dim \ker \mathcal{A} = s_1 + 2s_2 + \cdots + ns_n$. If the test succeeds, then the Cartan–Kähler theorem assures us of the local existence of integral manifolds parameterised by s_q functions of q variables, where s_q is the highest non-zero reduced character. That is, we have found an infinite-dimensional equivalence class characterised by the known invariants.

As a matter of fact, Cartan's test works even for the trivial case $G = \{e\}$ in which $s_1 = \cdots = s_n = 0$ and $\dim \ker \mathcal{A} = 0$, but the Cartan–Kähler theorem applies only to analytic problems, whereas the Frobenius theorem gives results in the smooth category, and so applies more widely.

If, on the other hand, Cartan's test fails, then (S, Ω) is not in involution. This is not a difficulty, it simply implies that there is more information in the equivalence relation (42) at the level of the second derivative, and hence not accessible through the structure equations. The extra information typically relates to the size of the equivalence class, but may also yield new scalar invariants.

As for the specific equivalence problem, the way forward is to construct the first prolonged exterior system (S^1, Ω^1) on the extended manifold $P^1 \times \bar{P}^1$ where $P^1 = P \times G^1$ and G^1 is abelian. The construction of the prolonged system is entirely analogous to that outlined in section 3, and will not be repeated here. Once again, we restart the process from the beginning, calculating intrinsic torsion, performing group reductions, collecting invariants and, if necessary, prolonging further. From the Cartan–Kuranishi theorem, we know that (always assuming sufficient regularity) we must achieve involution in finitely many steps.

This done, we have exhaustively analysed the structure equations for information about the equivalence relation. We have collected an initial set of scalar invariants (the remnant intrinsic torsion) characterising the equivalence classes and decided if the equivalence class is infinite-dimensional or finite-dimensional.

5   Characterising Equivalence Classes

After following the procedures outlined in the previous sections, reducing the equivalence group and prolonging as necessary, we can assume that the equivalence problem has been brought to a form where the action of G on the intrinsic torsion τ is trivial, and either the Frobenius condition or Cartan's test for involution is satisfied. The final 1-forms θ and π yield an *invariant* coframe whose structure equations are adapted in some way to the equivalence class structure. While this is very useful, we would really like to have a ready means of characterising equivalence classes, and the invariant coframe is not sufficient for this.

Besides the invariant coframe, we have also generated a set of scalar invariants of an equivalence class, the intrinsic torsion coefficients τ. By themselves, the torsion coefficients may not fully characterise the equivalence class either, but they can be used in conjunction with the final coframe to derive a set of invariant functions determining the equivalence class as far as possible, as we shall see in this section.

At this stage, it is useful to consider three possibilities for the final space P:

1. $P = M$, the original manifold, so that the original equivalence group has been completely reduced to the trivial group $\{e\}$ (equivalently, the

original G-structure has become an e-structure). Then $\boldsymbol{\theta}$ is a coframe for P, $\bar{\boldsymbol{\theta}}$ is a coframe for $\bar{P}$, and, granted equality of the invariants on P and $\bar{P}$, solutions to the final exterior system $\{\boldsymbol{\theta} - \bar{\boldsymbol{\theta}}\}$ are guaranteed by the Frobenius theorem.

2. $P = M \times G$ in which G is non-trivial, but $\ker \mathcal{A} = \{0\}$. There is thus no freedom in choosing integral elements of the final exterior system, and the prolonged system $\{\boldsymbol{\theta} - \bar{\boldsymbol{\theta}}, \boldsymbol{\pi} - \bar{\boldsymbol{\pi}}\}$ falls into class 1. Once more, the Frobenius condition guarantees local solutions to the exterior system.

3. $P = M \times G$ with $\ker \mathcal{A}$ non-trivial. The tableau $\boldsymbol{\pi}_{MC}$ for G satisfies Cartan's test for involutivity, so there is an infinite-dimensional set of equivalence transformations guaranteed by the Cartan–Kähler theorem.

For convenience, we will assume that we are in case 1, $P = M$ and $G = \{e\}$, where an immediate solution is possible. The modifications necessary for the other two cases will be discussed later.

In this case, the structure equations are simply

$$\mathrm{d}\theta^i = \tfrac{1}{2} T^i_{jk} \theta^j \wedge \theta^k, \tag{60}$$

and the intrinsic torsion $\boldsymbol{\tau} = \boldsymbol{T}$. As discussed earlier (cf. equation (51)), on integral manifolds f of $S = \{\boldsymbol{\theta} - \bar{\boldsymbol{\theta}}\}$, $\Omega = \{\boldsymbol{\theta}\}$, corresponding elements $F = T^i_{jk}$ and $\bar{F} = \bar{T}^i_{jk}$ of the intrinsic torsion on M and $\bar{M}$ must satisfy $f^*(F - \bar{F}) = 0$. As a trivial consequence, $f^*(\mathrm{d}F - \mathrm{d}\bar{F}) = 0$ as well. Defining the *coframe derivatives* $\partial_i F$ of a function F on M by

$$\mathrm{d}F = \partial_i F \, \theta^i, \tag{61}$$

(with a corresponding definition for $\bar{\partial}_i \bar{F}$), we have

$$f^*(\partial_i F - \bar{\partial}_i \bar{F}) = 0 \tag{62}$$

as a result of $f^*S = 0$ and the independence condition. The coframe derivatives of the intrinsic torsion provide a further set of invariant scalars characterising the equivalence class. Obviously, further derivatives can be taken indefinitely, yielding more and more invariant scalars, but since M is n-dimensional, at most n functions from the complete collection can be independent.

Put more formally, let $\mathfrak{F}_0 = \boldsymbol{T}$, and set

$$\mathfrak{F}_s = \{\boldsymbol{T}, \partial_{i_1}\boldsymbol{T}, \ldots, \partial_{i_1}\cdots\partial_{i_s}\boldsymbol{T} \mid 1 \leqslant i_1 \leqslant \cdots \leqslant i_s \leqslant n\} \qquad s = 1, \ldots, n. \tag{63}$$

Assuming that $\operatorname{rank} \mathrm{d}\mathfrak{F}_s$ is constant on M, we have $\operatorname{rank} \mathrm{d}\mathfrak{F}_s \leqslant \operatorname{rank} \mathrm{d}\mathfrak{F}_{s+1} \leqslant n$ for $s \geqslant 0$, so there is some value $k \geqslant 0$ at which the rank must stabilise: $\operatorname{rank} \mathrm{d}\mathfrak{F}_{k+1} = \operatorname{rank} \mathrm{d}\mathfrak{F}_k$. Setting $r = \operatorname{rank} \mathrm{d}\mathfrak{F}_k$, locally there exists a set $\{F^1, \ldots, F^r\}$ of independent functions in terms of which each element of $\mathfrak{F}_k$ can be written as a functional combination.

This yields the final solution to the equivalence problem in this case: the functions $\{F^1, \ldots, F^r\}$, and the way they generate $\mathfrak{F}_k$, completely characterise the equivalence classes, summarised in the following result from Cartan (the statement here follows that in [4]).

Theorem 5.1 *Let θ be a coframe on M with a maximal rank set $\mathfrak{F}_k$ of functions generating the intrinsic torsion τ and its coframe derivatives to all orders. Let $\bar{\theta}$ be a coframe on $\bar{M}$ with corresponding $\bar{\mathfrak{F}}_k$. Then there exists a diffeomorphism $\Phi \colon M \to \bar{M}$ in a neighbourhood of $\bar{p} = \Phi(p)$ satisfying*

$$\Phi^* \bar{\theta} = \theta \tag{64}$$

if and only if

(i) *$k = \bar{k}$ and* $\operatorname{rank} \mathrm{d}\mathfrak{F}_k = \operatorname{rank} \mathrm{d}\bar{\mathfrak{F}}_k$;

(ii) *if $\{F^1, \ldots, F^r\}$ is a set of $r = \operatorname{rank} \mathrm{d}\mathfrak{F}_k$ independent functions generating $\mathfrak{F}_k$, and $\{\bar{F}^1, \ldots, \bar{F}^r\}$ is the corresponding set for $\bar{\mathfrak{F}}_k$, then*

$$F^a(p) = \bar{F}^a(\bar{p}) \qquad a = 1, \ldots, r; \tag{65}$$

(iii) *for any element $h = H(F^1, \ldots, F^r) \in \mathfrak{F}_k$, the corresponding element $\bar{h} \in \bar{\mathfrak{F}}_k$ is $\bar{h} = H(\bar{F}^1, \ldots, \bar{F}^r)$.*

We have here a complete solution for the case $G = \{e\}$. The residual freedom in the equivalence problem is now contained in the self-equivalences $\Phi \colon M \to M$ with $\Phi^* \theta = \theta$. These self-equivalences form a Lie group of dimension $n - r$, as can be seen by pulling back the structure equations to a level set of $\{F^1, \ldots, F^r\}$.

In case 2, the elements θ and π of the prolonged system can be treated as an e-structure on $M \times G$ and we can repeat the same procedure as above. Note that since θ is semi-basic and the intrinsic torsion τ consists of functions on the base M, invariant under the action of G, we have

$$\mathrm{d}\tau = \partial_i \tau \theta^i \tag{66}$$

as before. However, since $\theta = \{\theta^i\}$ carries a representation of G, the coframe derivatives $\partial_i \tau$ carry the dual representation, and so $\mathrm{d}\partial_i \tau$, if non-zero, is not semi-basic. This raises the possibility of launching into a fresh round of group reductions and prolongations if the intrinsic torsion τ is not constant.

In case 3, similar remarks to those just made apply again. The added feature here is that the set of self-equivalences is an infinite-dimensional Lie pseudo-group. This is so because the general equivalence map solving (S, Ω) depends upon s_k arbitrary functions of k variables, where s_k is the highest non-zero character of the tableau π_{MC}.

This completes the description of the Cartan equivalence method. We have followed the consequences of the equivalence relation (1) right through to the stage where we now have an invariant coframe adapted to the equivalence class structure, and a (possibly incomplete) set of invariant functions characterising the equivalence classes. In addition, we know something about the size and nature of the group of self-equivalences. The only major item of the equivalence method which we have neglected is the actual construction of the equivalence maps Φ and γ for specific equivalent coframes. This is essentially an application of the Frobenius theorem and a reversal of the arguments used to establish the initial exterior system in section 3. The method is known as the *technique of the graph*, and the details can be found in papers already cited [2, 4, 9].

6 Examples

In this section, we will put the method of equivalence outlined above into practice by applying it to the two examples introduced in section 2: Lagrangian mechanics and Riemannian geometry. In both cases, we will study the classification problem, and so will focus on just one side of the full equivalence relation – the 1-forms $\boldsymbol{\theta}$ and their structure equations on the space $P = M \times G$. As indicated in earlier sections, we will follow the convention of re-cycling symbols as we move from space to space. Since there are a number of successive reductions or prolongations involved in a typical equivalence problem, this convention saves continually searching for new symbols or cluttering up the notation with extra indices.

6.1 Lagrangian Mechanics

In section 2, we established the Cartan form of the equivalence problem for one-dimensional mechanical Lagrangians under a change of variables. We will now use Cartan's method to complete the classification of such Lagrangians. The equivalence condition was

$$\Phi^* \bar{\boldsymbol{\omega}} = \gamma \, \boldsymbol{\omega} \tag{67}$$

where

$$\boldsymbol{\omega} = \begin{pmatrix} \omega^1 \\ \omega^2 \\ \omega^3 \end{pmatrix} = \begin{pmatrix} \mathrm{d}q - v \, \mathrm{d}t \\ L \, \mathrm{d}t \\ \mathrm{d}v \end{pmatrix} \qquad L \neq 0 \tag{68}$$

is a coframe for $M = J^1(\mathbb{R}, \mathbb{R})$, and

$$\gamma = \begin{pmatrix} \alpha & 0 & 0 \\ \beta & 1 & 0 \\ \rho & \sigma & \tau \end{pmatrix} \qquad \alpha \neq 0 \quad \tau \neq 0 \tag{69}$$

is a function on M taking values in a 5-parameter Lie group G.

Let

$$\boldsymbol{\theta} = \begin{pmatrix} \theta^1 \\ \theta^2 \\ \theta^3 \end{pmatrix} = \begin{pmatrix} \alpha & 0 & 0 \\ \beta & 1 & 0 \\ \rho & \sigma & \tau \end{pmatrix} \begin{pmatrix} \omega^1 \\ \omega^2 \\ \omega^3 \end{pmatrix} \tag{70}$$

be the matrix of lifted 1-forms $\boldsymbol{\theta} = g \boldsymbol{\omega}$ on $M \times G$. We take the structure equations one by one. An easy calculation gives

$$\mathrm{d}\theta^1 = \pi^1 \wedge \theta^1 + \frac{\alpha}{\tau L} \theta^2 \wedge \theta^3, \tag{71}$$

where

$$\pi^1 = \frac{1}{\alpha} \mathrm{d}\alpha - \frac{\rho}{\tau L} \theta^2 + \frac{\beta}{\tau L} \theta^3. \tag{72}$$

Since the final term contains no θ^1 factor, it is not possible to absorb it into the first term by redefining the complementary 1-form π^1. The coefficient $T^1{}_{23} = \alpha/(\tau L)$ is intrinsic torsion.

Likewise, the θ^2 structure equation is

$$\mathrm{d}\theta^2 = \pi^2 \wedge \theta^1 + \frac{\beta - L_v}{\tau L} \theta^2 \wedge \theta^3, \tag{73}$$

where

$$\pi^2 = \frac{1}{\alpha}\,\mathrm{d}\beta - \frac{\rho(\beta - L_v) + \tau L_q}{\alpha \tau L}\theta^2 + \frac{\beta(\beta - L_v)}{\alpha \tau L}\theta^3. \tag{74}$$

The coefficient $T^2{}_{23} = (\beta - L_v)/(\tau L)$ is also intrinsic torsion.

By inspection, the group G acts on $T^1{}_{23}$ by scaling, so we can normalise to, say, $T^1{}_{23} = 1$ by setting

$$\alpha = \tau L. \tag{75}$$

Likewise, G acts on $T^2{}_{23}$ through translation by β, allowing us to normalise it to zero by setting

$$\beta = L_v. \tag{76}$$

With the group reductions (75) and (76), the 1-form π^1 remains independent of θ (it has a term proportional to $\mathrm{d}\tau$), giving rise to no new intrinsic torsion. However, the 1-form π^2 becomes semi-basic,

$$\pi^2 = -\frac{J^2(\sigma L + \tau a)}{\tau^2 L^3}\theta^2 + \frac{J^2}{\tau^2 L^2}\theta^3, \tag{77}$$

in which

$$J^2 = L L_{vv}, \quad \text{and} \quad a = \frac{L_q - L_{tv} - v L_{qv}}{L_{vv}}. \tag{78}$$

The combination a is useful because the Euler–Lagrange equations for L can be written

$$\frac{\mathrm{d}v}{\mathrm{d}t} = a. \tag{79}$$

Note that this step assumes the usual regularity condition for a mechanical system that L_{vv} is nowhere zero and also requires sign $L_{vv} = $ sign L for real J.

The (reduced) θ^2 structure equation is thus pure intrinsic torsion:

$$\mathrm{d}\theta^2 = \frac{J^2(\sigma L + \tau a)}{\tau^2 L^3}\theta^1 \wedge \theta^2 - \frac{J^2}{\tau^2 L^2}\theta^1 \wedge \theta^3. \tag{80}$$

The torsion coefficients can be normalised to $T^2{}_{12} = 0$ and $T^2{}_{13} = -1$ by setting

$$\sigma = -\frac{\tau a}{L}, \quad \text{and} \quad \tau = \pm\frac{J}{L}. \tag{81}$$

We treat the branch $\tau = J/L$ (the other branch, and the complex case $T^2{}_{13} = +1$, are quite similar).

The θ^2 structure equation has now been exhausted:

$$\mathrm{d}\theta^2 = -\theta^1 \wedge \theta^3. \tag{82}$$

However, the second group reduction (81) has made π^1 semi-basic, so it is worth re-examining the θ^1 equation, which now reads

$$\mathrm{d}\theta^1 = \frac{\rho - J_2}{J}\theta^1 \wedge \theta^2 - \frac{J_3 + L_v}{J}\theta^1 \wedge \theta^3 + \theta^2 \wedge \theta^3. \tag{83}$$

Here, we can already make use of the coframe derivatives

$$\mathrm{d}J = J_1\theta^1 + J_2\theta^2 + J_3\theta^3 \tag{84}$$

since J is known to be a function on the base manifold M. Any further group reductions are performed by pullback, which commutes with the exterior derivative d, so that the definitions of J_1, J_2 and J_3 adjust themselves accordingly.

The θ^1 structure equation (83) gives us the final group reduction

$$\rho = J_2. \tag{85}$$

All group parameters have now been fixed in normalising intrinsic torsion. The remaining non-constant piece of torsion from this equation, $T^1{}_{13} = -(J_3 + L_v)/J$, yields our first scalar invariant of the equivalence class.

The final, invariant coframe is

$$\begin{pmatrix} \theta^1 \\ \theta^2 \\ \theta^3 \end{pmatrix} = \begin{pmatrix} J & 0 & 0 \\ L_v & 1 & 0 \\ J_2 & -aJ/L^2 & J/L \end{pmatrix} \begin{pmatrix} \mathrm{d}q - v\,\mathrm{d}t \\ L\,\mathrm{d}t \\ \mathrm{d}v \end{pmatrix}. \tag{86}$$

A simple calculation shows that, although the θ^2 coframe derivative J_2 appears in this expression, the specification of $\boldsymbol{\theta}$ is not circular. Explicitly,

$$\begin{aligned} J_2 &= \frac{\partial J}{\partial \omega^2} + \frac{a}{L}\frac{\partial J}{\partial \omega^3} \\ &= \frac{1}{L}\frac{\partial J}{\partial t} + \frac{v}{L}\frac{\partial J}{\partial q} + a\frac{\partial J}{\partial v}. \end{aligned} \tag{87}$$

It is worth noting that the final $\theta^2 = L_v\,\mathrm{d}q - (vL_v - L)\,\mathrm{d}t$ is the Cartan 1-form, central to a geometrical understanding of the calculus of variations (cf. G. Prince's chapter [10]). This adapted structure is exactly the kind of result which Cartan's method is bound to produce.

We have yet to examine the θ^3 structure equation. From the other two equations (82) and (83), we have $0 = \mathrm{d}^2\theta^2 = \theta^1 \wedge \mathrm{d}\theta^3$, so we know in advance that $T^3{}_{23} = 0$. The final structure equations are

$$\mathrm{d}\begin{pmatrix} \theta^1 \\ \theta^2 \\ \theta^3 \end{pmatrix} = \begin{pmatrix} -F^1\theta^1 \wedge \theta^3 + \theta^2 \wedge \theta^3 \\ -\theta^1 \wedge \theta^3 \\ -F^2\theta^1 \wedge \theta^2 + F^3\theta^1 \wedge \theta^3 \end{pmatrix}, \tag{88}$$

in which (after some slightly messy calculations)

$$\begin{aligned} F^1 &= \frac{J_3 + L_v}{J} = \frac{LL_{vvv} + 3L_vL_{vv}}{2L_{vv}\sqrt{LL_{vv}}} \\ F^2 &= \frac{J_{22}}{J} - 2\frac{J_2{}^2}{J^2} + \frac{J_2(L_t + vL_q + aL_v)}{JL^2} + \frac{Ja_1}{L^2} + \frac{L_v a_2}{L^2} \\ F^3 &= \frac{J_1 - J_{23}}{J} + \frac{J_2 F^1}{J} - \frac{L_q}{JL}. \end{aligned} \tag{89}$$

Applying the exterior derivative to both sides of the structure equations (88) yields the "Jacobi identities" for the structure functions F^a (only the first and third rows yield non-trivial relations):

$$\begin{aligned} F^1_2 &= -F^3 \\ F^3_2 - F^2_2 &= F^1 F^2. \end{aligned} \tag{90}$$

Since dim $M = 3$, there can be at most three independent functions among these invariants and their coframe derivatives. The maximal symmetry equivalence classes have all three F^a constant. By (90), this implies $F^3 = 0$ and, as long as $F^1 \neq 0$, $F^2 = 0$. Included in these classes are, for example, the free particle Lagrangian $L = \frac{1}{2}mv^2$, which has $J = \frac{1}{\sqrt{2}}mv$, and $a = 0$. The invariant coframe is

$$\begin{pmatrix} \theta^1 \\ \theta^2 \\ \theta^3 \end{pmatrix} = \begin{pmatrix} \frac{1}{\sqrt{2}}(mv\,dq - mv^2\,dt) \\ mv\,dq - \frac{1}{2}mv^2\,dt \\ \frac{\sqrt{2}}{v}\,dv \end{pmatrix}, \tag{91}$$

from which $J_3 = \frac{1}{2}mv$ while $J_1 = J_2 = 0$. Hence $F^1 = \frac{3}{\sqrt{2}}$ and $F^2 = F^3 = 0$, as required.

6.2   Riemannian Geometry

In section 2, the local isometry problem for n-dimensional Riemannian manifolds (M, ds^2) was formulated as a local equivalence problem for orthonormal coframes $\boldsymbol{\omega}$ under the orthogonal group $SO(n)$.

$$\Phi^* d\bar{s}^2 = ds^2 \qquad \Longleftrightarrow \qquad \Phi^* \bar{\boldsymbol{\omega}} = \gamma\boldsymbol{\omega}, \tag{92}$$

where γ is an $SO(n)$-valued function on M. To take this problem further, Gardner's implicit approach [2] will be used so that we can avoid introducing an explicit parameterisation of the equivalence group.

As it involves essentially no more work, but includes some physically interesting geometries, we will undertake a classification of pseudo-Riemannian or Lorentzian structures, so the orthonormal coframe obeys

$$ds^2 = \sum_{i=1}^{p} \omega^i \otimes \omega^i - \sum_{j=1}^{q} \omega^j \otimes \omega^j, \tag{93}$$

where $p + q = n$, and the equivalence group is $G = SO(p, q)$. It is convenient to introduce the matrix

$$\eta = (\eta_{ij}) = \begin{pmatrix} I_{p \times p} & 0 \\ 0 & -I_{q \times q} \end{pmatrix} = (\eta^{ij}) = \eta^{-1}, \tag{94}$$

in terms of which, $ds^2 = \boldsymbol{\omega} \otimes \eta\boldsymbol{\omega}$.

In order to treat the classification problem for Lorentzian structures, we lift the orthonormal coframe to the soldering form on the orthonormal frame bundle. Locally, let $P = M \times G$, and consider $\boldsymbol{\theta} = g\,\boldsymbol{\omega}$, where $g \in SO(p, q)$. As usual, the structure equations are

$$d\theta^i = \pi^i{}_j \wedge \theta^j + \frac{1}{2}T^i{}_{jk}\theta^j \wedge \theta^k, \tag{95}$$

but without a group parameterisation (and an explicit set of structure equations for the original orthonormal coframe $\boldsymbol{\omega}$), it is not possible to compute the torsion $\boldsymbol{T}$ explicitly. Nor is it possible to absorb the torsion as far as possible by an explicit change of coframe, since we don't have the 1-forms π^ρ at our disposal. Hence the means previously used to calculate the intrinsic torsion are not available.

Fortunately, all that is needed are the *defining relations* of the Lie algebra $\mathfrak{g}$ of the equivalence group G. These are the linear relations expressing the fact that for a proper subgroup of $GL(n)$, the components of $\boldsymbol{\pi}_{MC}$ are linearly dependent. Since $\pi^i{}_j = A^i{}_{\rho j}\pi^\rho$, the defining relations hold for each of the matrices A_ρ, and so any change of coframe to elements of the form $\pi^\rho - v_i^\rho \theta^i$ results in a new matrix of 1-forms (again denoted $(\pi^i{}_j)$) which still obey the defining relations. Conversely, replacing $\boldsymbol{\pi}_{MC}$ by another matrix $\boldsymbol{\pi} \simeq \boldsymbol{\pi}_{MC}$ (mod $\boldsymbol{\theta}$) of 1-forms obeying the defining relations corresponds to such a change in the coframe elements π^ρ. This is the Lie algebra-compatible absorption of torsion mentioned earlier in section 4. (Note that, in this section, the notation $\boldsymbol{\pi}$ is used for the matrix of linearly dependent 1-forms $(\pi^i{}_j)$, rather than the vector of independent 1-forms (π^ρ).) For $\mathfrak{so}(p,q)$, the defining relations are

$$\eta\boldsymbol{\pi} + (\eta\boldsymbol{\pi})^t = 0, \qquad \text{or} \qquad \eta_{ij}\pi^j{}_k + \eta_{kj}\pi^j{}_i = 0 \tag{96}$$

in components, and are satisfied by the Maurer–Cartan 1-forms $\boldsymbol{\pi}_{MC}$. The quasi-linear structure equations (95) can be re-written

$$d\boldsymbol{\theta} = \boldsymbol{\Delta} \wedge \boldsymbol{\theta}, \tag{97}$$

where $\boldsymbol{\Delta} \simeq \boldsymbol{\pi}_{MC}$ (mod $\boldsymbol{\theta}$). Defining

$$\boldsymbol{\pi} = \tfrac{1}{2}\left(\boldsymbol{\Delta} - \eta^{-1}(\boldsymbol{\Delta}\eta)^t\right) \tag{98}$$
$$\boldsymbol{\psi} = \tfrac{1}{2}\left(\boldsymbol{\Delta} + \eta^{-1}(\boldsymbol{\Delta}\eta)^t\right), \tag{99}$$

we see that $\boldsymbol{\pi} \simeq \boldsymbol{\pi}_{MC}$ (mod $\boldsymbol{\theta}$), and satisfies the defining relations (96) of $\mathfrak{g}$. In terms of these new 1-form matrices, we can write

$$d\boldsymbol{\theta} = \boldsymbol{\pi} \wedge \boldsymbol{\theta} + \boldsymbol{\psi} \wedge \boldsymbol{\theta}. \tag{100}$$

The matrix of 1-forms $\boldsymbol{\psi}$ obeys

$$\eta\boldsymbol{\psi} - (\eta\boldsymbol{\psi})^t = 0, \tag{101}$$

and its entries are called the *principal components*. They represent the torsion $\boldsymbol{T}$ of the structure equations in this implicit form, but not necessarily the intrinsic torsion.

As with an explicit parametric calculation, integral elements of the exterior system associated with the equivalence relation must annihilate additional 1-forms. Not knowing the explicit coframe elements, we can still write a linearly dependent set of additional 1-forms

$$\boldsymbol{\pi} - \boldsymbol{\phi}, \tag{102}$$

where $\boldsymbol{\phi} = (\phi^i{}_j)$ is a matrix of semi-basic 1-forms satisfying the defining relations of $\mathfrak{g}$. Using the structure equations (100), $\boldsymbol{\phi}$ must satisfy

$$\boldsymbol{\phi} \wedge \boldsymbol{\theta} = -\boldsymbol{\psi} \wedge \boldsymbol{\theta}. \tag{103}$$

This is the implicit form of the Grassmann variety conditions, $\boldsymbol{A}\boldsymbol{v} = \boldsymbol{T}$.

Before attempting to solve (103), we can establish the dimension of the solution space by solving the homogeneous equations $\boldsymbol{\phi} \wedge \boldsymbol{\theta} = 0$, or equivalently

$(\eta\phi)_{ij} \wedge \theta^j = 0$. This kind of equation is solved using the Cartan lemma, by which there exist functions $\alpha_{ijk} = \alpha_{ikj}$ such that $(\eta\phi)_{ij} = \alpha_{ijk}\theta^k$. But ϕ also satisfies the defining relations $(\eta\phi)^t = -(\eta\phi)$, from which it follows that $\alpha_{ijk} = -\alpha_{jik}$. Since the only rank three tensor with one symmetric and one antisymmetric pair of indices is the zero tensor, we have $\alpha_{ijk} = 0$. Thus $\ker \mathcal{A} = \{0\}$ and the solution to (103), if it exists, is unique.

The easiest way to see that (103) has a solution is to construct it. Using the Cartan lemma again, there exist functions $\beta_{ijk} = \beta_{ikj}$ such that $(\eta\phi + \eta\psi)_{ij} = \beta_{ijk}\theta^k$. Applying the defining relations, and using the symmetry (101) of the principal components, we find

$$2(\eta\psi)_{ij} = (\beta_{ijk} + \beta_{jik})\theta^k. \tag{104}$$

This relation is easily inverted, in precisely the same way that the explicit form of the Christoffel symbols is usually derived, to give

$$\beta_{ijk} = \psi_{ijk} + \psi_{kij} - \psi_{jki}, \tag{105}$$

where the components ψ_{ijk} are defined by $(\eta\psi)_{ij} = \psi_{ijk}\theta^k$.

Having fixed a matrix of 1-forms ϕ satisfying the defining relations of $\mathfrak{g}$ and solving (103) uniquely, we can adopt the new matrix of 1-forms $\pi' = \pi - \phi$ and (immediately discarding the dash on π') re-write the structure equations

$$d\theta = \pi \wedge \theta, \tag{106}$$

where, unlike Δ in (97), π satisfies the defining relations for $\mathfrak{g}$. Since the defining relations identify the coframe elements complementing θ on $M \times G$, this operation corresponds to making a change of coframe on $M \times G$ in which all the torsion has been absorbed. In fact, π is nothing other than the matrix of Levi-Civita connection 1-forms on the orthonormal frame bundle over M. With vanishing intrinsic torsion, there is no possibility of a group reduction for this equivalence problem. However, the equivalence group G is non-trivial, so we must apply Cartan's test for involution to see if the method has terminated. We have already established that $\dim \ker \mathcal{A} = 0$, so Cartan's test can be satisfied only if the Cartan characters all vanish: $s_1 = \cdots = s_n = 0$. However, given that the structure equations involve all the (implicitly defined) 1-forms π^ρ, it is apparent that $s_1 + \cdots + s_n = \dim G = \frac{1}{2}n(n-1)$. So Cartan's test fails and we must prolong the equivalence problem.

As discussed in section 3.3, the prolongation leads to the affine bundle $V_\Omega(S)$ over $M \times G$ of integral elements. However, as $\dim \ker \mathcal{A} = 0$, the fibres of this bundle are zero-dimensional, so the prolonged system resides on the original space $M^1 = M \times G$ and the prolonged equivalence group is trivial: $G^1 = \{e\}$. The prolonged exterior system can be expressed implicitly in terms of the linearly dependent set of 1-forms $\{\theta, \pi\}$ (and their counterparts on $\bar{M} \times G$).

Since the 1-forms $\{\theta, \pi\}$ for the prolonged equivalence problem span the cotangent spaces, we have finished the first stage of the equivalence method and can begin to calculate invariants and derived invariants. Before doing this, we must fill out the remaining structure equations for $d\pi$. In this case, these can be deduced from

$$0 = d^2\theta = (d\pi - \pi \wedge \pi) \wedge \theta. \tag{107}$$

Using a higher-degree analogue of the Cartan lemma, there exist 1-forms $\chi^i{}_{jk}$ on $M \times G$ with $\chi^i{}_{jk} \simeq \chi^i{}_{kj}$ (mod $\boldsymbol{\theta}$) such that

$$d\pi^i{}_j - \pi^i{}_k \wedge \pi^k{}_j = \chi^i{}_{jk} \wedge \theta^k. \tag{108}$$

However, the defining relations of $\mathfrak{g}$ imply that the left-hand side (with the indices all lowered using the matrix η) is antisymmetric, implying that $\chi_{ijk} \wedge \theta^k = -\chi_{jik} \wedge \theta^k$, or $\chi_{ijk} \simeq -\chi_{jik}$ (mod $\boldsymbol{\theta}$). Once again, the combination of symmetry and antisymmetry forces the result

$$\chi^i{}_{jk} \simeq 0 \quad (\text{mod } \boldsymbol{\theta}), \tag{109}$$

from which it follows that there are functions $R^i{}_{jkl}$ on $M \times G$ such that

$$d\pi^i{}_j - \pi^i{}_k \wedge \pi^k{}_j = R^i{}_{jkl}\theta^k \wedge \theta^l. \tag{110}$$

These functions are clearly (up to conventional signs and factors) the components of the Riemann tensor.

Thus the intrinsic torsion for the prolonged equivalence problem is given by the Riemann tensor components. Since the prolonged problem corresponds to a Frobenius-integrable system, there are no more group reductions or prolongations required. We have thus used Cartan's method to establish the well-known fact that locally isometric Lorentzian manifolds are classified by the components of their Riemann tensors and their derivatives. The precise statement of the result requires the additional qualifications stated in the theorem in section 5.

This example has been pursued further in a number of directions. One branch which is immediately accessible contains the rank zero coframes, for which the curvature components are constants. Simple algebraic analysis, which can be found in any book on Riemannian geometry, leads to the size of the symmetry group and other results. More intricate investigations have been made for Lorentzian metrics in four dimensions, the space–time of general relativity. Here the classification of coframes, derivation of invariants, and so on has been honed to a fine art [6] and even implemented in computer algebra packages [8] to determine the equivalence class of a given metric.

References

[1] R.L. Bryant, S.S. Chern, R.B. Gardner, H.L. Goldschmidt and P.A. Griffiths, *Exterior Differential Systems,* Springer-Verlag, New York, 1991.

[2] R.B. Gardner, *The Method of Equivalence and its Applications,* CBMS-NSF regional conference series in applied mathematics **58**, SIAM, Philadelphia, 1989.

[3] D. Hartley, Involution analysis for non-linear exterior differential systems, *Math. Computer Modelling* **25**, 51–62, 1997.

[4] N. Kamran, *Contributions to the Study of the Equivalence Problem of Élie Cartan and its Applications to Partial and Ordinary Differential Equations,* Mémoires de la classe des sciences, Académie Royale de Belgique **45**, Fac. 7, 1989.

[5] Niky Kamran's chapter in this volume.

[6] A. Karlhede, A review of the geometric equivalence of metrics in general relativity, *Gen. Rel. Grav.* **12**, 693–707, 1980.

[7] Ian Lisle and Greg Reid's chapter in this volume.

[8] M.A.H. MacCallum and J.E.F. Skea, SHEEP: A computer algebra system for general relativity, in *Algebraic Computing in General Relativity: Lecture Notes from the First Brazilian School on Computer Algebra*, eds. M.A.H. MacCallum *et al*, Oxford University Press, Oxford, 1994.

[9] P.J. Olver, *Equivalence, Invariants, and Symmetry*, Cambridge University Press, Cambridge, 1995.

[10] Geoff Prince's chapter in this volume.

[11] W. Seiler, *Applying AXIOM to Partial Differential Equations*, Internal Report 95-17. Fakultät für Informatik, Universität Karlsruhe, available from `http://iaks-www.ira.uka.de/iaks-calmet/werner/publ.html`

[12] K. Yang, *Exterior Differential Systems and Equivalence Problems*, Kluwer Academic Publishers, Boston, 1992.

The Inverse Problem in the Calculus of Variations and its Ramifications

GEOFF E. PRINCE

Keywords: Calculus of variations, inverse problem, Helmholtz conditions

1   The Problem and its History

The inverse problem in the calculus of variations involves deciding whether the solutions of a given system of second-order ordinary differential equations

$$\ddot{x}^a = F^a(t, x^b \dot{x}^b), \quad a, b = 1, \dots, n$$

are the solutions of a set of Euler–Lagrange equations

$$\frac{\partial^2 L}{\partial \dot{x}^a \partial \dot{x}^b} \ddot{x}^b + \frac{\partial^2 L}{\partial x^b \partial \dot{x}^a} \dot{x}^b + \frac{\partial^2 L}{\partial t \partial \dot{x}^a} = \frac{\partial L}{\partial x^a}$$

for some Lagrangian function $L(t, x^b, \dot{x}^b)$.

Because the Euler–Lagrange equations are not generally in normal form, the problem is to find a so-called multiplier matrix $g_{ab}(t, x^c, \dot{x}^c)$ such that

$$g_{ab}(\ddot{x}^b - F^b) \equiv \frac{d}{dt}\left(\frac{\partial L}{\partial \dot{x}^a}\right) - \frac{\partial L}{\partial \dot{x}^a}.$$

The most commonly used set of necessary and sufficient conditions for the existence of the g_{ab} are the so-called *Helmholtz conditions* due to Douglas [15] and put in the following form by Sarlet [42]:

$$g_{ab} = g_{ba}, \quad \Gamma(g_{ab}) = g_{ac}\Gamma^c_b + g_{bc}\Gamma^c_a, \quad g_{ac}\Phi^c_b = g_{bc}\Phi^c_a, \quad \frac{\partial g_{ab}}{\partial \dot{x}^c} = \frac{\partial g_{ac}}{\partial \dot{x}^b},$$

where

$$\Gamma^a_b := -\frac{1}{2}\frac{\partial F^a}{\partial \dot{x}^b}, \quad \Phi^a_b := -\frac{\partial F^a}{\partial x^b} - \Gamma^c_b\Gamma^a_c - \Gamma(\Gamma^a_b),$$

and where

$$\Gamma := \frac{\partial}{\partial t} + u^a\frac{\partial}{\partial x^a} + F^a\frac{\partial}{\partial u^a}.$$

In the introduction to his landmark 1941 paper [15], the Fields' medallist Jesse Douglas said that "*the problem ... is one of the most important hitherto unsolved problems of the calculus of variations*". Douglas goes on to solve the problem for the $n = 2$ case only: he does this by a painstaking study of four main cases and many subcases. These cases are distinguished by the vanishing or otherwise of quantities constructed from the F^a. Douglas expresses the necessary conditions to be satisfied by the multiplier matrix as partial differential equations and uses Riquier–Janet theory to solve them. In most of the cases

Douglas decides the existence question and gives all the possible Lagrangians; in the remaining cases, the problem becomes a question of the closure of a certain 1-form.

Douglas's method is not easily generalizable to higher dimension and for this reason the problem has not been solved in Douglas's sense for $n = 3$ or higher. This chapter describes the current situation, its development and the deep importance that the inverse problem still has for mathematics and mathematical physics.

1.1 Historical Note

(The reader is referred to the excellent Notes at the end of Chapter 5 of Olver's book [35] and the comprehensive first chapter of Anderson and Thompson [1] on the history of the variational bicomplex. This section is primarily intended to fill in the historical details of the local inverse problem for second order ordinary differential equations.)

Helmholtz [21] first raised the matter of deciding which systems of second order ordinary differential equations are Euler–Lagrange for a first-order Lagrangian (that is, one depending on velocities but not accelerations) *in the form presented*. Mayer [33] gave necessary and sufficient conditions for this to be true. Hirsch [24] is responsible for the *multiplier problem*, that is, the question of the existence of multiplier functions which convert second order o.d.e.'s in normal form into Euler–Lagrange equations. Hirsch produced certain self-adjointness conditions which, of themselves, are not particularly useful in classifying second order equations according to the existence and uniqueness of the corresponding multipliers.

This multiplier problem was completely solved by Douglas in 1941 [15] for two degrees of freedom, that is, a pair of second order equations on the plane. As described, Douglas's solution took the form of an exhaustive classification of all such equations in normal form. In each case, Douglas identified all (if any) Lagrangians producing Euler–Lagrange equations whose normal form is that of the equations in that particular case. His method studiously avoided Hirsch's self-adjointness conditions and he produced his own necessary and sufficient conditions, now known perversely as the *Helmholtz conditions*. These conditions are partly differential, partly algebraic and his approach was to generate a sequence of integrability conditions, solving these using Riquier–Janet theory.

Douglas's solution is so comprehensive that his conditions formed the basis for subsequent attempts on the three and higher degrees of freedom problems. His technique is so opaque that only recently has serious advancement been possible.

Following Douglas the next significant contribution to solving the Helmholtz conditions was that of the Henneaux [22] and Henneaux and Shepley [23]. Motivated by Wigner's question *"Do the equations of motion* (as opposed to the variational principle) *completely determine the corresponding quantum mechanics?"* Henneaux developed an algorithm for solving the Helmholtz conditions for **any given system of second order equations,** and, in particular, he solved the problem for spherically symmetric problems in dimension 3. In this fundamental case Henneaux and Shepley showed that a two-parameter family of

Lagrangians produce the same equations of motion, answering Wigner's question in the negative for the hydrogen atom. (Henneaux's Lagrangians produced inequivalent quantum mechanical hydrogen atoms. Further mathematical aspects of this case were elaborated by Crampin and Prince [10, 12].)

At around the same time Sarlet [42] showed that the part of the Helmholtz conditions which ensures the correct time evolution of the multiplier matrix could be replaced by a possibly infinite sequence of purely algebraic initial conditions. Like Henneaux's work, Sarlet's contribution made the solution of the problem for particular higher dimensional systems a real possibility.

Henneaux's prototype for geometrising Douglas's Helmholtz conditions inspired and eventually united a group of mathematicians (Cantrijn, Cariñena, Crampin, Ibort, Marmo, Prince, Sarlet, Saunders and Thompson) with interests in Lagrangian formulations in their work on the inverse problem. Using tangent bundle geometry (see Yano and Ishahara [49]) and the geometrical approach to second order equations of Klein and Grifone [19, 20, 26, 27], they reformulated the Helmholtz conditions for non-autonomous second order o.d.e.'s on a manifold in terms of the corresponding non-linear connection on its tangent bundle. This occurred in 1985 after a sequence of papers [8, 13, 42]. Over the next nine years, the group made considerable advances in the geometry of second order equations, notably the work of Sarlet [43], and collectively Martínez, Cariñena and Sarlet [30, 31, 32]. The work of these three on derivations along the tangent bundle projection produced the geometric machinery needed to reformulate geometrically Douglas's **solution** of the two degrees of freedom case opening the way to extension to arbitrary dimension. This was achieved in 1993 by Crampin, Sarlet, Martínez, Byrnes and Prince and is reported in [14]. Separately Anderson and Thompson [1] applied exterior differential systems theory to some special cases of the geometrised problem with considerable success. (Both of these advances are outlined in section 3.3.) In addition, the reader is directed to the review papers of Morandi et al. [34] and Anderson and Thompson [1] for accounts of the problem prior to these developments.

1.2   Scope of this Chapter

This chapter aims to give the reader an idea of the contributions to the local inverse problem made in the last fifteen years using the differential geometry of the tangent bundle, and in particular, to demonstrate how this work has led to an advance (the generalization of the Liouville–Arnol'd theorem) in the theory of second order ordinary differential equations. I make no attempt to discuss the global inverse problem for ordinary differential equations (which is essentially cohomological in character: see [25, 1]). I have not attempted a comprehensive review and I have included few proofs.

2   Existence and Uniqueness of Lagrangians

Naturally, it is the uniqueness and existence aspects of the inverse problem rather than its computational aspects that are of interest to physicists. However, actually tackling the problem from first principles for a given system is a very worthwhile exercise, providing insight into the sources of non-uniqueness and non-existence. So I begin this section with two of Douglas's examples (the elementary technique is mine: Douglas has far more sophisticated paraphernalia at hand from tackling the general problem). I suggest that you attempt the second example yourself after you've followed the first one.

2.1   Two Examples

Example 1. As an example of the non-existence of a Lagrangian, consider the pair

$$\ddot{x} = \dot{y}, \qquad \ddot{y} = y.$$

It is a simple matter to show that the Helmholtz conditions are not satisfied. The only non-zero components of Γ^a_b and Φ^a_b are:

$$\Gamma^1_2 = -1, \qquad \Phi^2_2 = -1.$$

The third of the Helmholtz conditions gives only $g_{12} = 0$ and the first then gives $g_{21} = 0$.
Now the second gives

$$\Gamma(g_{11}) = 0 = \Gamma(g_{22}), \qquad \Gamma(g_{12}) = -g_{11} = \Gamma(g_{21}).$$

Hence $g_{11} = 0$ and the multiplier matrix must be singular if it exists in violation of the initial assumption. It is illuminating to continue: the last Helmholtz condition gives

$$\frac{\partial g_{22}}{\partial \dot{x}} = 0$$

and the second gives

$$\frac{\partial g_{22}}{\partial t} + \dot{x}\frac{\partial g_{22}}{\partial x} + \dot{y}\frac{\partial g_{22}}{\partial y} + y\frac{\partial g_{22}}{\partial \dot{y}} = 0$$

from which we deduce

$$\frac{\partial g_{22}}{\partial x} = 0$$

by virtue of $\dfrac{\partial g_{22}}{\partial \dot{x}} = 0$. Now we know that

$$g_{11} = 0 = g_{12} = g_{21}; \; g_{22} = G(t, y, \dot{y}); \; \Gamma(G) = 0,$$

and so the expression

$$g_{ab}(\ddot{x}^b - F^b) \equiv \frac{d}{dt}\left(\frac{\partial L}{\partial \dot{x}^a}\right) - \frac{\partial L}{\partial \dot{x}^a}$$

degenerates to

$$G(\ddot{y} - y) = \frac{d}{dt}\left(\frac{\partial L}{\partial \dot{y}}\right) - \frac{\partial L}{\partial y}.$$

Example 2. As a tractable example of non-uniqueness consider:

$$\ddot{x} = y, \quad \ddot{y} = 0.$$

In this case,

$$\Gamma = \frac{\partial}{\partial t} + \dot{x}\frac{\partial}{\partial x} + \dot{y}\frac{\partial}{\partial y} + y\frac{\partial}{\partial \dot{y}};$$

$$\Gamma_b^a = 0; \quad \Phi_2^1 = -1; \quad \Phi_b^a = 0, \ a \neq 1, \ b \neq 2.$$

The symmetry condition $g_{ac}\Phi_b^c = g_{bc}\Phi_a^c$ leads immediately to $g_{11} = 0$ and the last of the Helmholtz conditions yields

$$0 = \frac{\partial g_{11}}{\partial \dot{y}} = \frac{\partial g_{12}}{\partial \dot{x}}; \quad \frac{\partial g_{21}}{\partial \dot{y}} = \frac{\partial g_{22}}{\partial \dot{x}}.$$

Using the first of these in the second condition, $\Gamma(g_{12}) = 0$, we obtain

$$\frac{\partial g_{12}}{\partial t} + \dot{x}\frac{\partial g_{12}}{\partial x} + \dot{y}\frac{\partial g_{12}}{\partial y} = 0,$$

and since $\dfrac{\partial g_{12}}{\partial \dot{x}} = 0$ we must have

$$\frac{\partial g_{12}}{\partial t} + \dot{y}\frac{\partial g_{12}}{\partial y} = 0, \quad \frac{\partial g_{12}}{\partial x} = 0.$$

Hence

$$g_{12} = G_1(\dot{y}, y - \dot{y}t).$$

It now remains to find g_{22} (since $g_{21} = g_{12}$ by the first Helmholtz condition). The cleanest way to proceed is to change coordinates: $(t, x, y, \dot{x}, \dot{y}) \mapsto (t, x, \dot{x}, u, v)$ where $u := y - \dot{y}t$, $v := \dot{y}$. In these coordinates

$$\Gamma = \frac{\partial}{\partial t} + \dot{x}\frac{\partial}{\partial x} + (u + vt)\frac{\partial}{\partial \dot{x}}$$

and

$$g_{12} = g_{21} = G_1(u, v).$$

The condition $\dfrac{\partial g_{21}}{\partial \dot{y}} = \dfrac{\partial g_{22}}{\partial \dot{x}}$ becomes

$$\frac{\partial G_1}{\partial v} - t\frac{\partial G_1}{\partial u} = \frac{\partial g_{22}}{\partial \dot{x}}$$

$$\Rightarrow g_{22} = \left(\frac{\partial G_1}{\partial v} - t\frac{\partial G_1}{\partial u}\right)\dot{x} + G_2(t, x, u, v).$$

Applying $\Gamma(g_{22}) = 0$ (from the second Helmholtz condition) yields

$$-\frac{\partial G_1}{\partial u}\dot{x} + \left(\frac{\partial G_1}{\partial v} - t\frac{\partial G_1}{\partial u}\right)(u + vt) + \frac{\partial G_2}{\partial t} + \dot{x}\frac{\partial G_2}{\partial t} + \dot{x}\frac{\partial G_2}{\partial x} = 0.$$

Equating coefficients of powers of $\dot{x}$ to zero gives

$$-\frac{\partial G_1}{\partial u} + \frac{\partial G_2}{\partial x} = 0, \quad \left(\frac{\partial G_1}{\partial v} - t\frac{\partial G_1}{\partial u}\right)(u + vt) + \frac{\partial G_2}{\partial t} = 0.$$

The first of these gives

$$G_2 = \frac{\partial G_1}{\partial u} x + G_3(t, u, v)$$

and the second can now be integrated with respect to t:

$$G_3 = -\frac{\partial G_1}{\partial v} ut + \frac{t^2}{2}\left(v\frac{\partial G_1}{\partial v} - u\frac{\partial G_1}{\partial u}\right) - \frac{t^3}{3}v\frac{\partial G_1}{\partial u} + G_4(u, v).$$

In summary,

$$g_{11} = 0, \quad g_{12} = g_{21} = G_1(u, v),$$

$$g_{22} = \left(\frac{\partial G_1}{\partial v} - t\frac{\partial G_1}{\partial u}\right)\dot{x}$$

$$+ \frac{\partial G_1}{\partial u}x - \frac{\partial G_1}{\partial v}ut + \frac{t^2}{2}\left(v\frac{\partial G_1}{\partial v} - u\frac{\partial G_1}{\partial u}\right)$$

$$- \frac{t^3}{3}v\frac{\partial G_1}{\partial u} + G_4(u, v),$$

$$u := y - \dot{y}t, \quad v := \dot{y}.$$

In this way, the system

$$\ddot{x} = y, \quad \ddot{y} = 0$$

is shown to be variational and the multiplier matrix (g_{ab}) is determined up to two arbitrary functions of two variables. To find all possible regular Lagrangians, it remains to integrate

$$g_{ab} = \frac{\partial^2 L}{\partial \dot{x}^a \partial \dot{x}^b}$$

which I leave as an exercise for the reader. A less strenuous exercise involves checking that the Lagrangians

$$L_1 := xy\dot{y} + \frac{1}{2}(x - \frac{t^2}{2}y)\dot{y}^2 - \frac{1}{2}t\dot{x}\dot{y}^2 + \frac{1}{36}t^3\dot{y}^3$$

and

$$L_2 := \frac{1}{2}\dot{x}\dot{y}^2 - \frac{1}{2}ty\dot{y}^2 + \frac{1}{4}t^2\dot{y}^3$$

both generate $\ddot{x} = y$, $\ddot{y} = 0$. (These Lagrangians result from choosing $G_1 := u$, $G_4 := 0$ and $G_1 := v$, $G_4 := 0$ respectively.)

2.2 Implications

The development of quantum mechanics, symplectic geometry, symplectic mechanics and geometric quantisation have focused mathematicians' attention on the Hamiltonian rather than Lagrangian formulation of classical mechanics. However, the questions of uniqueness and existence of variational formulations for mathematical models of reality are more easily dealt with in the Lagrangian setting. Henneaux's surprising result on the non-uniqueness of the quantum mechanical hydrogen atom highlights the longstanding importance of the inverse problem to physics and mathematicians have again become interested in second order ordinary differential equations and their possible variational formulations.

As we've seen, the pair of equations

$$\ddot{x} = \dot{y}, \quad \ddot{y} = y$$

is not derivable from a Lagrangian. The immediate consequence is the non-existence of a quantum mechanical analogue, at least according to current quantisation schemes. One might also suspect that the canonical "symmetry-first integral relationship" (Noether's theorem) of Lagrangian systems might be missing for this system, but as I show in section 4.1, this pair of equations enjoys just about every feature that we expect of an Euler–Lagrange pair. This is because all systems of second order ordinary differential equations (at least locally) satisfy all but one of the four Helmholtz conditions necessary for the existence of at least one Lagrangian!

This fact even guarantees that compact level sets of involutive first integrals of non-Lagrangian systems are tori (the pair above do not have integrals with compact level sets but section 4.3 contains an example which does).

So it appears that the only immediate consequence of the absence of a Lagrangian for a given system is a feeling of aesthetic let down on the part of the mathematical modeller. Even this may vanish if it becomes possible to quantise the equations of motion directly.

On the other hand, the surprising part of the inverse problem for physicists is the non-uniqueness of Lagrangians for a given, well-known Lagrangian system.[1] (The trivial non-uniqueness (gauge invariance) corresponding to the addition of a total time derivative and multiplication by a constant are, of course, well known.) For example, the classical, three-dimensional Kepler problem has the standard $T - V$ Lagrangian

$$L(\vec{x}, \vec{u}) := \frac{1}{2}\|\vec{u}\|^2 - \frac{\mu}{r}, \quad r := \|\vec{x}\|,$$

but Henneaux and Shepley [23] find a two-parameter family of degenerate Lagrangians which, when added to the standard Lagrangian, produce the same equations of motion. A particularly interesting one-parameter sub-family is

$$L_\gamma := L + \frac{\gamma J}{r^2}$$

where γ is a real parameter and J is the magnitude of the angular momentum, $J := \|\vec{x} \times \vec{u}\|$.

Henneaux and Shepley show that for $\gamma \neq 0$ these Lagrangians lead to a new quantisation of the hydrogen atom which doesn't display energy level degeneracy. In this sense *the equations of motion alone do not uniquely determine the commutation relations* and some sort of selection rule is desirable. Crampin and Prince took up this question and the matter of geometrically explaining the two-fold non-uniqueness in two papers in 1985 and 1988 [10, 12]. The key to our results lies in the observation that, upon the usual reduction from three dimensions to two,

$$\frac{\gamma J}{r^2} = \gamma(\dot{\theta}^2 - \dot{\phi}^2 \sin^2\theta)^{\frac{1}{2}} \mapsto \gamma|\dot{\theta}|$$

(in the usual coordinates), so that we needed to explain how to pullback trivial gauge freedom in dimension 2 to something nontrivial in dimension 3.

---

[1] See also the treatment of Lagrangian systems in Hartley's chapter in this volume.

We showed that the source of the non-uniqueness lay in the $SO(3)$ symmetry of the problem: an arbitrary 2-form on the two-sphere can be pulled back to a two-parameter family of degenerate Lagrangians by the smooth map

$$\hat{J}(\vec{x}, \vec{u}) := \frac{\vec{x} \times \vec{u}}{\|\vec{x} \times \vec{u}\|}$$

of $T\mathbb{E}_0^{3+}$ onto the unit two-sphere S^2. ($T\mathbb{E}_0^{3+}$ is the open submanifold of $T\mathbb{E}_0^3$ consisting of those points for which $\vec{x} \times \vec{u} \neq \vec{0}$.) $\hat{J}$ defines a fibration of $T\mathbb{E}_0^{3+}$ over S^2 in which the individual fibres are tangent bundles of 2-planes of constant direction of angular momentum and the lifted solution curves of the equations of motion lie in these fibres. Moreover, if f is any function on S^2, then $\hat{J}^*f$ is a constant of the motion. These facts express the usual reduction of three-dimensional central force problems to two dimensions. When applied to the two-parameter family of degenerate Lagrangians this reduction produces the usual gauge freedom.

We also showed that these problems held the promise of a selection rule for Lagrangians at the classical level by demonstrating that there was *a symmetry of the equations of motion* (see section 3.4) whose action by Lie differentiation on the family of Lagrangians had the correct Lagrangian as a fixed point. Specifically, this symmetry has generator

$$Z := t\frac{\partial}{\partial t} + \frac{2}{3}r\frac{\partial}{\partial r} - \frac{1}{3}\dot{r}\frac{\partial}{\partial \dot{r}} - \dot{\theta}\frac{\partial}{\partial \dot{\theta}} - \dot{\phi}\frac{\partial}{\partial \dot{\phi}},$$

a (projectable) vector field on $\mathbb{R} \times T\mathbb{E}_0^3$, and

$$\mathcal{L}_Z L_\gamma = -\frac{2}{3}L_{3\gamma/2}.$$

For the Kepler problem this (prolongation of a) point symmetry is related to the Runge–Lenz vector and for this case *the equations of motion, through their symmetry group, do completely determine the commutation relations.*

The relation between the non-uniqueness of Lagrangians and the symmetries of the second order differential equations is treated quite generally in [36, 11], where it is shown that the degeneracy is equivalent to a certain class of symmetries of the Euler–Lagrange equations. Nevertheless there are many open problems remaining in this degeneracy question (see Morandi et al. [34]).

3 Geometrical Formulations and Solutions

The advances of the last 15 years in the treatment of the inverse problem hinge on two developments. The first is the geometric formulation of arbitrary systems of second-order o.d.e's on a manifold M and the second is the geometric formulation of the ordinary problem of the calculus of variations on M.

The initial work on geometrising second order equations is due to Klein and his student Grifone [26, 27, 19, 20] The key feature of their work is the introduction of the *SODE* – a vector field on TM, or $\mathbb{R} \times TM$ in the non-autonomous case, whose integral curves are the solution curves of the second order o.d.e's lifted to TM or $\mathbb{R} \times TM$. Crampin [8] studied the tangent bundle TM with a view to generalizing the geodesic spray (itself a SODE) and its associated linear connection. As a result the interaction of the SODE and the canonical structure on TM (or $\mathbb{R} \times TM$) is now known to give rise to a non-linear connection and its associated non-involutory horizontal distribution which produces the curvature. Subsequent work by Sarlet [43], then Martínez, Cariñena and Sarlet [30, 31, 32] and finally, Sarlet, Vandecasteele, Cantrijn and Martínez [45] has deepened the geometric infrastructure induced on TM or $\mathbb{R} \times TM$ by a SODE to include a fundamentally important covariant derivative. Byrnes [4] provides further insight into the constructions of these authors. I recommend Crampin and Pirani [9] and Saunders [47] as a good introduction to the underlying differential geometry and Saunders [47] as a useful text on more advanced geometry of the tangent bundle.

3.1   Second Order o.d.e.

Here is a brief description of some of these features in the time-dependent case: Suppose that M is some smooth, n-dimensional, second countable, Hausdorff manifold with generic local coordinates (x^a). The *evolution space* is defined as $E := \mathbb{R} \times TM$, with projection onto the first factor being denoted by $t : E \to \mathbb{R}$ and bundle projection $\pi : E \to \mathbb{R} \times M$. E has adapted coordinates (t, x^a, u^a) associated with t and (x^a).

A system of second order differential equations with local expression

$$\ddot{x}^a = F^a(t, x^b, \dot{x}^b), \ a, b, = 1, \dots, n$$

is associated with a smooth vector field Γ on E given in the same coordinates by

$$\Gamma := \frac{\partial}{\partial t} + u^a \frac{\partial}{\partial x^a} + F^a \frac{\partial}{\partial u^a}.$$

Γ is called a *second order differential equation field* or *SODE*. Such vector fields are completely characterised by the requirement that

$$\pi_* \Gamma_{(t,x,u)} = \frac{\partial}{\partial t} + u.$$

The integral curves of Γ are just the parametrised and lifted solution curves of the differential equations.

$E := \mathbb{R} \times TM$ has a number of natural structures, the first is its *contact structure*. A curve σ on M defines a section of $\mathbb{R} \times TM \to \mathbb{R}$ by $t \mapsto (t, \sigma^a(t), \dot{\sigma}^a(t))$ in terms of our natural bundle coordinates. A section ζ of $\mathbb{R} \times TM \to \mathbb{R}$ will be related to a curve in this way if and only if

$$\zeta^* \theta^a = 0, \quad a = 1, \dots, n,$$

where θ^a are local contact forms with coordinate expression

$$\theta^a = dx^a - u^a dt.$$

Simply put, the distribution spanned by these forms annihilates vector fields whose integral curves are naturally lifted from $\mathbb{R} \times M$. In particular, the contact structure annihilates SODEs: $\theta^a(\Gamma) = 0$.

The second structure is the *vertical subbundle* $V(E)$ of TE. A vector at a point of $\mathbb{R} \times TM$ is said to be *vertical* if it is tangent to the fibres of π :

$\mathbb{R} \times TM \to \mathbb{R} \times M$. The vector fields $V_a := \frac{\partial}{\partial u^a}$ form a local basis for the *vertical subbundle* $V(E)$ consisting of these vertical vector fields. These two structures are combined in the so-called *vertical endomorphism, S,* on E. It is a $(1,1)$ tensor field on E defined by

$$S := V_a \otimes \theta^a.$$

It has the following properties which determine it uniquely:

(i) S annihilates $V(E)$ and second order differential equation fields.

(ii) $S(Z) \in V(E)$ for any Z on E.

(iii) $S\left(\frac{\partial}{\partial t}\right) = -\Delta$, the dilation field on $\pi : E \to \mathbb{R} \times M$.

One might expect that the $(1,1)$ tensor field $\mathcal{L}_\Gamma S$ is fundamental to the analysis of the SODE Γ on E. Indeed, it produces a (non-linear) connection on E, that is, a direct sum decomposition of each tangent space of E in the following way. It is shown in Crampin, Prince and Thompson [13] that $\mathcal{L}_\Gamma S$ (acting on vectors) has eigenvalues $0, +1$ and -1. The eigenspace at a point of E corresponding to the eigenvalue 0 is spanned by Γ, while the eigenspace corresponding to $+1$ is the *vertical subspace* of the tangent space. The remaining eigenspace (of dimension n) is called the *horizontal subspace*. Unlike the vertical subspaces, these eigenspaces are not integrable; their failure to be so is due to the curvature of the nonlinear connection (induced by Γ) with components

$$\Gamma^a_b := -\frac{1}{2}\frac{\partial F^b}{\partial u^a}.$$

The most useful basis for the horizontal eigenspaces has elements with local expression

$$H_a = \frac{\partial}{\partial x^a} - \Gamma^b_a \frac{\partial}{\partial u^b}$$

so that a local basis of vector fields for the direct sum decomposition of the tangent spaces of E is $\{\Gamma, H_a, V_a\}$ with corresponding dual basis $\{dt, \theta^a, \psi^a\}$ where

$$\psi^a = du^a - F^a dt + \Gamma^a_b \theta^b.$$

The facts that

$$dt(\Gamma) = 1, \quad \theta^a(\Gamma) = 0 = \psi^a(\Gamma)$$

are crucial to the geometric formulation of the inverse problem, as we will see.

The H_a do not form an integrable distribution as I indicated; the components of the curvature manifest themselves in the expression for the commutators of the horizontal fields:

$$[H_a, H_b] = R^d_{ab} V_d$$

where the curvature of the connection is defined by

$$R^d_{ab} := \frac{1}{2}\left(\frac{\partial^2 F^d}{\partial x^a \partial u^b} - \frac{\partial^2 F^d}{\partial x^b \partial u^a} + \frac{1}{2}\left(\frac{\partial F^c}{\partial u^a}\frac{\partial^2 F^d}{\partial u^c \partial u^b} - \frac{\partial F^c}{\partial u^b}\frac{\partial^2 F^d}{\partial u^c \partial u^a}\right)\right).$$

In the next section, it will be useful to have some other commutators:

$$[H_a, V_b] = -\frac{1}{2}(\frac{\partial^2 F^c}{\partial u^a \partial u^b})V_c = V_b(\Gamma^c_a)V_c = V_a(\Gamma^c_b)V_c = [H_b, V_a],$$

$$[\Gamma, H_a] = \Gamma^b_a H_b + \Phi^b_a V_b, \qquad [\Gamma, V_a] = -H_a + \Gamma^b_a V_b,$$

and, of course, $[V_a, V_b] = 0$.

The case where Γ is a geodesic spray (so that the corresponding differential equations are the geodesic equations of a linear connection on M) is discussed in detail in [40, 41]. The horizontal fields are just

$$H_a := \frac{\partial}{\partial x^a} - \Gamma^b_{ac} u^c \frac{\partial}{\partial u^c}$$

and

$$[H_a, H_b] = -R^c_{dab} u^d V_c.$$

The reader is referred to Crampin's paper [7] on nonlinear connections associated with SODEs for a discussion of curvature, covariant differentiation, etcetera.

3.2   The Helmholtz Conditions

Now we turn to the geometric formulation of the ordinary problem in the calculus of variations. The extremals of the variational problem with regular Lagrangian $L \in C^\infty(E)$ are the base integral curves (on M) of the Euler–Lagrange equation, now represented by a SODE Γ, called the Euler–Lagrange field. The primary object of the Cartan–Hamilton formulation is the Cartan 1-form , θ_L, of the Lagrangian L:

$$\theta_L := L dt + dL \circ S.$$

In coordinates,

$$\theta_L = L dt + \frac{\partial L}{\partial u^a} \theta^a.$$

There is a brief discussion of the definition of θ_L at the end of this section. The key result is (see Goldschmidt and Sternberg [18] and Sternberg [46]):

Proposition 3.1. *If L is a regular Lagrangian (so that the matrix whose entries are $\dfrac{\partial^2 L}{\partial u^a \partial u^b}$ is everywhere nonsingular), then there is a unique vector field Γ on E such that*

$$\Gamma \lrcorner \, d\theta_L = 0 \quad and \quad dt(\Gamma) = 1.$$

This vector field is a SODE, and the equations satisfied by its integral curves are the Euler–Lagrange equations for L.

The coefficients F^a in the expression for the Euler–Lagrange field Γ are determined by the equation

$$\frac{\partial^2 L}{\partial u^a \partial u^b} F^b = \frac{\partial L}{\partial x^a} - \frac{\partial^2 L}{\partial u^a \partial x^b} u^b - \frac{\partial^2 L}{\partial u^a \partial t}.$$

If we use the basis $\{dt, \theta^a, \psi^a\}$, we obtain a particularly simple expression for the Cartan 2-form $d\theta_L$:

$$d\theta_L = \frac{\partial^2 L}{\partial u^a \partial u^b} \psi^a \wedge \theta^b$$

and this points the way to obtaining the Helmholtz conditions in a geometric form.

Notice first of all that $\dfrac{\partial^2 L}{\partial u^a \partial u^b}$ must satisfy all the conditions on the multiplier matrix g_{ab}, and secondly that, in the basis $\{dt, \theta^a, \psi^a\}$, $d\theta_L$ is completely determined by $\dfrac{\partial^2 L}{\partial u^a \partial u^b}$. These two facts indicate that we should look for a closed 2–form of maximal rank amongst 2-forms in $Sp\{\theta^a \wedge \psi^b\}$. The following theorem from [13] gives the most transparent geometric version of the Helmholtz conditions.

Theorem 3.2. *Given a SODE Γ, the necessary and sufficient conditions for there to be a Lagrangian for which Γ is the Euler–Lagrange field is that there should exist a 2-form Ω, of maximal rank, such that*

$$\Omega(V_1, V_2) = 0, \quad \forall V_1, V_2 \in V(E)$$

$$\Gamma \lrcorner \, \Omega = 0$$

$$d\Omega = 0$$

Ω is of maximal rank.

(The second and third conditions imply that $\mathcal{L}_\Gamma \Omega = 0$ and the fourth condition means that Ω has a one-dimensional kernel which the second condition shows is spanned by Γ.)

Necessity is obvious and the proof of sufficiency is a simple matter of starting with an arbitrary 2-form $\Omega \in Sp\{\theta^a \wedge \theta^b, \psi^a \wedge \theta^b\}$ and showing that

$$\Omega = g_{ab}\psi^a \wedge \theta^b$$

where g_{ab} satisfies Douglas's Helmholtz conditions (given at the beginning of this chapter).

The simplest way to see how the Helmholtz conditions arise from Theorem 3.2 is to put $\Omega := g_{ab}\psi^a \wedge \theta^b$ and compute $d\Omega$:

$$\begin{aligned}
d\Omega = {}& (\Gamma(g_{ab}) - g_{cb}\Gamma_a^c - g_{ac}\Gamma_b^c)dt \wedge \psi^a \wedge \theta^b \\
& + (H_d(g_{ab}) - g_{cb}V_a(\Gamma_d^c))\psi^a \wedge \theta^b \wedge \theta^d \\
& + V_c(g_{ab})\psi^c \wedge \psi^a \wedge \theta^b \\
& + g_{ab}\psi^a \wedge \psi^b \wedge dt \\
& + g_{ca}\Phi_b^c \theta^a \wedge \theta^b \wedge dt \\
& + g_{ca}H_b(\Gamma_d^c)\theta^a \wedge \theta^b \wedge \theta^d.
\end{aligned}$$

The four Helmholtz conditions are

$$\begin{aligned}
d\Omega(\Gamma, V_a, V_b) = 0, \qquad & d\Omega(\Gamma, V_a, H_b) = 0, \\
d\Omega(\Gamma, H_a, H_b) = 0, \qquad & d\Omega(H_a, V_b, V_c) = 0.
\end{aligned}$$

The remaining conditions arising from $d\Omega = 0$, namely

$$d\Omega(H_a, H_b, V_c) = 0 \quad \text{and} \quad d\Omega(H_a, H_b, H_c) = 0,$$

can be shown to be derivable from the first four (notice that this last condition is void in dimension 2).

At this point it is worthwhile identifying the separate sources of the algebraic and differential conditions. Recall the identity

$$d\Omega(X, Y, Z) = X(\Omega(Y, Z)) + Y(\Omega(Z, X)) + Z(\Omega(X, Y))$$
$$- \Omega([X, Y], Z) - \Omega([Y, Z], X) - \Omega([Z, X], Y)$$

for an arbitrary 2-form Ω and vector fields X, Y, Z. The first three terms involve the derivatives of Ω and the last three do not. In our case, using

$$\{X, Y, Z\} := \{\Gamma, H_a, H_b\}, \{\Gamma, V_a, V_b\}\{H_a, H_b, H_c\},$$

in turn makes the corresponding conditions $d\Omega(X, Y, Z) = 0$ purely algebraic. Any other choices produce differential conditions (except $d\Omega(V_a, V_b, V_c) = 0$ which is identically satisfied).

Example 1. Applying this theorem to the first example of section 2.1

$$\ddot{x} = \dot{y}, \qquad \ddot{y} = y,$$

we have the following (in coordinates (t, x, y, u, v) for $E := \mathbb{R} \times T\mathbb{R}^2$):

$$\Gamma := \frac{\partial}{\partial t} + u^a \frac{\partial}{\partial x^a} + v \frac{\partial}{\partial u} + y \frac{\partial}{\partial v},$$
$$\theta^x := dx - v dt, \ \theta^y := dy - v dt,$$
$$\psi^x := -\frac{1}{2}(dy + v dt) + du, \ \psi^y := dv - y dt,$$
$$\Omega := g_{ab} \psi^a \wedge \theta^b.$$

The algebraic conditions corresponding to the first and third Helmholtz conditions are

$$d\Omega(\Gamma, V_x, V_y) = g_{21} - g_{12} = 0, \qquad d\Omega(\Gamma, H_x, H_y) = -g_{21} = 0,$$

respectively.

The differential conditions corresponding to the second and fourth Helmholtz conditions are

$$d\Omega(\Gamma, H_x, V_x) = -\Gamma(g_{11}) = 0,$$
$$d\Omega(\Gamma, H_y, V_y) = -\Gamma(g_{22}) - \frac{1}{2}(g_{12} + g_{21}) = 0,$$
$$d\Omega(\Gamma, H_x, V_y) = -\Gamma(g_{21}) - \frac{1}{2}g_{11} = 0,$$
$$d\Omega(\Gamma, H_y, V_x) = -\Gamma(g_{12}) - \frac{1}{2}g_{11} = 0,$$

and

$$d\Omega(H_x, V_x, V_y) = \frac{\partial g_{21}}{\partial u} - \frac{\partial g_{11}}{\partial v} = 0,$$
$$d\Omega(H_y, V_x, V_y) = \frac{\partial g_{22}}{\partial u} - \frac{\partial g_{12}}{\partial v} = 0,$$

respectively.

The additional (redundant) conditions are

$$d\Omega(H_x, H_y, V_x) = \frac{1}{2}\left(\frac{\partial g_{11}}{\partial u} + 2\frac{\partial g_{11}}{\partial y} - 2\frac{\partial g_{12}}{\partial x}\right) = 0,$$

$$d\Omega(H_x, H_y, V_y) = \frac{1}{2}\left(\frac{\partial g_{21}}{\partial u} + 2\frac{\partial g_{21}}{\partial y} - 2\frac{\partial g_{22}}{\partial x}\right) = 0.$$

3.3 Recent Progress

The current decade has seen progress in solving the inverse problem on both fronts, that is, solving the Helmholtz conditions for a particular problem, and solving the problem in fixed or arbitrary dimension in the sense of Douglas.

In 1992, Anderson and Thompson [1] treated examples of both aspects using exterior differential systems theory.[2] They pose the inverse problem as that of finding all the closed 2-forms in $Sp\{\psi^a \wedge \theta^b\}$. We have already seen the algebraic and differential conditions which result; the big advantage in using EDS to treat these conditions in both concrete and abstract cases is that it answers the existence question and computes the degeneracy in the multiplier using the Cartan–Kähler theorem and the Cartan test for involution. This happens whether the multiplier can be explicitly calculated or not. In concrete cases the EDS algorithm produces a solution in which the degeneracy is manifest (of course, the user has to be able to integrate the differential conditions). On the abstract side, Anderson and Thompson show that in arbitrary dimension n, Douglas's Case I (Φ is a multiple of the identity) is always variational and that the multiplier is determined up to n functions of $n + 1$ variables.

The EDS technique does nothing more than the Riquier–Janet theory which Douglas used (although there is a technical difference in the way the degeneracy information is presented). However, EDS does appear to involve fewer integrability conditions and it is phrased in the language of exterior calculus.

The advance of Crampin et al. in 1993 [14] noted in section 1.1 comes from further geometrisation of the evolution space E and is of a structural nature rather than the computational nature of Anderson and Thompson's contribution. Specifically, we used a more highly refined calculus of forms which makes the generalization of Douglas's case-by-case analysis to higher dimension relatively transparent.

The details of this calculus are beyond the scope of the chapter, but it is based on the theory of sections of pullback bundles: for any map $f : N \to M$, the pullback of TM by f is a bundle and a section of this bundle is called *a vector field along f* (see, for example, [47]). Its application here is a restricted version of the theory of derivations of geometric objects defined along the projection $\pi : \mathbb{R} \times TM \to \mathbb{R} \times M$, a restriction which makes it identifiable with a calculus along $\tau : \mathbb{R} \times TM \to M$. (The corresponding calculus for the autonomous case, that is, for geometric objects defined along the projection $\tau : TM \to M$, has been developed by Martínez, Cariñena and Sarlet [30, 31, 45] and applied successfully by these authors in [32] to the solution of a related problem, namely the determination of necessary and sufficient conditions for a system of second-order differential equations to be separable into independent one-dimensional equations. The extension to the non-autonomous case can be found in [45].)

---

[2]Or, EDS for short. See for example, [3] and the chapter by Kamran in this volume.

The matrix Φ_b^a is now central to our discussion of the inverse problem. It appears as a type $(1,1)$ tensor field, Φ, along τ called the *Jacobi endomorphism* and our discussion of different subcases of the inverse problem is related in one way or another to the different possible algebraic normal forms of Φ. It is then natural to express the unknown multiplier g (which appears as a symmetric type $(0,2)$ tensor field along τ) in terms of a basis of 1-forms along the projection τ dual to a basis of vector fields adapted to Φ – a basis of eigenvectors of Φ, for example.

Using this calculus of derivations of geometric objects defined along τ to perform a normal form analysis on Φ we think we now have large parts of the Douglas paper well under control. Specifically, by using a frame adapted to the algebraic structure of Φ with respect to which Φ is in Jordan canonical form, we can write down the most general form of the multiplier which satisfies the algebraic conditions, and derive the detailed differential equations for the free components of the multiplier resulting from the invariance and closure conditions, in terms of certain "structure functions" of the frame.

Douglas's explicit calculations make more sense when interpreted in these terms: for example, we can explain how the classification into cases and subcases works, and we can show where expressions which play key roles in his analysis, but which are otherwise quite mysterious, come from. We are also able to see that some of his results are accidents of dimension, that is, occur only for $n = 2$.

Since the publication of this paper we have been working on the dimension n version of Douglas's Case I and Case IIa, that is, when Φ is a multiple of the identity and when both Φ is diagonalizable with distinct eigenvalues and $\mathcal{L}_\Gamma \Phi$ is a linear combination of Φ and the identity respectively.

3.4 Symmetries and First Integrals

The inverse problem for a given SODE is intimately related to the symmetries and first integrals of the second order equation(s). While the eigenvectors of Φ do not appear explicitly in the Helmholtz conditions, in many of the cases arising from the classification according to the Jordan normal form of Φ, they produce symmetries. Furthermore, in the Liouville–Arnol'd theorem and its generalization, the concept of complete integrability is central and the well-known Noether's theorem is indispensible in the proofs so I have collected some useful results on symmetries and first integrals for later use.[3] A *first integral* of a SODE Γ is a smooth function f on (some open subset of) the evolution space E satisying $\Gamma(f) = 0$. As a result f takes a constant value on any integral curve of Γ. The exterior derivatives of first integrals are important examples of a wider class of forms known as Γ-*basic forms*.

A Γ-*basic form* (or *invariant form* of Cartan [6]) is a form α on E which satisfies any one of three equivalent conditions:

1. $\alpha(\Gamma) = 0$ and $\Gamma \lrcorner \, d\alpha = 0$.

2. $\mathcal{L}_{f\Gamma}\alpha = 0$ for any smooth function f on E.

3. α is the pullback of a form from the quotient of E by the action of Γ.

---

[3]See also the treatment in Lisle and Reid's chapter in this volume.

The closed, Γ-basic 1-forms are locally just the exterior derivatives of first integrals of Γ.

A *symmetry* of the differential equation is a local one-parameter Lie group action on E which permutes the integral curves of Γ and, by projection, the solution curves on the base M (although this induced action will not in general be that of a Lie group on M). If an action is generated by a vector field X on E then it is a symmetry of Γ if and only if $\mathcal{L}_X \Gamma = \lambda \Gamma$ for some $\lambda \in C^\infty(E)$. Consequently, the symmetries of Γ do not form a module over $C^\infty(E)$, but over the first integrals of Γ and this module is closed under the Lie bracket. However, because any $C^\infty(E)$-multiple of Γ is a trivial symmetry of Γ, addition of such multiples to generators of symmetries produces further symmetry generators. The images of the integral curves of Γ under the original and modified actions differ only by parametrisation. For this reason we define a *transverse field*, $[X]$, to be the equivalence class of symmetries differing from X by the addition of a $C^\infty(E)$-multiple of Γ. As an example of selection from the equivalence class, we could choose $\hat{X} \in [X]$ so that $\hat{X}(t) = 0$ by setting $\hat{X} = X - X(t)\Gamma$, with X any element of $[X]$. There is an important one-to-one correspondence between transverse fields on E and vector fields on the quotient of E by the action of Γ. This results from the identification of $[X]$ with the representative $X \in [X]$ which passes to the quotient (this X generates an action leaving the parametrisation of the integral curves of Γ unchanged because $\mathcal{L}_X \Gamma = 0$).

Now we turn to a discussion of symmetries in the Euler–Lagrange case. Suppose that we have a regular, non-zero Lagrangian on E. (Given a regular Lagrangian, bounded below on E, we can always add a constant so as to make it non-zero.) It is shown in Proposition 1 of [39] that symmetries of $d\theta_L$ are also symmetries of Γ (but the converse is not true). The generators of such symmetries satisfy $\mathcal{L}_X d\theta_L = 0$ as do all other elements of their transverse field. It will be important for us to consider those elements X with $\theta_L(X) = 0$ (Proposition 1 of [39] guarantees the existence of this element). Because L is regular, $d\theta_L$ has maximal rank and the 2-form $d\theta_L$ provides a bijection $X \mapsto X \lrcorner\, d\theta_L$ between those vector fields on E satisfying $\mathcal{L}_X d\theta_L = 0$, $\theta_L(X) = 0$ and closed, Γ-basic 1-forms. This well-known result is a special case of the Noether–Cartan theorem (see [39] for a proof).

Theorem 3.3 (Noether–Cartan). *The map* $[X] \mapsto X \lrcorner\, d\theta_L$ *is a bijection of symmetries of* $d\theta_L$ *(that is,* $\mathcal{L}_X d\theta_L = 0$*) to closed basic 1-forms for* Γ*.*

We will restrict ourselves to the bijection $\Theta_L : X \mapsto X \lrcorner\, d\theta_L$ between symmetries of $d\theta_L$ with $\theta_L(X) = 0$ and closed Γ-basic 1-forms. We can use this map to define involutive first integrals in a natural way which corresponds to the Poisson bracket involution of Hamiltonian mechanics: two smooth first integrals f, g on E with $df \wedge dg \neq 0$ on E are said to be *in involution with respect to* L if $d\theta_L(X_f, X_g) = 0$. Here $X_f := \Theta_L^{-1}(df)$ and $X_g := \Theta_L^{-1}(dg)$, that is, for example, $\mathcal{L}_{X_f} d\theta_L = 0, \theta_L(X_f) = 0$ and $df = X_f \lrcorner\, d\theta_L$. We remark that the map $(f, g) \mapsto -d\theta_L(X_f, X_g)$ is not a Poisson bracket on E as it is only well defined on first integrals. However, while the map doesn't make E a Poisson manifold, it does turn the quotient of E by Γ into a symplectic manifold.

The important properties of involutive first integrals are given in the following propositions from [39]:

Proposition 3.4. *The following conditions are equivalent:*

(a) f *and* g *are involutive first integrals with respect to* L,

(b) $X_f(g) = 0 = X_g(f)$,

(c) $[X_f, X_g] = 0$.

This result shows that each such involutive pair produces a Lagrangian submanifold of $d\theta_L$ of dimension 3 whose tangent spaces are spanned by X_f, X_g and Γ.

Proposition 3.5. *If* f *and* g *are involutive first integrals and if* $Y_f \in [X_f]$ *and* $Y_g \in [X_g]$ *are defined by*

$$Y_f := X_f - X_f(t)\Gamma \ and \ Y_g := X_g - X_g(t)\Gamma$$

then

$$[Y_f, Y_g] = 0.$$

We can now define a Lagrangian system to be *completely integrable with respect to* L if there exist n first integrals f^a with $df^1 \wedge \cdots \wedge df^n \neq 0$ which are in involution with respect to L. We will see in section 4.1 how this concept can be applied to arbitrary SODEs and how it opens the way to the generalization of the Liouville–Arnol'd theorem.

3.5 Construction of θ_L

(This section can be skipped according to the taste of the reader.)
The heading is misleading because I'm just going to explain the coordinate free definitions of the Legendre transformation $\mathcal{L} : TM \to T^*M$ and the symplectic two form $d\omega$ on T^*M. The Cartan 2-form in this autonomous setting is then $\mathcal{L}^* d\omega$. The non-autonomous case is rather more complicated and even Sternberg skips some technicalities in his book [46]. I hope the reader will get an idea of the intrinsic definition of θ_L on E from what follows.

Suppose now that L is an autonomous Lagrangian function on TM, and consider the vector space T_xM. Then $L : T_xM \to \mathbb{R}$ induces a map

$$\mathcal{L} : T_xM \to T_x^*M$$

defined by

$$\mathcal{L}(u_x) = dL_{u_x} \circ \ell_u^{-1}$$

where ℓ_u is the canonical isomorphism

$$\ell_u(x, u, w) = (x, w)$$

which identifies the two vector spaces T_xM and $V_{(x,u)}TM$. This construction gives a coordinate-free definition of the Legendre transformation $\mathcal{L} : TM \to T^*M$. The local coordinates (x^a, u^b) for TM induce coordinates (x^a, p_b) on T^*M via $\mathcal{L}$:

$$x^a \circ \mathcal{L} = x^a, \ p_b \circ \mathcal{L} = \frac{\partial L}{\partial u^b}.$$

Now $\mathcal{L}$ is an immersion if and only if L is regular and in this case $\mathcal{L}^{-1}$ can be derived from the Hamiltonian function, $H : T^*M \to \mathbb{R}$, defined by

$$\mathcal{L}^* H = L.$$

Now we construct the canonical 1-form ω on T^*M whose coordinate expression is the familiar $p_a\,dx^a$ ($p_a\,dq^a$ in the usual mechanics coordinates).

$$
\begin{array}{ccc}
& & TT^*M \\
& \nearrow\, \pi_* & \Big\downarrow \pi_1 \\
TM & \xrightarrow[\;\mathcal{L}\;]{} & T^*M \\
& & \Big\downarrow \pi \\
& & M
\end{array}
$$

Let $v_{(x,p)} \in T_{(x,p)}T^*M$, then $\pi_1(v_{(x,p)}) = (x,p)$ and $\pi_*v_{(x,p)} = (x,v)$ so that $\pi_1(v)(\pi_*v)$ is linear in v. So we define the 1-form ω on T^*M by

$$\omega(v) := \pi_1(v)(\pi_*v).$$

In coordinates, $\omega = p_a dx^a$. $d\omega$ is the natural symplectic form on T^*M. We pull ω back by the Legendre transformation to get

$$\theta_L = \mathcal{L}^*\omega = dL \circ S$$

where, in the autonomous case, $S = V_a \otimes dx^a$, and so in coordinates

$$\theta_L = \frac{\partial L}{\partial u^a} dx^a.$$

(Compare these expressions with those given earlier for θ_L in the non-autonomous setting.)

$$
\begin{array}{ccccc}
TT^*M & & & & TTM \\
\Big\downarrow \pi_1 & & \pi_* \searrow & & \Big\downarrow \tau_1 \\
T^*M & & \xleftarrow[\;\mathcal{L}\;]{} & & TM \\
& \searrow \pi & & \nearrow \tau & \\
& & M & &
\end{array}
$$

Of course what remains is the construction of the variational principle whose integrand is θ_L: I refer you again to the references above the proposition at the start of this section.

4 The Liouville–Arnol'd Theorem on Contact Manifolds

As I indicated in section 2, every second order differential equation satisfies three of the four Helmholtz conditions. As a result, I will show in this section that each SODE Γ possesses a 1-form analogous to the Cartan 1-form and it is this fact that allows us to generalize the Liouville–Arnol'd theorem. This 1-form is called a Γ-contact form for the equation. The study of such forms is of major independent interest and I will mention only a few salient points.

4.1   Γ–contact Forms

A *contact form* on a $(2n + 1)$-dimensional manifold P is a smooth 1-form ω such that $\omega \wedge (d\omega)^n$ is nowhere zero. A contact form uniquely determines a vector field V, called the *Reeb field*, on the manifold such that $\omega(V) = 1$ and $\mathcal{L}_V \omega = 0$. It is apparent that an equivalent pair of conditions is $\omega(V) = 1$ and $V \lrcorner \, d\omega = 0$. Less obvious is that a contact form ω for V exists if and only if there is a Riemannian metric on P which makes the orbits of V geodesic and V have unit length (see Gluck [17]). The flow of V is called the *contact flow*.

A manifold which admits a contact form is called a *contact manifold*. A survey of results relating contact forms and Riemannian geometry is available in, for example, Blair [2]. Under certain conditions, local existence of contact forms is straightforward (as we shall see below) but global existence remains an important open problem. The classification of contact structures on a given manifold is also an active area of research: see, for example, Eliashberg [16] for a recent classification of contact structures on S^3.

It should be clear that any regular Lagrangian system provides a contact form for E for which Γ is the Reeb field by virtue of the properties of $d\theta_L$, namely $\omega := \theta_L - dG + dt$ where G satisfies $\Gamma(G) = L$. Furthermore, the following proposition shows that *every* second order equation field Γ has a contact form associated with it. We call such a form a *Γ-contact form*. Finally, don't confuse these Γ-contact forms with the contact forms associated with the contact structure on E.

Proposition 4.1. *Every second order differential equation field Γ defines a local contact flow on its evolution space E.*

Proof. Suppose that, as a result of the local rectification theorem, we have on some open domain U of E, $2n$ functionally independent first integrals of our system, $h^1, \ldots, h^{2n}$, then the form $\omega := h^1 dh^2 + \cdots + h^{2n-1} dh^{2n} + dt$ is a Γ-contact form. That is, $d\omega = dh^1 \wedge dh^2 + \cdots + dh^{2n-1} \wedge dh^{2n}$ is of maximal rank on U, $\omega(\Gamma) = 1$ and $\Gamma \lrcorner \, d\omega = 0$. □

This settles the existence question for (local) Γ-contact forms (but of course we might hope to have such a form without having to know $2n$ first integrals). The immediate consequence of this result is that for an arbitrary system of second order differential equations there is only one obstacle to the existence of a Lagrangian. This is because the exterior derivative of a Γ-contact form satisfies all the Helmholtz conditions except the one concerning the vertical fields. It is interesting to see what this condition turns out to be in terms of a local basis constructed from first integrals.

Proposition 4.2. *Suppose that ω is a Γ-contact form so that $\omega = k_\alpha dh^\alpha + dt$ in a basis $\{dh^\alpha, dt\}$ of exact 1-forms, where $\Gamma(h^\alpha) = 0$ and $\det \left(\frac{\partial k_\alpha}{\partial h^\beta} \right) \neq 0$. Then Γ is the Euler field of a Lagrangian L if and only if the coefficient functions k_α satisfy*

$$\frac{\partial h^\alpha}{\partial u^a} \frac{\partial h^\beta}{\partial u^b} \left(\frac{\partial k_\beta}{\partial h^\alpha} - \frac{\partial k_\alpha}{\partial h^\beta} \right) = 0 \quad a, b = 1, \ldots, n.$$

Proof. By the assumptions, $d\omega = \frac{\partial k_\alpha}{\partial h^\beta} dh^\beta \wedge dh^\alpha$. Consequently, $d\omega$ is closed, of maximal rank and $\Gamma \lrcorner \, d\omega = 0$. The remaining Helmholtz condition for the existence of a Lagrangian is

$$d\omega(V_1, V_2) = 0 \quad \text{for any vertical } V_1, V_2,$$

or equivalently,

$$d\omega(\frac{\partial^\cdot}{\partial u^a}, \frac{\partial}{\partial u^b}) = 0 \quad a, b = 1, \dots, n.$$

So in terms of the coefficient functions k_β this becomes

$$\frac{\partial h^\alpha}{\partial u^a} \frac{\partial h^\beta}{\partial u^b} \left(\frac{\partial k_\beta}{\partial h^\alpha} - \frac{\partial k_\alpha}{\partial h^\beta} \right) = 0 \quad a, b = 1, \dots, n,$$

where we have used $\frac{\partial}{\partial u^a} = \frac{\partial h^\alpha}{\partial u^a} \frac{\partial}{\partial h^\alpha}$ (note that $\frac{\partial k_\beta}{\partial h^\alpha} - \frac{\partial k_\alpha}{\partial h^\beta} \neq 0$ for some α, β because $d\omega \neq 0$).   □

Let us now suppose that we have a contact form ω for our second order differential equation field Γ on E. The 2-form $d\omega$ is of maximal rank and so, like the Lagrangian case, we have a Noether's theorem: there is a bijection $\mathcal{N}oether : X \mapsto X \lrcorner \, d\omega$ between vector fields satisfying $\mathcal{L}_X d\omega = 0, \omega(X) = 0$ and closed 1-forms annihilating Γ (that is, the exterior derivatives of first integrals of the differential equations). The conditions $\mathcal{L}_X d\omega = 0, \omega(X) = 0$ are compatible and further they imply that $\mathcal{L}_X \Gamma = 0$ so that X generates a symmetry of the differential equations. We will denote by X_f the pre-image of an exact 1-form df which annihilates Γ.

By analogy with the Lagrangian case, two first integrals f, g of Γ with $df \wedge dg \neq 0$ on E will be said to be *in involution with respect to* ω if $d\omega(X_f, X_g) = 0$. We then have a result identical to Proposition 3.5.

Proposition 4.3. *The following conditions are equivalent:*

(a) f *and* g *are involutive first integrals with respect to* ω,

(b) $X_f(g) = 0 = X_g(f)$,

(c) $[X_f, X_g] = 0$.

4.2   The Generalised Liouville–Arnol'd Theorem

Recent work by Byrnes, Aldridge, Sherring, Godfrey and Prince shows that, in addition to a Noether–Cartan theorem and the idea of involutivity, other important paraphenalia of classical mechanics are available to arbitrary second order equations via the associated contact manifold structure. This includes generalizations of the Hamilton–Jacobi equation (and powerful theorems on its separability) [38, 37] and the famous Liouville–Arnol'd theorem [39] which I will now describe.

The classical Liouville theorem in Lagrangian mechanics is a corollary of a more general result, obtained by showing that if ω is a Γ-contact form, then $\omega = d_D G$ for some function G on E where D is spanned by vector fields $X_{f^1}, \dots, X_{f^n}$ and Γ. (This exterior derivative d_D is the *foliated exterior derivative* associated with the integrable distribution D. I have included a brief description of it at the end of this section.) The partial derivatives of G with respect to these known integrals (in an appropriate chart) constitute the remaining integrals. I emphasise that local Γ-contact forms are guaranteed to exist (unlike Cartan one-forms θ_L), and may be found, at least theoretically, by solving the conditions $\omega(\Gamma) = 1$, $\Gamma \lrcorner\, d\omega = 0$ with the requirement that $d\omega$ has maximal rank. However, this may be a highly non-trivial task, even if the known integrals are to be involutary with respect to the Γ-contact form.

The proof of the extension of the part of the theorem due to Arnol'd is based on a proof given in Libermann and Marle [28], adapted to the contact form case and to the non-autonomous setting. I will only be looking at Liouville's part of the theorem in this chapter; the full proof is in [39].

It will be useful to have the following submanifolds defined: For a given set of involutive first integrals $f^1, \ldots, f^n$ and $(c, \tau) \in \mathbb{R}^n \times \mathbb{R}$,

$$E^\tau := \{v \in E \,:\, t(v) = \tau\};$$

$$E_c := \{v \in E \,:\, f^a(v) = c^a, \quad a = 1, \ldots, n\};$$

$$E_c^\tau := E_c \cap E^\tau.$$

Theorem 4.4 (Extended Liouville–Arnol'd). *Suppose we have a system of n second order differential equations (able to be put in normal form) and defined by Γ on $\mathbb{R} \times TM$, with Γ-contact form ω. If there are n first integrals $f^1, \ldots, f^n$, in involution with respect to ω and such that the differentials $df^1, \ldots, df^n$ are independent at every point of E, then the following are true:*

(1) a further n involutive first integrals $g_1, \ldots, g_n$ can be constructed by quadrature;

if a connected component N of E_c^τ is compact, then

(2) N is diffeomorphic to the n torus T^n;

(3) there exist action-angle variables (I, σ) on some open subset $U \subset E$, and a map

$$(\sigma, I, t) : U \longrightarrow T^n \times W \times S,$$

where $W \subset \mathbb{R}^n$ and $S \subset \mathbb{R}$, such that

(a) $(\sigma, I) : E^\tau \cap U \to T^n \times W$ is a symplectomorphism for all $\tau \in S$ when $T^n \times W$ is provided with the symplectic form $d\gamma_a \wedge dw^a$, where $w^1, \ldots, w^n$ are the canonical coordinates for $\mathbb{R}^n$ and $\gamma_1, \ldots, \gamma_n$ are the angular coordinates induced on T^n from $\mathbb{R}$ by the identification

$$T^n := \mathbb{R}^n / \mathbb{Z}^n;$$

(b) σ is a differentiable submersion $U \to T^n$;

(c)

$$I^{-1}(0) \cap E^\tau = N$$

and there exists an affine transformation $\chi : \mathbb{R}^n \to \mathbb{R}^n$ such that $I = \chi \circ f$ (where f is the map $U \to \mathbb{R}^n$ defined by the first integrals $f^1, \ldots, f^n$);

(d) W is open in $\mathbb{R}^n$, $0 \in W$ and S is open in $\mathbb{R}$.

In order to prove part 1 of the theorem, we needed to use the Poincaré lemma for the foliated derivative to find a function G, which plays a role similar to a solution of the Hamilton–Jacobi equation in the Lagrangian case. This is done by first constructing complementary distributions D and $D^\perp$ with respect to which ω is of type $(0, 1)$.

Lemma 4.5. *Let ω be a Γ-contact form on E. Let $f^1, \ldots, f^n$ be n involutive first integrals with respect to ω and let $X_{f^1}, \ldots, X_{f^n}$ be vector fields satisfying $X_{f^a} \lrcorner\, d\omega = df^a$ and $\omega(X_{f^a}) = 0$ on E.*

Let D be the distribution $Sp\{\Gamma, X_{f^1}, \ldots, X_{f^n}\}$ and choose a complementary distribution $D^\perp$ to be $\mathrm{Sp}\{X_{\bar{f}^1}, \ldots, X_{\bar{f}^n}\}$ where these vector fields are the images of a further n first integrals $\bar{f}^1, \ldots, \bar{f}^n$ with $df^1 \wedge \cdots \wedge df^n \wedge d\bar{f}^1 \wedge \cdots \wedge d\bar{f}^n \neq 0$ on E. These additional first integrals need not be in involution.

Then

(a) *D is integrable of dimension $n + 1$.*

(b) *$\omega = d_D G$ for some function G on an open subset V of E.*

Proof. First of all note that the integrals $\bar{f}^1, \ldots, \bar{f}^n$ are guaranteed locally by the usual existence theorem: the rectification lemma guarantees $2n$ local first integrals h^α. The given functions f^a can be written as functions of the h^α and it is then possible to construct a further n independent functions of the h^α, the $\bar{f}^a$. We need them to ensure that ω is type $(0, 1)$, which is to say that $D^\perp$ is in the annihilator of ω.

(a) In the discussion preceding Proposition 4.3 we noted that the X_{f^a} were symmetries of the equations so that $[\Gamma, X_{f^a}] = 0$ and since Proposition 4.3 ensures that X_{f^a} commute among themselves, D is integrable. D has dimension $(n+1)$ because $df^1 \wedge \cdots \wedge df^n \neq 0$ on E guarantees the pointwise linear independence of $\{\Gamma, X_{f^1}, \ldots, X_{f^n}\}$.

(b) ω takes zero values on $D^\perp$ by definition of the $X_{\bar{f}^a}$. Hence ω is of type $(0, 1)$ with this choice of $D, D^\perp$. Now $\Gamma \lrcorner\, d\omega = 0$ and $d\omega(X_{f^a}, X_{f^b}) = 0$ by assumption, so the $(0, 2)$ component of $d\omega$ is zero. Since ω is type $(0, 1)$, $d_D\omega$ is the $(0, 2)$ component of $d\omega$ and is thus zero. Hence $\omega = d_D G$ for some function G on an open subset V of E. $\qquad\square$

Proof of part 1 of the theorem. Utilizing the machinery of the preceding lemma, we have $\omega = d_D G$ where ω is of type $(0, 1)$. Computing G is the quadrature of the theorem. Now

$$dG = d_{01}G + d_{10}G = \omega + d_{10}G$$
$$\Rightarrow d\omega = -d(d_{10}G)$$
$$= -d\left(\frac{\partial G}{\partial f^a} df^a\right)$$
$$= -d\left(\frac{\partial G}{\partial f^a}\right) \wedge df^a$$

in a chart (t, x^a, f^a) for $V' \subset E$ related by local inversion of the expressions for the f^a to the chart (t, x^a, u^a). Now $\Gamma \lrcorner \, d\omega = 0$, $\Gamma(f^a) = 0$ and $d\omega$ has maximal rank so that

$$\Gamma \left(\frac{\partial G}{\partial f^a} \right) = 0$$

and

$$d \left(\frac{\partial G}{\partial f^1} \right) \wedge \cdots \wedge d \left(\frac{\partial G}{\partial f^n} \right) \wedge df^1 \wedge \cdots \wedge df^n \neq 0$$

on V'. Furthermore, if the vector fields Z_a are given by $Z_a \lrcorner \, d\omega = d \left(\frac{\partial G}{\partial f^a} \right)$ and $\omega(Z_a) = 0$, then $Z_b \lrcorner \, d\omega = -Z_b \left(\frac{\partial G}{\partial f^a} \right) df^a + Z_b(f^a) d \left(\frac{\partial G}{\partial f^a} \right)$ implies that $Z_b \left(\frac{\partial G}{\partial f^a} \right) = 0$ and $Z_b(f^a) = \delta_b^a$. This in turn means that $d\omega(Z_a, Z_b) = 0$ so that the integrals $\frac{\partial G}{\partial f^a}$ are in involution. This completes the proof of part 1 of the theorem.   □

In the case where the leaves of the foliation generated by the distribution $D := Sp\{\Gamma, X_{f_1}, \ldots, X_{f_n}\}$ at a fixed time are tori, it is natural to be concerned about the completeness of these vector fields. The key to this completeness lies in the fact that a vector field of compact support is complete. For the vector fields at hand, it remains to construct suitable vector fields of compact support which agree with the given fields on the tori. This is a technical task which involves using part of the foliation by tori as the compact support and the construction of a suitable bump function with which to multiply our given fields so that they are zero off the support. In this way, the completeness should appear as a corollary to the extended Liouville–Arnol'd theorem. I am not aware that this has been done and I thank Peter Vassiliou for raising the question.

4.3   Examples

We will now look at two examples of non–Lagrangian systems in the plane which illustrate respectively the two parts of the generalized Liouville–Arnol'd theorem.

Example 1. We return to the first example of section 2.1 and apply the generalization of Liouville's part of the original theorem.

So consider again the differential equations on $\mathbb{R}^2$:

$$\ddot{x} = \dot{y}, \quad \ddot{y} = y.$$

We use the contact forms, $\theta^x := dx - u\,dt$ and $\theta^y := dy - v\,dt$, modified force forms $\psi^x := -\frac{1}{2}(dy + v\,dt) + du$ and $\psi^y := dv - y\,dt$, and dt as before to form a convenient basis for T^*E. It is reasonably straightforward to find a Γ-contact form for this system, for example,

$$\omega := (u - y)\theta^x + \frac{1}{2}x\theta^y + (v - x)\psi^x + (\frac{1}{2}y - u)\psi^y + dt.$$

Relative to our basis, $d\omega$ is free of dt terms:

$$d\omega = \theta^x \wedge \theta^y + 2\psi^x \wedge \theta^x - 2\psi^x \wedge \psi^y.$$

It is a simple matter to check that solving $\Gamma \lrcorner \, \omega = 1$, $\Gamma \lrcorner \, d\omega = 0$ and $d\omega \wedge d\omega \neq 0$ in this case involves finding one first integral, here this integral is $u - y$.

The system has two obvious (point) symmetries generated by $\frac{\partial}{\partial t}$ and $\frac{\partial}{\partial x}$ (regarded as vector fields on E). These fields commute and $\mathcal{L}_{\frac{\partial}{\partial t}} d\omega = 0 = \mathcal{L}_{\frac{\partial}{\partial x}} d\omega$ so that $\frac{\partial}{\partial t} \lrcorner \, d\omega$ and $\frac{\partial}{\partial x} \lrcorner \, d\omega$ produce involutory first integrals, namely $f := u^2 + \frac{1}{2}(y^2 + v^2) - 2uy$ and $g := 2(y - u)$ respectively. Note that X_f, X_g differ from $\frac{\partial}{\partial t}, \frac{\partial}{\partial x}$ by multiples of Γ.

Now we wish to find a further two integrals by utilising $\omega = d_D G$, $D := \mathrm{Sp}\{\Gamma, \frac{\partial}{\partial t}, \frac{\partial}{\partial x}\}$; a straightforward calculation shows that

$$G = -\frac{1}{2}gx + (1 - f)t + F(f, g, y)$$

where $F(f, g, y)$ is a tedious composite of f, g and y. Computing $\frac{\partial G}{\partial f}, \frac{\partial G}{\partial g}$ and returning to the original chart, we obtain the integrals

$$k := \log(|y + v|) - t, \quad \ell := v - x + (u - y)\log(|y + v|)$$

up to arbitrary functions of f and g. The corresponding vector fields, up to a multiple of Γ, are

$$X_k = \frac{1}{y + v}\left(\frac{\partial}{\partial x} - \frac{\partial}{\partial y} - \frac{\partial}{\partial u} + \frac{\partial}{\partial v}\right)$$

and

$$X_\ell = \left(\frac{u - y}{v + y}\right)\left(\frac{\partial}{\partial x} - \frac{\partial}{\partial y} - \frac{1}{2}\frac{\partial}{\partial u} + \frac{\partial}{\partial v}\right) - \frac{1}{2}\left(\log|y + v|\frac{\partial}{\partial x} + \left(\frac{u + v}{v + y}\right)\frac{\partial}{\partial u}\right).$$

The patient reader can verify that $d\omega(X_k, X_\ell) = 0$ so that k and ℓ are involutive with respect to $d\omega$. As a last remark to this example, we point out that Douglas claims that this system has no Lagrangian because it is of his type IV. In fact it is easily shown that it is actually of type IIIb, for which the result still holds.

Example 2. Now we turn to an example of the Arnol'd part of the theorem. An alternative approach to this example, which is due to James Sherring, can be found in [39]. We will construct a pair of second order equations which are not Euler–Lagrange but which have compact first integral surfaces which are tori. We begin with

$$\ddot{x} = -x$$

with solutions

$$x = A\sin(t + \epsilon)$$

and first integrals (one global, one local)

$$A^2 := x^2 + \dot{x}^2 \quad \text{and} \quad \epsilon := \tan^{-1}\left(\frac{x}{\dot{x}}\right) - x.$$

Now we augment this differential equation with another so that the level sets of A^2 will contribute to 2-tori.

If we choose

$$y = \frac{B}{A^2} \sin(A^2 \sinh(t) + \sigma)$$

away from the zeros of A^2, differentiation leads to

$$\ddot{y} = -A^4 \cosh^2(t)y + \tanh(t)\dot{y}.$$

In this way, we achieve another pair of first integrals (again one global, one local), namely

$$B^2 := A^4 y^2 + \text{sech}^2(t)\dot{y}^2$$

and

$$\sigma := \tan^{-1}\left(\frac{A^2 \cosh(t)y}{\dot{y}}\right) - A^2 \sinh(x).$$

In making this construction (and this is part of the reason for the hyperbolic functions), we have created a Douglas Case IIIb example and so there is no Lagrangian. Notice that A^2 and B^2 are global first integrals while ϵ and σ are not but that $dA^2 \wedge dB^2 \wedge d\epsilon \wedge d\sigma$ has global meaning even though $d\epsilon$, $d\sigma$ are themselves not defined on the whole domain of this 4-form. We can also construct a global contact form on E:

$$\omega := A^2 d\epsilon + B^2 d\sigma + dt.$$

Notice that $d\omega$ has maximal rank on the complement of the zero set of A^2.

Now the intersection of the level sets of $A^2 (\neq 0)$ and B^2 at time τ, $E^\tau_{(A^2, B^2)}$, is clearly closed and bounded and hence compact. It is the intersection of a pair of cylinders which have been oriented so that this intersection can be identified as the cartesian product of subsets of $\mathbb{R}^2/\{(0,0)\}$:

$$\{(a,b) : a^2 + b^2 = R^2\} \times \{(c,d) : R^4 c^2 + \text{sech}^2(\tau)d^2 = P^2\} \cong S^1 \times S^1.$$

To see this from the point of view of the generalized Liouville–Arnol'd theorem we need to check that A^2 and B^2 are in involution with respect to $d\omega$. If we use the bijection described in the second paragraph preceding Proposition 4.3 we see that

$$X_{A^2} \lrcorner\, d\omega = dA^2 \Rightarrow X_{A^2}(B^2) = 0,$$
$$X_{B^2} \lrcorner\, d\omega = dB^2 \Rightarrow X_{B^2}(A^2) = 0,$$

so that

$$d\omega(X_{A^2}, X_{B^2}) = 0$$

and so the first integrals are involutive. It is left as an exercise to compute the vector fields X_{A^2}, X_{B^2} and show that they are complete.

4.4 The Foliated Derivative

I close this section with a brief description of the foliated exterior derivative on E. Given a foliation of E, this derivative coincides with the usual exterior derivative on the leaves of the foliation.

Let D be an integrable m-dimensional distribution of smooth vector fields on $V \subset E$ and let $D^\perp$ be a $(2n + 1) - m$ dimensional complement. A form α on V is said to be of *type* (p, q) if it has degree $(p + q)$ and for vectors W_a in D or $D^\perp$, $\alpha(W_1, \ldots, W_{p+q}) = 0$ except for at most the case where p of the W_a are in $D^\perp$ and q of them in D.

Every form σ of degree $(p + q)$ has a unique decomposition into (p, q) forms as follows: the *component of type* (p, q) of the form σ is the form σ_{pq} which has the same values as σ when p of its arguments are in $D^\perp$ and q in D and zero values otherwise.

Vaisman [48] proves the following proposition: "Let α be a type (p, q) form. Then $d\alpha$ has only non-zero components of the types $(p, q + 1), (p + 1, q), (p + 2, q - 1)$." In this way the exterior derivative can be decomposed as

$$d = d_{01} + d_{10} + d_{2,-1}$$

in an obvious notation. d_{01} is called the *foliated exterior derivative* and we will denote it d_D, it extends by linearity and skew-symmetry to all differential forms on $V \subset E$, not just those of type (p, q).

In fact d_D is just the restriction to D of the conventional exterior derivative d. Since D is assumed to be integrable, the foliated derivative d_D inherits the properties of d:

(a) $d_D(\alpha \wedge \beta) = d_D \alpha \wedge \beta + (-1)^{\deg \alpha} \alpha \wedge d_D \beta$

(b) $d_D^2 \alpha = 0$

(c) if $d_D \alpha = 0$ then locally $\alpha = d_D \beta$.

If, instead, d is restricted to $D^\perp$, we obtain d_{10}. Because $D^\perp$ is not assumed integrable, d_{10} is not guaranteed to satisfy the above conditions. Moreover, if $D^\perp$ is non-integrable there exist $X, Y \in D^\perp$ such that $[X, Y] \in D$. This gives rise to the $d_{2,-1}$ component.

5 The Future

I would be foolish to claim that a solution (in Douglas's sense) of the inverse problem is imminent, even in dimension 3. However, I believe that it is a real possibility within five years or so. What is certainly true is that the inverse problem in the calculus of variations is stimulating more research than at any time since Douglas's landmark paper.

On the one hand, there is the direct attack on the problem using differential geometric methods [14]. On the other, there is the application of these new methods to the related topic of second order differential equations in the areas of separability [32], contact flows [39], Hamilton–Jacobi separability for geodesic sprays [38] and so on. On top of this comes work on higher-order differential equations [5] and references therein.

Over the last 15 years, connection theory for second order ordinary differential equations has shown us that the geodesic case and all its geometric content is not so special as we thought. When a metric (Lagrangian) is responsible for the connection, it is certainly distinguished, but from a connection viewpoint it is just an example of the more general theory in which objects such as curvature and torsion have a natural meaning.

Connection theory has also shown that the non-metric Euler–Lagrange case is not so special either. We now know that arbitrary second order equations enjoy almost all its features, such as a Noether–Cartan theorem, a Liouville–Arnol'd theorem and a Hamilton–Jacobi equation. In general, only the Lagrangian itself is missing.

The fundamental role of the inverse problem is to classify the connections associated with second order equations into those which are linear and metric, linear but not metric, non-linear and Euler–Lagrange and nonlinear but not Euler–Lagrange. This viewpoint will appeal to the mathematician looking for an all-encompassing theory which subsumes the special, but beautiful, cases of metric differential geometry and Lagrangian mechanics without losing too much fine structure.

Acknowledgements

I have borrowed freely from papers which I have co-authored and I humbly acknowledge the long-term assistance and collaboration of Mike Crampin and Willy Sarlet who have both made contributions to the field far greater than my own. I also thank Jon Aldridge for sharing his ideas on the Helmholtz conditions. Finally, I thank Mike Crampin and Jon Aldridge for their careful reading of various versions of this chapter and Peter Vassiliou and Ian Lisle for their careful and patient editorship.

References

[1] I. Anderson and G. Thompson, The inverse problem of the calculus of variations for ordinary differential equations, *Memoirs Amer. Math. Soc.* **98**, No. 473 (1992).

[2] D.E. Blair. Contact manifolds in Riemannian geometry, *Lecture Notes in Mathematics* **509** Springer-Verlag, Berlin, 1976.

[3] R.L. Bryant, S.S. Chern, R.B. Gardner, H.L. Goldschmidt and P.A. Griffiths. *Exterior Differntial Systems*. Springer-Verlag, Berlin, 1991.

[4] G B. Byrnes, A complete set of Bianchi identities for tensor fields along the tangent bundle projection, *J. Phys. A: Math. Gen.* **27**, 6617–6632 (1993).

[5] G.B. Byrnes, A linear connection for higher-order ordinary differential equations, *J. Phys. A: Math. Gen.* **29**, 1685–1694 (1996).

[6] É. Cartan, *Leçons sur les Invariants Intégraux,* Hermann, Paris, 1922.

[7] M. Crampin, Generalized Bianchi identities for horizontal distributions, *Math. Proc. Camb. Phil. Soc.* **94**, 125–132 (1983).

[8] M. Crampin, Tangent bundle geometry for Lagrangian dynamics, *J. Phys. A: Math. Gen.* **16**, 3755–3772 (1985).

[9] M. Crampin and F.A.E. Pirani, *Applicable Differential Geometry*, CUP, 1987.

[10] M. Crampin and G.E. Prince, Generalizing gauge freedom for spherically symmetric potentials, *J. Phys. A: Math. Gen.* **18**, 2167–2175 (1985).

[11] M. Crampin and G.E. Prince, Equivalent Lagrangians and dynamical symmetries: some comments, *Phys. Lett.* **108A**, 191–194 (1985).

[12] M. Crampin and G.E. Prince, Alternative Lagrangians for spherically symmetric potentials, *J. Math. Phys.* **29**, 1551–1555 (1988).

[13] M. Crampin, G.E. Prince and G. Thompson, A geometric version of the Helmholtz conditions in time dependent Lagrangian dynamics, *J. Phys. A: Math. Gen.* **17**, 1437–1447 (1984).

[14] M. Crampin, W. Sarlet, E. Martínez, G.B. Byrnes and G.E. Prince, Toward a geometrical understanding of Douglas's solution of the inverse problem in the calculus of variations, *Inverse Problems* **10**, 245–260 (1994).

[15] J. Douglas, Solution of the inverse problem of the calculus of variations, *Trans. Am. Math. Soc.* **50**, 71–128 (1941).

[16] Y. Eliashberg, Contact 3-manifolds twenty years since J. Martinet's work, *Ann. Inst. Fourier* **42**, 165–192 (1992).

[17] H. Gluck, Global theory of dynamical systems, *Lecture Notes in Mathematics* **819** Springer-Verlag, Berlin, 1980.

[18] H. Goldschmidt and S. Sternberg, The Hamilton–Cartan formalism in the calculus of variations, *Ann. Inst. Fourier* **23**, 203–267 (1973).

[19] J. Grifone, Structure presque tangente et connexions I, *Ann. Inst. Fourier* **22**(1), 287–334 (1972).

[20] J. Grifone, Structure presque tangente et connexions II, *Ann. Inst. Fourier* **22**(3), 291–338 (1972).

[21] H. Helmholtz, Über der physikalische Bedeutung des Princips der kleinsten Wirkung, *J. Reine Angew. Math.* **100**, 137–166 (1887).

[22] M. Henneaux, On the inverse problem of the calculus of variations, *J. Phys. A: Math. Gen.* **15**, L93–L96 (1982).

[23] M. Henneaux and L.C. Shepley, Lagrangians for spherically symmetric potentials, *J. Math. Phys.* **23**, 2101–2107 (1988).

[24] A. Hirsch, Die Existenzbedingungen des verallgemeinerten kinetischen Potentials, *Math. Ann.* **50**, 429–441 (1898).

[25] L.A. Ibort and C. López-Lacasta, On the existence of local and global Lagrangians for ordinary differential equations, *J. Phys. A: Math. Gen.* **23**, 4779–4792 (1990).

[26] J. Klein, Espaces variationels et mécanique, *Ann. Inst. Fourier* **12**, 1–124 (1962).

[27] J. Klein, Structures symplectique ou *J*–symplectiques homogènes sur l'espace tangent à une variété, *Symp. Math.* **14**, 181–192 (1974).

[28] P. Libermann and C.-M. Marle, *Symplectic Geometry and Analytical Mechanics,* Reidel, Dordrecht, 1987.

[29] J. Liouville, *Journal de Math.* **20**, 137 (1855).

[30] E. Martínez, J.F. Cariñena and W. Sarlet, Derivations of differential forms along the tangent bundle projection, *Diff. Geom. Applic.* **2**, 17–43 (1992).

[31] E. Martínez, J.F. Cariñena and W. Sarlet, Derivations of differential forms along the tangent bundle projection. Part II, *Diff. Geom. Applic.* **3**, 1–29 (1993).

[32] E. Martínez, J.F. Cariñena and W. Sarlet, A geometric characterization of separable second-order differential equations, *Math. Proc. Camb. Phil. Soc.* **113**, 205–224 (1993).

[33] A. Mayer, Die Existenzbedingungen eines kinetischen Potentiales, *Berich. Verh. Konig. Sach. Gesell. wissen. Leipzig, Math. Phys. Kl.* **84**, 519–529 (1896).

[34] G. Morandi, C. Ferrario, G. Lo Vecchio, G. Marmo and C. Rubano, The inverse problem in the calculus of variations and the geometry of the tangent bundle, *Phys. Rep.* **188**, 147–284 (1990).

[35] P.J. Olver, *Applications of Lie Groups to Differential Equations*, 2nd Edition, Springer-Verlag, Berlin, 1993.

[36] G.E. Prince, A complete classification of dynamical symmetries in classical mechanics, *Bull. Austral. Math. Soc.* **32**, 299–308 (1985).

[37] G.E. Prince, J. Aldridge and G.B. Byrnes, A universal Hamilton–Jacobi equation for second order ODEs, *J. Phys. A* **32**, 827–844 (1999).

[38] G.E. Prince, J. Aldridge, S.E. Godfrey and G.B. Byrnes, The separation of the Hamilton–Jacobi equation for the Kerr metric, *J. Aust. Math. Soc. Series B* (1999), to appear.

[39] G.E. Prince, G.B. Byrnes, J. Sherring and S.E. Godfrey, A generalization of the Liouville–Arnol'd theorem, *Math. Proc. Camb. Phil. Soc.* **117**, 353–370 (1995).

[40] G.E. Prince and M. Crampin, Projective differential geometry and geodesic conservation laws in general relativity, I: projective actions, *Gen. Rel. Grav.* **16**, 921–942 (1984).

[41] G.E. Prince and M. Crampin, Projective differential geometry and geodesic conservation laws in general relativity, II: conservation laws, *Gen. Rel. Grav.* **16**, 1063–1075 (1984).

[42] W. Sarlet, The Helmholtz conditions revisited. A new approach to the inverse problem of Lagrangian dynamics, *J. Phys. A: Math. Gen.* **15**, 1503–1517 (1982).

[43] W. Sarlet, Geometrical structures related to second-order equations, in D. Krupka and A. Svec, eds., *Differential Geometry and its Applications*, Proc. Conf. 1986 Brno, Czechoslovakia, (J.E. Purkyně University, Brno 1987) 279–288.

[44] W. Sarlet, G.E. Prince and M. Crampin, Adjoint symmetries for time dependent second order equations, *J. Phys. A: Math. Gen.* **23**, 1335–1347 (1990).

[45] W. Sarlet, A. Vandecasteele, F. Cantrijn, and E. Martínez, Derivations of forms along a map: the framework for time-dependent second-order equations, *Diff. Geom. Applic.* **5**, 171–203 (1995).

[46] S. Sternberg, *Differential Geometry*, 2nd edition, Chelsea, New York, 1983.

[47] D.J. Saunders, *The Geometry of Jet Bundles,* CUP, 1989.

[48] I. Vaisman, *Cohomology and Differential Forms*, Marcel Dekker, New York, 1973.

[49] K. Yano and S. Ishahara, *Tangent and Cotangent Bundles,* Marcel Dekker, New York, 1973.

Twistor Theory

MICHAEL K. MURRAY

Keywords: Mini-twistors, holomorphic line bundles, Penrose transform, monopoles

1   Introduction

Twistor theory began with the work of Roger Penrose who introduced the powerful techniques of complex algebraic geometry into general relativity. Loosely speaking, it is the use of complex analytic methods to solve problems in real differential geometry. In most cases, the emphasis is on the geometry of the problem rather than the analysis. My lectures are not designed to be a survey of all of twistor theory or even the most important parts but will concentrate instead on those areas with which I am most familiar.

I will first describe the so-called mini-twistor space of $\mathbb{R}^3$ and how it can be used to find harmonic functions, or functions in the kernel of the Laplacian, by an integral transform. This integral transform is, in fact, a classical result given in Whittaker and Watson's famous book [21]. Also in there is an integral formula for solutions of the wave equation on $\mathbb{R}^4$. I will show how this also fits into the context of twistor theory by giving a general twistor integral transform – called a Penrose transform – for the solution of any constant coefficient homogeneous linear differential equation. To present the general result requires a certain amount of complex geometry, in particular, the theory of complex line bundles. This will be explained as I go along. A good reference is [9].

Twistor theory can also be used to solve nonlinear diferential equations which are related to the self-duality equations that describe instantons in $\mathbb{R}^4$. I will present a brief account of the theory of the Bogomolny equations. These are essentially time-invariant instantons and the twistor correspondence uses the mini-twistor space I described at the beginning.

There are many other topics in twistor theory. For example, there are twistor methods for defining hyper-Kähler manifolds [2] and applications of twistor theory to the representation theory of Lie groups [4]. Many short notes on twistor theory have traditionally appeared in the informal "Twistor Newsletter". Various compendiums of these are available and I have listed them in the bibliography [13, 16, 17]

2   Mini-twistors

It has been known for a long time that problems in real differential geometry can often by simplified by using complex coordinates. For example, in the plane $\mathbb{R}^2$ we can write $z = x + iy$ and thereby identify $\mathbb{R}^2$ with $\mathbb{C}$. We then discover that a function $f \colon \mathbb{R}^2 \to \mathbb{R}$ is harmonic if and only if we can write it as

$$f = \psi + \bar{\psi}$$

where $\psi\colon \mathbb{C} \to \mathbb{C}$ is a holomorphic (complex-analytic) function. So we have related harmonic, real-valued functions – a problem in real differential geometry on the plane – to holomorphic functions of one-variable – a problem in complex analysis.

If we try the same technique in $\mathbb{R}^3$ we discover the unfortunate fact that it has odd dimension and hence cannot be identified with some $\mathbb{C}^n$! We can, however, form another space closely associated to the geometry of $\mathbb{R}^3$ that is intrinsically complex. This is the space Z of all oriented lines in $\mathbb{R}^3$ known to twistor theorists as *mini-twistor space*. The word "mini" here refers to the fact that Penrose's original twistor theory was defined in dimension four. Notice that any oriented line ℓ in $\mathbb{R}^3$ is determined uniquely by giving the unit vector u parallel to the line in the direction of the orientation and the shortest vector v joining the origin to ℓ. The vector v is well known to be determined by the fact that it is orthogonal to u. We then have

$$\ell = \{v + tu \mid t \in \mathbb{R}\}.$$

This shows that

$$Z = \{(u, v) \mid u, v \in \mathbb{R}^3, \|u\|^2 = 1, \langle u, v \rangle = 0\} \subset S^2 \times \mathbb{R}^3.$$

The mini-twistor space Z is readily seen to be TS^2 the tangent bundle of the two-sphere, that is the space formed by taking the union of all the tangent planes to the two-sphere or

$$TS^2 = \cup_{u \in S^2} T_u S^2$$

where

$$T_u S^2 = \{v \in \mathbb{R}^3 \colon \langle u, v \rangle = 0\}$$

is the tangent plane to the two-sphere at u. Notice that there is a projection map $Z \to S^2$ sending a line to its direction or a pair (u, v) to v.

Whereas $\mathbb{R}^3$ is three dimensional, the mini-twistor space Z is four dimensional and now we have a chance of identifying it with something complex. Of course, it is not a Euclidean space but a manifold so we have to consider complex manifolds.

Recall that a (real) manifold M of dimension n is essentially a set which can be covered by coordinate charts in such a way that if coordinates $x = (x^1, \ldots, x^n)$ and $y = (y^1, \ldots, y^n)$ have overlapping domains then on that overlap we have

$$x^i(p) = F^i(y(p))$$

for some $F(y^1, \ldots, y^n)$ which is a smooth (that is infinitely differentiable) function of n real variables. We call F the change of coordinates map. To remove the "essentially" and give a precise definition I would only need to be a little more careful about the domain of definition of x, y and F. I refer the reader to [10] or any other book on differential geometry.

A complex manifold is the obvious generalisation of this definition. It has complex valued coordinates and the corresponding coordinate change maps like F are required to be holomorphic. Being holomorphic is a stronger condition than infinite differentiability and the theory of complex manifolds has a rich structure because of this. A good reference is [9]. Instead of making the definition anymore precise, I will illustrate it with an example. The example is the simplest, non-trivial complex manifold – the two-sphere.

Example 2.1 (The two-sphere). Define two open subsets U_0 and U_1 of S^2 by removing the north and south poles of the two-sphere. That is

$$U_0 = S^2 - \{(0,0,1)\} \quad \text{and} \quad U_1 = S^2 - \{(0,0,-1)\}.$$

These will be the domain of our coordinates. The union of U_0 and U_1 clearly covers all of S^2. We define complex co-ordinates on U_0 by

$$\xi_0(x,y,z) = \frac{x+iy}{1-z}$$

and on U_1 by

$$\xi_1(x,y,z) = \frac{x-iy}{1+z}.$$

Geometrically, the first set of coordinates is defined by stereographic projection from $(0,0,1)$. That is if we draw a line through $(0,0,1)$ and (x,y,z) then it meets the XY plane at $(x/(1-z), y/(1-z))$. The second set is defined by stereographic projection from $(0,0,-1)$ with a change of sign in the y coordinate. The change of sign is so that the coordinate change map is holomorphic. If we do not change the sign we find that the coordinate change map is anti-holomorphic, that is of the form $w \mapsto \bar{F}(w)$ where F is holomorphic. We calculate the coordinate change map to be

$$\xi_0(x,y,z) = \frac{1}{\xi_1(x,y,z)} = F(\xi_1(x,y,z))$$

and see that it is the holomorphic function $F(w) = 1/w$.

In the case of $\mathbb{R}^2$, harmonic functions h are generated by holomorphic functions ψ by the simple device of writing

$$h(x,y) = \psi(x+iy) + \bar{\psi}(x+iy).$$

This exploits the fact that the map $(x,y) \mapsto x+iy$ identifies the space $\mathbb{R}^2$ with the space $\mathbb{C}$. To construct harmonic functions on $\mathbb{R}^3$ we need to do something more complicated because the spaces $\mathbb{R}^3$ and Z cannot be identified. There are lots of reasons for this, for example they do not have the same topology; $\mathbb{R}^3$ can be contracted to a point and Z can only be contracted to a two-sphere, but it is simplest to note that they have different real dimensions, three and four respectively. To understand the relationship between $\mathbb{R}^3$ and Z, we have to explore their geometry a little further.

By definition, the points of Z are oriented lines in $\mathbb{R}^3$. Moreover, any point in $\mathbb{R}^3$ defines a two-sphere's worth of lines, all the oriented lines going through that point. If the point in question is p then the set of all lines through p is the set of all (u,v) satisfying

$$v = p - \langle p, u \rangle u.$$

We shall call this subset a real section of Z and denote it by X_p. I need to explore further the geometry of these real sections.

Firstly, let me explain why they are called sections. Notice that the map

$$\rho_p: \quad \begin{array}{ccc} S^2 & \to & Z \\ u & \mapsto & (u, v = p - \langle p, u \rangle u) \end{array}$$

defines a section of the projection $Z \to S^2$. (Recall that a section of a projection $\pi \colon X \to Y$ is a map $\rho \colon Y \to X$ with $\pi(\rho(y)) = y$ for all $y \in Y$.) The image of this section is X_p. I am, of course, abusing language here as I am calling the X_p a real section rather than the image of a real section – I will continue to do this.

Next, we need a map

$$\tau \colon Z \to Z,$$

called the real structure. It is defined to be the involution that sends a line with orientation to the same line with opposite orientation, that is $\tau(u, v) = (-u, v)$. The real structure fixes the set X_p because

$$\tau(u, p - \langle p, u \rangle u) = (-u, p - \langle p, u \rangle u) = (-u, p - \langle p, -u \rangle (-u)).$$

This is the reason the section X_p is called real.

Finally, the map ρ_p is holomorphic, in fact given by a quadratic polynomial. To see this, we need to consider the complex structure of Z which we have avoided up to now. The complex coordinates we have defined above on the two-sphere can be used to define complex coordinates (η, ζ) on Z. Let $\widetilde{U}_0$ be the set of all $(u, v) \in Z$ where $u \in U_0$. Then we define

$$\zeta(u, v) = \frac{u_1 + i u_2}{1 - u_3}. \tag{1}$$

Notice that this is just the coordinates on S^2 applied to u, the point v has played no role yet. To obtain complex coordinates, we differentiate the map ζ. We obtain

$$\eta(u, v) = \frac{v_1 + i v_2}{1 - u_3} + \frac{(u_1 + i u_2) v_3}{(1 - u_3)^2}. \tag{2}$$

If $p = (x, y, z)$ is a point in $\mathbb{R}^3$ then we have defined X_p to be the set of all (u, v) that correspond to lines through p. This is the set

$$X_p = \{(u, p - \langle u, p \rangle u) \mid u \in S^2\}.$$

If we substitute $v = p - \langle u, p \rangle u$ into the equation for η and simplify we see that the equation defining X_p as a subset of Z is

$$\eta = \frac{1}{2}((x + iy) + 2z\zeta - (x - iy)\zeta^2) \tag{3}$$

where (x, y, z) are the coordinates of p.

Hence, in local coordinates, the map ρ_p is

$$\rho_p(\zeta) = (\frac{1}{2}((x + iy) + 2z\zeta - (x - iy)\zeta^2), \eta). \tag{4}$$

We call a section that can be written locally as a holomorphic function as in equation (4) a holomorphic section. It is possible to show (see section 5) that all holomorphic sections $S^2 \to Z$ take the form

$$\zeta \mapsto (\zeta, a + b\zeta + c\zeta^2)$$

in coordinates, where a, b and c are complex numbers. With our choice of coordinates and the definition of the real structure, it is an easy exercise to see that if (η, ζ) are the coordinates of p then $(-\bar{\eta}/\bar{\zeta}^2, -1/\bar{\zeta})$ are the coordinates of $\tau(p)$. So τ is an anti-holomorphic map. A section is therefore real if the equation

$$\eta = a + b\zeta + c\zeta^2$$

defines the same subset of Z as the equation

$$-\frac{\bar{\eta}}{\bar{\zeta}^2} = a + b\frac{-1}{\bar{\zeta}} + c\frac{1}{\bar{\zeta}^2}.$$

Simplifying we see that this is true if and only if $a = -\bar{c}$ and b is real. Hence the real sections ρ_p defined by points in $\mathbb{R}^3$ are precisely all the real sections of Z. So we have a bijection between points of $\mathbb{R}^3$ and real holomorphic sections of Z.

The correspondence we have now established between $\mathbb{R}^3$ and Z is completely symmetric; points in Z define special subsets (oriented lines) in $\mathbb{R}^3$ and points in $\mathbb{R}^3$ define special subsets (holomorphic real sections) in Z.

Consider a differential form $\omega = g(\eta, \zeta)d\eta$ on Z. We can use this to define a function on $\mathbb{R}^3$ by

$$\phi(x, y, z) = \int g(((x + iy) + 2z\zeta - (x - iy)\zeta^2)/2, \zeta)d\zeta. \tag{5}$$

It is easy to differentiate through the integral sign in (5) and see that

$$\frac{\partial^2 \phi}{\partial x^2} + \frac{\partial^2 \phi}{\partial y^2} + \frac{\partial^2 \phi}{\partial z^2} = 0.$$

So the function ϕ is harmonic. This result is a classical integral formula contained in Whittaker and Watson [21]. To obtain their formulation, we replace the integral in η by a contour integral around the circle and let $\eta = -\exp(iw)$ to obtain

$$\phi(x, y, z) = \int_{-\pi}^{\pi} h(z + ix\cos(w) + iy\sin(w), w)dw$$

where $h(v) = g(\exp(iw)v$.

Notice that instead of performing an integral in equation (5) over S^2, we can think of this as being first the restriction of ω to X_p and second its integral over X_p so that we have

$$\phi(p) = \int_{X_p} \omega.$$

3   A Classical Result

Whittaker and Watson also give an integral formula for the solutions of the wave equation on $\mathbb{R}^4$. It is

$$\phi(x, y, z, t) =$$
$$\int_{-\pi}^{\pi} g(x\sin(w)\cos(w') + y\sin(w)\sin(w') + z\cos(w) + t, w, w')dwdw'.$$

Again it is easy to check that this satisfies

$$\frac{\partial^2 \phi}{\partial x^2} + \frac{\partial^2 \phi}{\partial y^2} + \frac{\partial^2 \phi}{\partial z^2} - \frac{\partial^2 \phi}{\partial t^2} = 0.$$

I want to explain how these two classical examples form part of a general Penrose transform or twistor correspondence. This Penrose transform will give rise to integral formulae for the solutions of any homogeneous real-valued polynomial differential equation, that is any differential equation for functions ϕ on $\mathbb{R}^{n+1}$ of the form

$$D_f \phi = f(\frac{\partial}{\partial x^0}, \frac{\partial}{\partial x^1}, \dots, \frac{\partial}{\partial x^n}) \phi(x^0, \dots, x^n) = 0 \tag{6}$$

where f is a homogeneous real-valued polynomial of degree $k > 1$ and $n > 1$. I will show that there is a twistor space Z, a vector space $H^{n-1}(Z, \mathcal{O}(-n-1+k))$ of Dolbeault cohomology classes (think of them for now as differential forms) and a twistor correspondence

$$T: H^{n-1}(Z, \mathcal{O}(-n-1+k)) \to C^\omega(\mathbb{R}^{n+1}, \mathbf{C})$$

where C^ω denotes the real-analytic functions. This is an injective map, defined by integrating the differential form, onto the space of all the real-analytic functions in the kernel of D_f if $k \leq n$.

The flavour of this correspondence is as follows. We start with a polynomial f and $\mathbb{R}^{n+1}$. There is a twistor space Z which depends on f, and has a family of submanifolds X_p indexed by elements p in $\mathbb{R}^{n+1}$. In the mini-twistor space case, we integrated differential forms over the X_p. In the more general case, we integrate so-called differential forms ρ "of type $(0, n-1)$ with values in the line bundle $\mathcal{O}(-n-1+k)$", or their cohomology classes. The result then is a function

$$T(\rho)(p) = \int_{X_p} \omega$$

on $\mathbb{R}^{n+1}$.

To explain this correspondence in more detail, I have to explain the differential geometry of Z, why these differential forms ρ are the correct things to integrate over X_p and what the cohomology space $H^{n-1}(Z, \mathcal{O}(-n-1+k))$ is. The spaces X_p are all isomorphic to a subvariety X of complex projective space defined by f so we start then with the geometry of complex projective space.

4 Complex Projective Space

Recall that n-dimensional complex projective space is the n-dimensional complex manifold $\mathbb{C}P_n$ of all lines through the origin in $\mathbb{C}^{n+1}$. If z is a non-zero vector in $\mathbb{C}^{n+1}$ then we denote by $[z]$ the set of all non-zero multiples of z, that is

$$[z] = \{\lambda z \mid \lambda \in \mathbb{C} - \{0\}\}.$$

Notice that $[z]$ is the line through z with the origin or 0 removed. Clearly we can identify the line through z with $[z]$ so we identify the elements of $\mathbb{C}P_n$ with the set of all $[z]$. Indeed, I will often refer to $[z]$ as a line. It follows from this definition that

$$[z^0, \ldots, z^n] = [\lambda z^0, \ldots, \lambda z^n]$$

for any non-zero complex number λ. If $[z]$ is a line then the numbers $(z^0, \ldots, z^n)$ are called the homogeneous coordinates of the line. Note that the homogeneous coordinates of a line are not unique although we can uniquely specify things like ratios of the z^i when these are finite.

We will need to know below that $\mathbb{C}P_n$ is compact as a topological space. To see this, notice that the $2n-1$ sphere in $\mathbb{C}^n = \mathbb{R}^{2n}$ is the set of all z such that $\sum_{i=1}^{n} |z^i|^2 = 1$. If we map each such unit vector to the line containing it we define a continous, onto map

$$S^{2n-1} \to \mathbb{C}P_n.$$

The sphere is, of course, compact and hence $\mathbb{C}P_n$ is compact.

We make $\mathbb{C}P_n$ a complex manifold by considering the open subsets U_i where the ith homogeneous coordinate is non-zero. On these, we can define coordinates

$$[z^0, \ldots, z^n] \mapsto \left(\frac{z^0}{z^i}, \ldots, \frac{z^{i-1}}{z^i}, \frac{z^{i+1}}{z^i}, \ldots, \frac{z^n}{z^i} \right) \tag{7}$$

taking values in $\mathbb{C}^n$.

Example 4.1 ($\mathbb{C}P_1$). We have already met one-dimensional complex projective space. It is just the two-sphere. An explicit diffeomorphism between S^2 and $\mathbb{C}P_1$ is given by the map

$$\begin{array}{ccc} S^2 & \to & \mathbb{C}P_1 \\ (x, y, z) & \mapsto & [x + iy, 1 - z]. \end{array}$$

We leave it as an exercise for the reader to check that the coordinates defined in section 2 are essentially the same as the coordinates defined in equation (7) when composed with this diffeomorphism.

If f is a homogeneous polynomial of degree k then we can apply it to the homogeneous coordinates of a line and we have, of course

$$f(\lambda z^0, \ldots, \lambda z^n) = \lambda^k f(z^0, \ldots, z^n).$$

It follows that although f applied to a line $[z]$ is not well defined, we can uniquely define the subset X of $\mathbb{C}P_n$ where f vanishes. We just define this to be the set of $[z]$ where $f(z) = 0$. In general, X is a projective algebraic variety, for convenience we will assume that it is a smooth submanifold of $\mathbb{C}P_n$.

Example 4.2 (Mini-twistors). For the case of the Laplacian in $\mathbb{R}^3$, we are interested in the polynomial

$$f(x, y, z) = x^2 + y^2 + z^2.$$

The variety X in $\mathbb{C}P_2$ is the quadric defined to be the set of all $[z^0, z^1, z^2]$ satisfying

$$(z^0)^2 + (z^1)^2 + (z^2)^2 = 0.$$

This is isomorphic to $\mathbb{C}P_1$ or S^2 via the map

$$w \mapsto [i(w^2 + 1), (w^2 - 1), 2w].$$

On the open set U_i, the set where f vanishes is the zero set of the polynomial

$$f(\frac{z^0}{z^i}, \ldots, 1, \ldots, \frac{z^n}{z^i})$$

but this polynomial does not extend to a nice function on all of $\mathbb{C}P_n$. There is a good reason for this which I wish to discuss next as it is also the reason that we have to introduce holomorphic line bundles in the next section.

If we have a complex manifold M then it makes sense to talk about holomorphic functions $\chi \colon M \to \mathbb{C}$, just as for the case of smooth functions on a real manifold. We just define a function $\chi \colon M \to \mathbb{C}$ to be holomorphic if for every set of coordinates $z^1, \ldots, z^n$ it can be written as

$$\chi(m) = g(z^1(m), \ldots, z^n(m))$$

for some holomorphic function $g(z^1, \ldots, z^n)$. However if $\chi \colon M \to \mathbb{C}$ is a holomorphic function on a compact complex manifold then it must be constant. This is because of the maximum modulus theorem. Indeed the function $|\chi|^2$ is a continous function on the compact space M so has a maximum at some point m. But we can now choose an open set about m and introduce complex coordinates to make that set look like an open subset of $\mathbb{C}^n$. But a holomorphic function whose modulus has a maximum in the interior of an open set is constant. It follows that, to obtain a useful function theory on compact complex manifolds, we have to do something more than just look at globally defined functions. There are two generalisations that are possible, the first is to sections of holomorphic line bundles and the second is to sheaves. We shall need only to consider holomorphic line bundles for the purposes of these lectures and we will do that in the next section, 5.

The integral transform we wish to construct takes the form:

$$T(\omega)(p) = \int_{X_p} \omega$$

and I have been claiming that ω is a type of differential form. I now want to consider precisely what kind of differential form it is.

Recall that if M is an oriented real manifold of dimension m then we can integrate differential forms of degree m over it. In the case that M is a complex manifold of dimension n with coordinates $z^1, \ldots, z^n$ we can integrate complex differential forms of the type:

$$g(z, \bar{z})dz^1 \wedge \cdots \wedge dz^n \wedge d\bar{z}^1 \wedge \cdots \wedge d\bar{z}^n. \tag{8}$$

The reason for this is that if we let $z^j = x^j + iy^j$ we obtain coordinates (x^j, y^j) for M thought of as an $m = 2n$-dimensional real manifold. If we then use $dz^j = dx^j + idy^j$ and $d\bar{z}^j = dx^j - idy^j$ and expand out equation (8) becomes a differential form which is a multiple of

$$dx^1 \wedge \cdots \wedge dx^n \wedge dy^1 \wedge \cdots \wedge dy^n$$

and this is something we can integrate.

More general differential forms on a complex manifold can also be decomposed into products of dz's and $d\bar{z}$'s. We say a general form is of type (p, q) if it has p dz's and q $d\bar{z}$'s. We will be particularly interested in forms of type $(0, n)$. Indeed, we will adopt the rather perverse notion that the forms like

$$g(z, \bar{z})dz^1 \wedge \cdots \wedge dz^n \wedge d\bar{z}^1 \wedge \cdots \wedge d\bar{z}^n$$

should be written as

$$[g(z, \bar{z})dz^1 \wedge \cdots \wedge dz^n] \, d\bar{z}^1 \wedge \cdots \wedge d\bar{z}^n$$

and thought of, not as forms of type (n, n), but as forms of type $(0, n)$ with values in the *holomorphic line bundle* of all forms of type $(n, 0)$. We denote by K_m the space of all forms at $m \in M$ of type $(n, 0)$. In other words, K_m is the highest exterior power of the cotangent space $T_p M^*$. It follows that K_m is one dimensional. The union of all these one-dimensional spaces as m varies is called the *canonical line bundle* of M. The moral of the story then is that on a complex manifold of dimension n we can integrate $(0, n)$ forms with values in the canonical line bundle K.

More generally, we will be interested in other holomorphic line bundles over a complex manifold M so in the next section we will discuss the theory of these.

5 Holomorphic Line Bundles

We have seen that on a compact complex manifold the only holomorphic functions are the constants. We want to introduce the concept of a holomorphic section of a holomorphic line bundle to give us a larger collection of function-like objects to work with. To do this, we associate to every point m of a complex manifold M a one-dimensional complex vector space L_m. Denote by L the union of all these spaces. Note that, in principle, the L_m's are all different but not very different. A one-dimensional complex vector space is *nearly* the same as $\mathbb{C}$. Indeed, if we pick any non-zero vector v in L_m then it defines a basis of L_m and hence a linear isomorphism

$$\begin{array}{ccc} \mathbb{C} & \to & L_m \\ z & \mapsto & xv \end{array} \, .$$

Instead of holomorphic functions, we want to consider holomorphic sections of L. A section is a function

$$\psi \colon M \to L = \cup_{m \in M} L_m$$

with the property that $\psi(m) \in L_m$ for every m. Notice that because L_m is a vector space we can add sections by defining

$$(\psi + \chi)(m) = \psi(m) + \chi(m)$$

and multiply them by scalars by defining

$$(z\psi)(m) = z\psi(m).$$

So the collection of all sections forms a vector space and they behave in many ways like functions.

We want to consider holomorphic sections and for that we need a notion of a family of vector spaces L_m varying holomorphically with m. To make sense of this, we need to consider the precise definition of a holomorphic line bundle.

Definition 5.1 (Holomorphic line bundle). A holomorphic line bundle over a manifold M is a holomorphic manifold L with a holomorphic map $\pi \colon L \to M$ such that:

1. each fibre $L_m = \pi^{-1}(m)$ is a one-dimensional complex vector space, and

2. we can cover M with open sets U_α such that there is a bi-holomorphic map $\psi_\alpha \colon \pi^{-1}(U_\alpha) \to U_\alpha \times \mathbb{C}$ with the property that for all $m \in U_\alpha$ we have $\psi_\alpha(L_m) \subset \{m\} \times \mathbb{C}$ and moreover $\psi_{\alpha|L_m}$ is a linear isomorphism.

In the second part of this definition which requires that

$$\psi_{\alpha|L_m} \colon L_m \to \{m\} \times \mathbb{C}$$

is a linear isomorphism, we make $\{m\} \times \mathbb{C}$ into a vector space by defining

$$\lambda(b, z) + \mu(b, w) = (b, \lambda z + \mu w).$$

A bi-holomorphic map is just an invertible holomorphic map whose inverse is also holomorphic. We call the set $L_m = \pi^{-1}(m)$ the fibre of L over m.

We can now define

Definition 5.2 (Holomorphic section). If $L \to M$ is a holomorphic line bundle then a holomorphic section of L is a holomorphic map $\psi \colon M \to L$ such that $\psi(m) \in L_m$ for all $m \in M$.

Example 5.1 (The trivial bundle). The simplest example of a holomorphic line bundle is the *trivial* bundle $L = M \times \mathbb{C}$. In this case, it is easy to see that a section is just a holomorphic function. Indeed, a section must have the form $\psi(m) = (m, \chi(m))$ for some holomorphic map $\chi \colon M \to \mathbb{C}$.

It is often useful to have a "local" description of sections. To do this, we note from the definition that we can cover M with open sets $\{U_\alpha\}$ such that there are local non-vanishing holomorphic sections $\psi_\alpha \colon U_\alpha \to L$. At any point $m \in U_\alpha \cap U_\beta$, we now have two non-zero elements $\psi_a(m)$ and $\psi_\beta(m)$ in the one-dimensional vector space L_m. These must differ by a scalar $g_{\alpha\beta}(m)$ and hence there are holomorphic maps

$$g_{\alpha\beta} \colon U_\alpha \cap U_\beta \to \mathbb{C} - \{0\}$$

(called transition functions) such that

$$\psi_\alpha = \psi_\beta g_{\alpha\beta} \tag{9}$$

on $U_\alpha \cap U_\beta$.

Notice that to be completely precise we should write equation (9) as

$$\psi_{\alpha|U_\alpha \cap U_\beta} = \psi_{\beta|U_\alpha \cap U_\beta} g_{\alpha\beta}$$

but we shall usually drop the notation that indicates the restriction of functions to subsets such as $U_\alpha \cap U_\beta$.

Any section ξ of L can then be written at points of each U_α as

$$\xi = \xi_\alpha \psi_\alpha.$$

The section ξ is holomorphic if and only if each of the

$$\xi_\alpha : U_\alpha \to \mathbb{C}$$

is holomorphic. The ξ_α must satisfy

$$\xi_\alpha g_{\alpha\beta} = \xi_\beta$$

on the intersection $U_\alpha \cap U_\beta$. A converse to this result is also true. If we can find holomorphic functions $\xi_\alpha : U_\alpha \to \mathbb{C}$ such that

$$\xi_\alpha g_{\alpha\beta} = \xi_\beta$$

on the intersection $U_\alpha \cap U_\beta$ then defining $\xi = \xi_\alpha \psi_\alpha$ on each open set U_α defines a global holomorphic section of L.

Example 5.2 (The tautological bundle). We now define a natural holomorphic line bundle H over complex projective space called the *tautological line bundle*. The fibre of H over a point $[z] \in \mathbb{C}P_n$ is just the line $[z]$. To see how all the fibres fit together we define

$$H = \{([z], w) \mid w = \lambda z, \lambda \in \mathbb{C} - \{0\}\} \subset \mathbb{C}P_n \times \mathbb{C}^{n+1}.$$

The projection map $H \to \mathbb{C}P_n$ is just the restriction of the obvious projection from $\mathbb{C}P_n \times \mathbb{C}^{n+1}$, that is $([z], w) \mapsto [z]$. We define local sections $\psi_i : U_i \to H$ for each $i = 0, \ldots, n$ by

$$\psi_i([z]) = ([z], (\frac{z_0}{z_i}, \ldots, 1, \ldots, \frac{z_n}{z_i})).$$

Note that they do have image in H because

$$(\frac{z_0}{z_i}, \ldots, 1, \ldots, \frac{z_n}{z_i}) = \frac{1}{z_i}(z_0, \ldots, z_n).$$

From the definition of the ψ_i, we have

$$\psi_i([z]) = \frac{1}{z_i}(z) = \frac{z_j}{z_i}\psi_j([z]).$$

and hence the transition functions are

$$g_{ij} = \frac{z_j}{z_i}.$$

Example 5.3 (The tangent bundle to S^2). In the case of $\mathbb{C}P_1$, the tangent space at any point is a one-dimensional complex vector space and hence is a complex line bundle. Notice that elementary topology tells us that it cannot be a trivial line bundle. If it were it would have a non-vanishing (holomorphic) section and therefore TS^2 would have a non-vanishing section. But this means that the two-sphere would have a non-vanishing continuous tangent vector field and this is not possible [18].

Operations on vector spaces such as taking duals and forming tensor products can also be applied, fibre by fibre, to a line bundle. So we define L^* the dual of the line bundle L by $(L^*)_m = (L_m)^*$ and $L \otimes J$, the tensor product of the line bundles L and J by $(L \otimes J)_m = L_m \otimes J_m$. If v is a non-zero element of a one-dimensional vector space, it naturally defines an element v^{-1} of the dual space by the requirement that $v^{-1}(v) = 1$. Non-vanishing local sections ψ_α of L hence give rise to non-vanishing sections ψ_α^{-1} of L^*. Similarly local sections of L and J can be tensored together to give local sections of $L \otimes J$. Notice that if $g_{\alpha\beta}$ are transition functions for L and $h_{\alpha\beta}$ are transition functions for J then the transition functions for L^* are $g_{\alpha\beta}^{-1}$ and those for $L \otimes J$ are $g_{\alpha\beta}h_{\alpha\beta}$.

We can now define

Definition 5.3 (Isomorphism of line bundles). Two line bundles L and J are said to be *isomorphic* if at each point m of M we have a linear isomorphism

$$\alpha_m \colon L_m \to J_m.$$

such that the induced section of $L^* \otimes J$ is holomorphic.

Example 5.4 (The bundles $\mathcal{O}(p)$). Returning to the line bundle H over $\mathbb{C}P_n$ we can now define a family of line bundles over $\mathbb{C}P_n$ denoted by $\mathcal{O}(p)$ for any integer p as follows. For p a negative integer, we define $\mathcal{O}(p)$ to be the $-p$th tensor power of H with itself and for p a positive integer, we define $\mathcal{O}(p)$ to be the pth tensor power of H^* with itself. These line bundles satisfy: $\mathcal{O}(-1) = H$, $\mathcal{O}(p) \otimes \mathcal{O}(q) = \mathcal{O}(p + q)$, $\mathcal{O}(p)^* = \mathcal{O}(-p)$ and $\mathcal{O}(0)$ is the trivial line bundle.

The reason for the choice of sign here is that the line bundle H has no holomorphic sections except for the zero section. Let us see why this is true in the case of $\mathbb{C}P_1$. There we have two open sets U_0 and U_1 and a holomorphic section of H consists of a pair of functions $\xi_0 \colon U_0 \to \mathbb{C}$ and $\xi_1 \colon U_1 \to \mathbb{C}$ such that

$$\xi_0([z]) = \frac{z_1}{z_0}\xi_1([z]).$$

To understand this, let us translate it into coordinates. Let $\zeta([z]) = z_0/z_1$, then we require two holomorphic functions ξ_0 and ξ_1, defined on all of $\mathbb{C}$, such that

$$\xi_0(\zeta) = \frac{1}{\zeta}\xi_1(\frac{1}{\zeta}).$$

If we Taylor expand both sides of this equation in ζ and $1/\zeta$ we deduce that $\xi_0 = \xi_1 = 0$ is the only solution. Similarly, if we try and find a holomorphic section of $\mathcal{O}(p)$ we seek holomorphic functsions ξ_0 and ξ_1 defined on all of $\mathbb{C}$ such that

$$\xi_0(\zeta) = \zeta^p \xi_1(\frac{1}{\zeta}).$$

Again Taylor expanding shows that if p is negative there are no non-zero sections and if p is positive there is a vector space of sections of dimension $p + 1$.

Notice that this gives us a simple interpretation of the number p in the case of a line bundle $\mathcal{O}(p)$ over $\mathbb{C}P_1$ with $p \geq 0$. That is, it is the number of zeroes (counted with multiplicity) of a holomorphic section of the line bundle. It is in fact true (see [9]) that every holomorphic line bundle over $\mathbb{C}P_1$ is isomorphic to some $\mathcal{O}(p)$. This means that if we have a holomorphic line bundle over $\mathbb{C}P_1$ that has a holomorphic section which has p zeroes then the line bundle has to be isomorphic to $\mathcal{O}(p)$. We shall use this result later.

In the case of $\mathcal{O}(p)$ defined on $\mathbb{C}P_n$, a similar result holds. If $p < 0$ then there are no sections and if $p > 0$ then the space of sections has the same dimension as the space of all homogeneous polynomials in $n + 1$ variables of degree p. To see why this space of polynomials arises, recall that $\mathcal{O}(1) = H^*$ is the dual bundle of $\mathcal{O}(-1)$. We can define sections of $\mathcal{O}(1)$ by noting that, for any $i = 1, \ldots, n+1$, we can define a linear map on $\mathbb{C}^{n+1}$ and hence, by restriction on any line in $\mathbb{C}^{n+1}$, by picking out the ith component of a vector. Call this holomorphic section of $\mathcal{O}(1)$, ζ^i. Notice that it is non-vanishing on U_i. These sections in fact span the space $H^0(\mathbb{C}P_n, \mathcal{O}(1))$ of all holomorphic sections. We can multiply or tensor these sections together p times to get a section of $\mathcal{O}(p)$ and so we find that $H^0(\mathbb{C}P_n, \mathcal{O}(-p))$ has the same dimension as the space of all homogeneous polynomials of degree p.

Example 5.5 (The section of $\mathcal{O}(k)$ defined by f). If we take the homogeneous polynomial $f(z)$ then $f(\zeta) = f(\zeta^0, \ldots, \zeta^n)$ defines a section of $\mathcal{O}(k)$ vanishing precisely on X. This then is the solution to the problem raised in the previous section of extending the functions

$$f(\frac{z^0}{z^i}), \ldots, 1, \ldots, \frac{z^n}{z^i})$$

defined on the set U_i to all of $\mathbb{C}P_n$. We can but only if we intepret the result of doing so as a section of the line bundle $\mathcal{O}(k)$.

Example 5.6 (The bundle $\mathcal{O}(2)$ over Z). We have seen that S^2 and $\mathbb{C}P_1$ are the same as complex manifolds and that Z is the same real manifold as TS^2. It is also true that Z is isomorphic as a complex manifold, with the complex structure we have defined, to $T\mathbb{C}P_1$. Moreover, as we have noted, this is a line bundle on $\mathbb{C}P_1$. The complex holomorphic sections are just those described in section 2. Hence the space of all holomorphic sections of Z has the same dimension as the space of all homogeneous polynomials of degree two in two variables. So it is a copy of $\mathbb{C}^3$. The real sections are a real subspace of this $\mathbb{R}^3$ as we discussed in section 2.

6　The Canonical Bundle of X

We want to integrate things over X. For this purpose we need to know what its canonical bundle is. It turns out that it is isomorphic to the bundle $\mathcal{O}(-n-1+k)$ restricted to X. I will sketch the proof of this result.

As X is a submanifold of $\mathbb{C}P_n$, the tangent space to X at any point of X is a subspace of the tangent space to all of $\mathbb{C}P_n$. The quotient of these is the normal space to X denoted by

$$N = T\mathbb{C}P_n/TX. \tag{10}$$

Notice that in Riemannian geometry we would use an inner product actually to pick out a normal. We have to resist the temptation to do this in complex geoemtry as it would involve using a hermitian inner product to pick out a normal subspace of $T\mathbb{C}P_n$ to X and that would involve taking conjugates and spoil the "holomorphicness" of N. Note also that elementary linear algebra shows that $N^* \subset T\mathbb{C}P_n^*$, in fact it is the subspace of all linear functions on $T\mathbb{C}P_n$ which vanish on TX.

We have defined the canonical bundle of X to be the highest exterior power of TX^*. The highest exterior power of a vector space V is often denoted $\det(V)$

because if $X : V \to V$ is a linear map and V is, say, n dimensional then X induces a linear map from the one-dimensional space $\bigwedge^n(V)$ to itself. Such a linear map must be multiplication by a complex number which is actually $\det(X)$. The quotient (10) induces a natural isomorphism $N \otimes \det(TX) = \det(T\mathbb{C}P_n)$. Note that N is one dimensional so that $N = \det(N)$. Hence if K is the canonical bundle of X we have

$$K = \det(TX)^* = \det(T\mathbb{C}P_n)^* \otimes N.$$

Now we only need to calculate N and the canonical bundle of $\mathbb{C}P_n$. This is actually a step forward!

Recall that f defines a holomorphic section of $\mathcal{O}(k)$ vanishing on X, $\mathcal{O}(k)$ has non-vanishing sections χ_i say over each of the open sets U_i and hence over these

$$f = f_i \chi_i$$

for some holomorphic functions $f_i : U_i \to \mathbb{C}$. On a complex manifold, the usual exterior derivative

$$d = \sum_{i=1}^{n} \frac{\partial}{\partial x^i} dx^i$$

has a holomorphic analogue

$$\partial = \sum_{i=1}^{n} \frac{\partial}{\partial z^i} dz^i.$$

Define

$$\partial f = (\partial f_i) \chi_i.$$

As f is constant on X, this is zero when applied to vectors tangent to X and hence ∂f_i is a section of N^*. It follows that ∂f is a section, over U_i of

$$N^* \otimes \mathcal{O}(k).$$

It is, in fact, a non-vanishing section. The reason for this is a little technical. If it vanished then that would amount to f having a multiple root somewhere on X and that would contradict the assumption that X is a smooth submanifold of $\mathbb{C}P_n$. I do not wish to dwell on this point but refer the interested reader instead to [9].

Notice that if we change from U_i to U_j then $\chi_i = g_{ij}\chi_j$ and thus $f_i = g_{ji}f_j$ where g_{ij} is a non-vanishing holomorphic function. Differentiating gives

$$\partial f_i = \partial(g_{ij})f_j + g_{ij}\partial f_j$$

and evaluating at any point on X where f_j vanishes gives

$$(\partial f_i)\chi_i = (\partial f_j)\chi_j$$

on X. Hence ∂f is a non-vanishing section of $N^* \otimes \mathcal{O}(k)$ over all of X. So we conclude that $N = \mathcal{O}(k)$.

Finally, we calculate $\det(T\mathbb{C}P_n)$ or the canonical bundle of projective space. The tangent space to projective space at a line $[z]$ can be shown to be isomorphic to all the linear maps from the line $[z]$ to the quotient $\mathbb{C}^{n+1}/\mathcal{O}(-1)_{[z]}$. In other words

$$TCP_n = (\mathcal{O}(-1)^* \otimes \mathbb{C}^{n+1})/\mathcal{O}(-1).$$

When we apply det to a tensor product, it behaves like taking the determinant of a Kronecker product of matrices. The latter multiplies and the same is true for spaces. That is $\det(\mathbb{C}^m \otimes V) = \det(V)^{\otimes m}$ so

$$\det(TCP_n)^* = \mathcal{O}(-n) \otimes \det(\mathbb{C}^{n+1})^* \otimes \mathcal{O}(-1) = \mathcal{O}(-n-1)$$

and finally

$$K = \mathcal{O}(-n-1+k).$$

We can now begin to explain the twistor correspondence we are interested in. For the time being, we ignore the question of defining a twistor space. We use the basis $\zeta^0, \ldots, \zeta^n$ of $H^0(\mathbb{C}P_n, \mathcal{O}(1))$ to define an explicit isomorphism with $\mathbb{C}^{n+1}$ by

$$(z^0, \ldots, z^n) \mapsto \sum_{i=0}^{n} z^i \zeta^i.$$

It is possible to show that the restriction map

$$H^0(\mathbb{C}P_n, \mathcal{O}(1)) \to H^0(X, \mathcal{O}(1))$$

is an isomorphism but this needs some of the theory of sheaves and would divert us too far from the main task. The point is that sections of $\mathcal{O}(1)$ vanish on linear subspaces of $\mathbb{C}P_n$, whereas the subspace X is not linear because $k > 1$, hence we cannot have a section of $\mathcal{O}(1)$ vanishing on all of X. For further details look at [9]. Let us just note then that

$$\mathbb{C}^{n+1} = H^0(\mathbb{C}P_n, \mathcal{O}(1)) = H^0(X, \mathcal{O}(1)).$$

Choose some differential forms $\omega_0, \ldots, \omega_m$ on X of type $(0, n-1)$ where ω_l has values in $\mathcal{O}(-n-1+k-l)$. Then each

$$\omega_l(\sum_{i=0}^{n} z^i \zeta^i)^l$$

is a differential form with values in $\mathcal{O}(-n-1+k)$ and hence so is their sum. We can then integrate to define

$$\phi(z) = \int_X \sum_{i=0}^{m} \omega_l(\sum_{i=0}^{n} z^i \zeta^i)^l. \tag{11}$$

This is a function on $\mathbb{C}^{n+1}$. If we apply a monomial differential operator

$$\frac{\partial}{\partial z^{i_1}} \frac{\partial}{\partial z^{i_2}} \cdots \frac{\partial}{\partial z^{i_k}}$$

to $\phi(z)$ we obtain

$$\int_X \sum_{i=0}^{m} \omega_l l(l-1) \cdots (l-k+1)(\sum_{i=0}^{n} z^i \zeta^i)^{l-k} \zeta^{i_1} \zeta^{i_2} \cdots \zeta^{i_k}.$$

Hence, if we apply D_f to $\phi(z)$ we obtain

$$\int_X \sum_{i=0}^m \omega_l l(l-1)\cdots(l-k+1)(\sum_{i=0}^n z^i\zeta^i)^{l-k} f(\zeta^0,\zeta^1,\ldots,\zeta^k)$$

which vanishes because f vanishes on X.

So far, we have a transformation from collections of differential forms $\omega_0,\ldots,\omega_m$ to functions in the kernel of the differential operator D_f. Clearly an infinite sum of ω_l's would also work if the sum was convergent. We might also expect that sometimes the function ϕ may be the zero function. To understand these details, we need to define a twistor space Z for this problem. Before doing that, it is interesting to note that something similar to a Fourier transform is going on here in as much as differentiation is being turned into multiplication. Eastwood has made this analogy precise by using the *twistor transform* [7].

7   Twistor Space

To understand the construction introduced in the previous section we need to construct an analogue of mini-twistor space. We shall also call it Z in this case. We define Z to be the line bundle $\mathcal{O}(1)$ restricted to X. Notice that, in as much as X depends on f, the twistor space Z also depends on f.

Example 7.1 (Mini-twistors). In the case of the Laplacian on $\mathbb{R}^3$ we have seen that $X = S^2 = \mathbb{C}P_1$ and sits inside $\mathbb{C}P_2$ as the subvariety

$$(z^0)^2 + (z^1)^2 + (z^2)^2 = 0$$

via the map

$$w \mapsto [i(w^2+1),(w^2-1),2w].$$

We have noted that Z is the line bundle $\mathcal{O}(2)$ over $\mathbb{C}P_1$ whereas I am now suggesting that it is the line bundle $\mathcal{O}(1)$ over $\mathbb{C}P_2$ restricted to X. This is not a contradiction because the line bundle $\mathcal{O}(1)$ on $\mathbb{C}P_2$ restricted to X and thought of as line bundle over $\mathbb{C}P_1$ is actually the line bundle $\mathcal{O}(2)$. To see this, recall from section 5 that we can find out which $\mathcal{O}(p)$ a line bundle over $\mathbb{C}P_1$ is by choosing a holomorphic section and counting its zeroes. In this case, we can take a holomorphic section of $\mathcal{O}(1)$ given by the linear function z^0. This vanishes on the hyperplane $z^0 = 0$ and the restriction of this to the subvariety $X = \mathbb{C}P_1$ is the pair of points $[0,1,i]$ and $[0,-1,i]$. Hence it vanishes at two places and so the line bundle in question is isomorphic to $\mathcal{O}(2)$.

The relationship between the twistor space Z and $\mathbb{R}^{n+1}$ revolves around the following *double fibration* where we identify $\mathbb{R}^{n+1}$ with a real subspace of $H^0(X,\mathcal{O}(1))$.

$$\mathbb{R}^{n+1} \times X$$
$$\swarrow \qquad\qquad \searrow \qquad\qquad (12)$$
$$\mathbb{R}^{n+1} \qquad\qquad\qquad\qquad Z$$

If $(\sum_{i=0}^n x^i\zeta^i, x) \in \mathbb{R}^{n+1} \times X$ then the left-hand map sends this to $x = (x^0,\ldots,x^n)$ and the right-hand arrow sends it to $\sum_{i=0}^n \zeta^i(x)$. If we fix a point

x in $\mathbb{R}^{n+1}$ then the image of the section

$$\sum_{i=0}^{n} x^i \zeta^i \colon X \to Z$$

is a subvariety X_z in Z which is identified with X by the projection $\pi\colon Z \to X$. This is the analogue of the real section determined by the point x in the mini-twistor case.

Example 7.2 (Mini-twistors). Recall from section 2 the construction of the mini-twistor space. We can introduce the double fibration

$$\begin{array}{ccc}
 & \mathbb{R}^3 \times S^2 & \\
 \swarrow & & \searrow \\
 \mathbb{R}^3 & & Z
\end{array} \qquad (13)$$

Here the left-hand arrow is the map $(x, u) \mapsto v$ and the right-hand arrow is the map $(v, u) \mapsto (v- <v, u> u, u)$ or in complex coordinates

$$(v, \zeta) \mapsto \left(\frac{1}{2} \big((x+iy) + 2z\zeta - (x-iy)\zeta^2 \big), \zeta \right).$$

We will need various line bundles on Z. We can use the projection $\pi\colon Z \to X$ to *pull-back* line bundles from X. That is, if L is a line bundle on X, we define a line bundle $\pi^{-1}L$ on Z by $(\pi^{-1}(L))_z = L_{\pi(z)}$. We adopt the convention of denoting $\pi^{-1}(\mathcal{O}(p))$ by just $\mathcal{O}(p)$. Notice that a peculiar thing happens for the bundle $\mathcal{O}(1)$. Here if z is an element of Z then because Z is itself the line bundle $\mathcal{O}(1)$ we have $z \in \mathcal{O}(1)_{\pi(z)}$ and hence $z \in \pi^{-1}(\mathcal{O}(1))_z$. We denote this section of $\mathcal{O}(1)$, $z \mapsto z$ by η. Notice that the variety X_z, in this notation, is the subset of Z where $\eta = \sum_{i=0}^{n} \zeta^i z^i$. Finally, we remark that differential forms also pull back from X to Z. This is a standard piece of differential geometry, see for example [9, 10].

Recall the expression for ϕ in equation (11):

$$\phi(z) = \int_X \sum_{i=0}^{m} \omega_l (\sum_{i=0}^{n} z^i \zeta^i)^l$$

where each of the ω_l is a $(0, n-1)$ form on X with values in $\mathcal{O}(-n-1+k-l)$ respectively. Consider the $(0, n)$ form on Z with values in $\mathcal{O}(-n-1+k)$ defined by

$$\omega = \sum_{i=1}^{m} \pi^{-1}(\omega_l) \eta^l.$$

Because each X_z is a copy of X, we can restrict ω to X_z and integrate. We get

$$\int_{X_z} \omega = \int_{X_z} \sum_{i=1}^{m} \pi^{-1}(\omega_l) \eta^l$$

$$= \int_{X_z} \sum_{i=1}^{m} \pi^{-1}(\omega_l) (\sum_{i=0}^{n} z^i \zeta^i)^l$$

$$= \int_{X} \sum_{i=1}^{m} \omega_l (\sum_{i=0}^{n} z^i \zeta^i)^l.$$

Hence

$$\phi(z) = \int_{X_z} \omega.$$

We can in fact define such a function ϕ by this formula for any differential form ω on the twistor space Z of type $(0, n-1)$ with values in $\mathcal{O}(-n-1+k)$. Some of these, when integrated over any X_z, give zero. The situation is analogous to that of a real manifold M of dimension n. In that case, a differential n form can be integrated and an application of Stoke's theorem tells us that we will get zero if the differential form is of the form $d\mu$ for μ an $n-1$ form. The quotient of the space of all n forms by those of the form $d\mu$ is $H^n(M, \mathbb{R})$, the nth de Rham cohomology group of M. In the complex situation the appropriate analogue of d is

$$\bar{\partial} = \sum_{i=1}^{n} \frac{\partial}{\partial \bar{z}^i} d\bar{z}^i$$

which maps $(0, p)$ forms to $(0, p+1)$ forms. The space of all $(0, n-1)$ forms with values in $\mathcal{O}(-n-1+k)$ modulo those of the form $\bar{\partial}$ of something is denoted by

$$H^{n-1}(Z, \mathcal{O}(-n-1+k))$$

and called the $n-$ 1st Dolbeault cohomology of Z with values in $\mathcal{O}(-n-1+k)$. This is an infinite dimensional vector space (it would be finite dimensional if Z was compact). Finally, we can define the twistor correspondence:

$$T \colon H^{n-1}(Z, \mathcal{O}(-n-1+k)) \to H^0(\mathbb{C}^{n+1}, \mathcal{O})$$

by $T(\omega)(z) = \int_{X_z} \omega$.

The theorem proved in [19] is that this map T is a bijection onto the kernel of D_f. The method of proof is to note that the expansion in terms of powers of η of a differential form ω corresponds to the power series expansion of $T(\omega)$. In algebraic geometry, this expansion in powers of η corresponds to expanding the form ω normal to the subvariety X_0. It then remains essentially to compare terms in the power series.

8 The Examples Revisited

Let us revisit the examples in Whittaker and Watson [21]. The twistor theory tells us that on $\mathbb{R}^3$ a function in the kernel of the Laplacian is given by

$$\phi(x, y, z) = \int_{S^2} \sum g_l(w)(x\zeta^1 + y\zeta^2 + z\zeta^3)^l dw.$$

Restricting w to the circle and using a contour integral, this becomes

$$\phi(x, y, z) = \int_{\pi}^{-\pi} \sum g_l(w) 2^l \exp(ilw)(xi\cos(w) + yi\sin(w) + z)^l dw.$$

Assuming the sum converges and defining

$$g(v, w) = \sum g_l(w) 2^l \exp(ilw) v^l,$$

we recover the formula in section 3.

In the second case, if the polynomial is

$$f(x, y, z, t) = x^2 + y^2 + z^2 - t^2$$

and X is again a quadric in $\mathbb{C}P_3$. In this case, it is known that this quadric is a product of two spheres, that is $X = \mathbb{C}P_1 \times \mathbb{C}P_1$. We can define "coordinates" on this (actually a two-to-one map) by

$$(w, w') \mapsto [(1 + w^2)(w'^2 - 1), -i(w^2 - 1)(w'^2 - 1), 2i(1 + w'^2)^2, 4iww']$$

and a similar calculation yields the formula of Whittaker and Watson.

9   Correspondences

The double fibration in equation (12) is an example of a *correspondence*. More generally, we say that a correspondence is a submanifold of $X \times Z$ such that the induced maps $W \to X$ and $W \to Y$ are fibrations. We summarise this data by considering the double fibration

$$
\begin{array}{ccc}
 & W & \\
\swarrow & & \searrow \\
X & & Z
\end{array}
\tag{14}
$$

Notice that a correspondence is a generalisation of the idea of having a map from X and Z or vice versa. For example if $F \colon X \to Z$ then the graph of F inside $X \times Z$ defines a correspondence; similarly for a map from $Y \to X$.

Another way of obtaining a correspondence is to consider a Lie group G with two subgroups H and K. Then we have the diagram of homogeneous spaces:

$$
\begin{array}{ccc}
 & G/(H \cap K) & \\
\swarrow & & \searrow \\
G/K & & G/H
\end{array}
\tag{15}
$$

Because many representations arise as spaces of sections of bundles over homogeneous spaces, we are naturally led to Penrose transforms between representations. Many *integral intertwining operators* arise in this way [4].

The classical Radon transform also fits into this picture if we let $X = \mathbb{R}^3$, Z be the space of planes in $\mathbb{R}^3$ and W the subset of $\mathbb{R}^3 \times Z$ where the point in $\mathbb{R}^3$ lies on the plane [10].

Given a correspondence we can try and transform objects such as differential forms, sections of bundles and bundles backwards and forwards between Z and X via W. To do this, we need to be able to pull objects back from Z and W and then push them down from W to X. A quite general machinery for this has been developed [6].

10   The Weierstrass Representation

Examples of nonlinear twistor transforms really go back to Weierstrass' work on minimal surfaces, although Weierstrass, of course, was not thinking about twistors. Let me explain briefly how the classical Weierstrass representation of a minimal surface [14] fits into the context of mini-twistor space.

If Σ is an oriented surface in $\mathbb{R}^3$ then for any point p of Z there is a well-defined oriented normal line ℓ_p. Hence we have a map

$$\begin{array}{ccc} \Sigma & \to & Z \\ p & \mapsto & \ell_p \end{array}$$

whose image defines a submanifold $\widetilde{\Sigma}$ of Z. Weierstrass' result is that if the surface Σ is minimal then $\widetilde{\Sigma}$ is a holomorphic submanifold of Z. There is also a reverse correspondence that constructs a minimal surface out of a holomorphic submanifold of Z.

Note that if we compose with the projection from Z to S^2 that maps a line to the unit normal in its direction then we obtain the classical Gauss map which associates to a point on the surface its unit normal in S^2. For further details on the Weierstrass representation from this point of view, the reader should look at the paper of Hitchin [11].

11   Monopoles

The original work in this area was the Atiyah–Ward transform for the self-duality equations [3] on $\mathbb{R}^4$. A good reference is the Pisa lecture notes of Atiyah [1]. These can be hard to find so it is worth noting that they are in his collected works. As I am more interested in monopoles and how they fit into the mini-twistor space setting we have developed here, let me say something briefly about them. They are, in any case, the time invariant version of the self-duality equations. The best place to learn about monopoles is the book by Atiyah and Hitchin [2] and the references therein.

Monopoles are a gauge theory on $\mathbb{R}^3$. To define a monopole we start with a pair (A, ϕ) consisting of a connection 1-form A on $\mathbb{R}^3$ with values in $LSU(2)$, the Lie algebra of $SU(2)$, and a function ϕ (the Higgs field) from $\mathbb{R}^3$ into $LSU(2)$. The Yang–Mills–Higgs energy on this pair is

$$\mathcal{E}(A, \phi) = \int_{\mathbb{R}^3} (|F_A|^2 + |\nabla_A \phi|^2) d^3 x$$

where $F_A = dA + A \wedge A$ is the curvature of A, $\nabla_A \phi = d\phi + [A, \phi]$ is the covariant derivative of the Higgs field, and we use the usual norms on 1-forms and 2-forms and the standard inner product on $LSU(2)$. The energy is minimized by the solutions of the Bogomolny equations

$$\star F_A = \nabla_A \phi \tag{16}$$

where $\star$ is the Hodge star on forms on $\mathbb{R}^3$. These equations, and the energy, are invariant under gauge transformations, where the gauge group $\mathcal{G}$ of all maps g from $\mathbb{R}^3$ to $SU(2)$ acts by

$$(A, \phi) \mapsto (gAg^{-1} - dgg^{-1}, g\phi g^{-1}).$$

Finiteness of the energy, and the Bogomolny equations, imply certain boundary conditions at infinity in $\mathbb{R}^3$ on the pair (A, ϕ) which are spelt out in detail in [2]. In particular, $|\phi| \to c$ for some constant c which is usually taken to be 1.

A monopole, then, is a gauge equivalence class of solutions to the Bogomolny equations subject to these boundary conditions. In some suitable gauge, there is a well-defined Higgs field at infinity

$$\phi^\infty \colon S^2_\infty \to S^2 \subset LSU(2)$$

going from the two-sphere of all oriented lines through the origin in $\mathbb{R}^3$ to the unit two-sphere in $LSU(2)$. The degree of ϕ^∞ is a positive integer k called the magnetic charge of the monopole.

Because the Bogomolny equations are nonlinear, the set of all solutions is not a linear space. After we quotient it by the gauge group, it is a manifold M_k of dimension $4k$ called the moduli space of charge k monopoles. In the case that $k = 1$, there is a spherically symmetric monopole called the Bogomolny–Prasad–Sommerfeld (BPS) monopole, or unit charge monopole. Its Higgs field has a single zero at the origin, and its energy density is peaked there so it is reasonable to think of the origin as the centre or location of the monopole. The Bogomolny equations are translation invariant so this monopole can be translated about $\mathbb{R}^3$ and also rotated by the circle of constant diagonal gauge transformations. This, in fact, generates all of M_1 which is therefore diffeomorphic to $S^1 \times \mathbb{R}^3$. The coordinates on M_1 specify the location of the monopole and what can be thought of as an internal phase. More generally, there is an asymptotic region of the moduli space M_k consisting of approximate superpositions of k unit charge monopoles located at k widely separated points and with k arbitrary phases.

Hitchin [11, 12] developed the Atiyah–Ward correspondence for monopoles and this transforms a monopole into a holomorphic vector bundle on Z. The vector bundle is easy to describe. If γ is an oriented line we can restrict the connection and Higgs field to the line and consider the ordinary one variable differential equation

$$(\nabla_{\dot\gamma} - \Phi)\psi = 0. \tag{17}$$

The space of solutions to this is a two-dimensional complex vector space E_γ depending on γ. As such, it forms a vector bundle $E \to Z$. The real trick now is to make this a *holomorphic* vector bundle. To do this, the original connection and Higgs field have to satisfy the Bogomolny equations. Conversely, given this bundle we can recover the monopole and the solution of the Bogomolny equation.

Hitchin uses this Atiyah–Ward formalism to prove a number of interesting results. Let me conclude by mentioning the *spectral curve* of a monopole. Generally, the equation (17) has no solutions that decay at both ends of the line. The set of all lines for which it does have such solutions form a special subset of Z called the *spectral curve*. This is determined by an equation of the form

$$\eta^k + a_1(\zeta)\eta^{k-1} + \cdots + \alpha_k(\zeta) = 0$$

where each of the a_i is a polynomial of degree $2i$. So, in particular, the spectral curve is holomorphic. This sort of result would be enormously difficult to prove with just analysis in $\mathbb{R}^3$. Hitchin goes on to show how to recover the monopole from its spectral curve [11] and to show precisely which curves correspond to monopoles [12]. This is just the beginning of an exciting story. We refer the reader to the book of Atiyah and Hitchin [2] and references therein for the details.

References

[1] M.F. Atiyah, *Geometry of Yang–Mills Fields*, Lezione Fermione, Pisa, 1979. (Also appears in Atiyah's collected works.)

[2] M.F. Atiyah and N.J. Hitchin, *The Geometry and Dynamics of Magnetic Monopoles*, Princeton University Press, Princeton, 1988.

[3] M.F. Atiyah and R.S. Ward, Instantons and algebraic geometry, *Commun. Math. Phys.* **55**, 117–124 (1977). (Also appears in Atiyah's collected works.)

[4] R.J. Baston and M.G. Eastwood, *The Penrose Transform, its Interaction with Representation Theory*, Oxford University Press, Oxford, 1989.

[5] M.G. Eastwood and R.O. Wells, Cohomology and massless fields, *Commun. Math. Phys.* **78**, 305–351 (1981).

[6] M.G. Eastwood, The generalised Penrose–Ward transform, *Math. Proc. Camb. Phil. Soc.* **97**, 165–187 (1985).

[7] M.G. Eastwood, On Michael Murray's twistor correspondence, in [16].

[8] M.G. Eastwood, Introduction to the Penrose Transform, *Contemporary Mathematics* **154**, 71–75 (1993).

[9] P. Griffiths and J. Harris, *Principles of Algebraic Geometry*, Wiley, New York, 1978.

[10] S. Helgason, *Differential Geometry, Lie Groups and Symmetric Spaces*, Academic Press, New York, 1978.

[11] N.J. Hitchin, Monopoles and geodesics, *Commun. Math. Phys.* **83**, 579–602, (1982).

[12] N.J. Hitchin, On the construction of monopoles, *Commun. Math. Phys.* **89**, 145–190, (1983).

[13] L.P. Hughston and R.S. Ward eds., *Advances in Twistor Theory*, Pitman Research Notes in Math. **37** (1979).

[14] J. Lucas, M. Barbosa and A. Gervasio Colares, *Minimal surfaces in* $\mathbb{R}^3$, Lecture Notes in Mathematics **1195**, Springer-Verlag, New York, 1986.

[15] N.J. Hitchin, N.S. Manton and M.K. Murray, Symmetric monopoles, *Nonlinearity* **8**, 661–692 (1995).

[16] L.J. Mason and L.P. Hughston eds., *Further Advances in Twistor Theory*, Pitman Research Notes in Math. **231** (1990).

[17] L.J. Mason, L.P. Hughston and P.Z. Kobak eds., *Further Advances in Twistor Theory Volume II*, Pitman Research Notes in Math. **232**, (1995).

[18] J.R. Munkres, *Topology A First Course*. Prentice-Hall, New Jersey, 1975.

[19] M.K. Murray, A twistor correspondence for homogeneous polynomial differential operators, *Math. Ann.* **272**, 99–115 (1985).

[20] R.O. Wells, *Differential Analysis on Complex Manifolds*, Springer-Verlag, New York, 1973.

[21] E.T. Whittaker and G.N. Watson, *A Course in Modern Analysis*, Cambridge University Press, Cambridge, 1950.

Index

anharmonic lattice model, 30
  Bäcklund transformation, 32
Atiyah–Ward correspondence, 221
Atiyah–Ward transform, 220

Bäcklund transformation, 3, 4, 16–
  47, 86–94
  anharmonic lattice model, 32
  for Heisenberg model, 87
  for sine-Gordon equation, 18–
    24
  for vortex filament equation, 89
  infinitesimal, 44–46
  nonlinear Schrödinger equation,
    37
  Weingarten system, 33
Bäcklund, A.V., 16–18
  photograph, 48
Baker–Akhiezer function, 66–72
bi-Hamiltonian structure, 84
Bianchi diagram, 25, 26
Bianchi lattice, 26, 27
Bianchi permutability theorem, 24–
  26
Bianchi surface
  figure, 35
Bianchi system, 21
Bogomolny equations, 201, 220, 221
Bogomolny–Prasad–Sommerfeld
  monopole, 221

calculus of variations, 171
canonical bundle, 209, 213
Cartan
  1-form, 165, 181, 187–188
  2-form, 181
  characters, 111
    reduced, 111, 160
  lemma, 168
  system, 104
Cartan's involutivity test, 108–115,
  151, 160, 161
Cartan–Kähler theorem, 108, 110,
  151, 161
Cartan–Kuranishi theorem, 155
Cauchy–Riemann equations, 121, 134
Christoffel symbols, 19

circulation, 56
coframes, 146, 147
  classification of, 155
  derivatives, 161
  invariant, 146, 160, 165, 166
  orthonormal, 148, 166
commutator, 180
completely integrable system, 2, 4,
  16–47, 60–62, 86–87, 94–
  96
complex line bundles, 206, 211
complex manifolds, 202
complex projective space, 206, 207
conformal map, 121
connection, 77
connection form, 77, 220
contact
  flow, 189
  forms, 106, 125, 179, 189, 195
  manifold, 189
  structure, 7, 149, 179
  transformation, 10
correspondence, 219
critical points, 91
curvature, 180
  total, 20

de Rham cohomology, 218
defining system, 117–123, 125–133
  infinitesimal, 134, 136–141
    initial data for, 137–141
derivative
  parametric, 120, 137, 140
  principal, 120
derived flag, 104
differential equations manifold, 10
differential forms of type (p, q), 208
Dini surface, 27
  figure, 28
Dolbeault cohomology, 218
  classes, 206
double fibration, 216, 219
dual bundle, 213

e-structure, 159, 162
equivalence group, 147
equivalence problem, 146–169

Gardner's approach, 158, 159
  overdetermined, 147
  prolongation of, 151
  reduction of, 151
  specific, 149, 155
  under-determined, 147
Euler–Lagrange equations, 171, 172, 181
Euler–Lagrange field, 181, 182
evolution space, 179
exterior derivative, 214
  foliated, 190, 196
exterior differential system, 100–115, 184
  in involution, 111
  prolongation of, 153
  quasilinear, 112, 151, 159
  reduction, 153
  torsion of, 152
  with independence condition, 109

first fundamental form, 19
first integrals, 185–187, 192, 194
  in involution, 187, 194
  in involution w.r.t. ω, 190, 192, 195
  in involution w.r.t. L, 186
Floquet eigenfunction, 74
Floquet exponent, 74
Frobenius theorem, 103

G-structures, 156
Γ-basic form, 185
Γ-contact form, 189–190, 192, 193
Gardner's approach to equivalence problem, 158, 159
gauge theory, 220
gauge transformation, 64, 69, 70, 80
Gauss equations, 19, 21
Gauss map, 220
Goursat normal form, 106

harmonic functions, 203, 205
Hasimoto transformation, 62, 63
heat equation, 117
Heisenberg model, 90
  continuous, 64
Helmholtz conditions, 171–177, 181–184, 189
Higgs field, 221

hodograph system, 43
holomorphic line bundle, 209, 210
holomorphic section, 204, 210, 214
  of a holomorphic line bundle, 209
homogeneous polynomials, 213
horizontal subspace, 180
horizontal vector field, 181

infinite Lie pseudogroup, *see* Lie pseudogroup, infinite
instantons, 201
integral element, 108–111, 150–152
  admissible, 110
  Kähler regular, 109, 110
  ordinary, 110
integral manifold, 101–102
  admissible, 110
  Kähler regular, 109
integral transform, 208
invariant form, 125–127, 129, 130, 132
  mod ω^i, 128
invariant, scalar, *see* scalar invariant
involutive tableau, 112
isotropy algebra, 134, 136
  linear, 134–140
isotropy group, 121, 128, 131, 134
  linear, 128, 135

Jacobi endomorphism, 185
jet bundle, 6

Kadomtsev–Petviashvili equation, 142
Kepler problem, 177, 178
Korteweg–de Vries equation, 3
Korteweg–de Vries equation, 31
  modified, *see* mKdV equation

Lagrangian, 12, 171–182
Lagrangian mechanics, 148, 163
Lagrangian system
  completely integrable, 187
Laplace transformation, 1, 2
Laplacian, 207, 216, 218
Lax pair, 61, 64
Levi-Civita connection 1-forms, 168
Lie algebra, 134, 135
  conformal, 134

defining relations, 167
  orthogonal, 134
Lie group
  linear, 131, 135
  one-parameter, 134
Lie pseudogroup, 116–143
  Cartan structure, 123–133
    nonconstant, 131–133
  definition, 119
  finite, 120, 121
  infinite, 116–143, 162
  intransitive, 131, 132
  isomorphism, 124, 125, 133
  similar, 124, 140
  transitive, 131
line bundles
  isomorphism of, 212
linear isotropy algebra, *see* isotropy
    algebra, linear
linear isotropy group, *see* isotropy
    group, linear
Liouville equation, 117, 141
Liouville–Arnol'd theorem, 188–195
LKR system, 44–46
locally solvable, 123
Loewner transformation, 42, 43
loop space, 58
Lorentzian structures, 169

Mainardi–Codazzi equations, 19
Marsden–Weinstein Poisson bracket,
    82, 85
Marsden–Weinstein symplectic form,
    59
Maurer–Cartan 1-forms, 128, 151,
    167
method of moving frames, 11
mini-twistor space, 201, 202, 219,
    220
mini-twistors, 207, 216, 217
minimal surface, 219
mKdV equation, 31
  one-soliton surface, 33
  figure, 34, 35
monopoles, 220
  spectral curve of, 221
multiple points, 91
multiplier matrix, 171, 176
multiplier problem, 172

N-phase curve, 72, 73
N-phase solution, 65, 66
natural curvatures, 80
natural frame, 80
NLS equation, *see* nonlinear Schröd-
    inger equation
non-Lagrangian system, 193–195
nonlinear Schrödinger equation, 36–
    38, 86
  auto-Bäcklund transformation,
    37
  focussing cubic, 60, 62, 63
  one-soliton surface, 38
  figure, 38
  periodic, 84

oriented singular knot, 58

parametric derivative, *see* derivative,
    parametric
Penrose transform, 201, 206, 219
periodic and antiperiodic points, 91
Pfaff problem, 104
Pfaffian system, 101–107
Poisson bracket, 82
polar space, 109
principal components, 167
principal derivative, *see* derivative,
    principal
prolongation, 152
  isomorphic, 124, 133
  of a Lie algebra, 152
  of pseudogroup, 124
pseudo-sphere, 27
  figure, 28
pseudo-spherical surface, 21–29
  motion, 29–34
  single soliton, 26
pseudogroup, 118
  Lie, *see* Lie pseudogroup
  not Lie type, 121, 122

Radon transform, 219
real section, 204, 217
real structure, 204
Reeb field, 189
Riemannian geometry, 147
Riquier–Janet theory, 120, 171, 172
Runge–Lenz vector, 178

scalar invariant, 123, 133

second fundamental form, 19
second order differential equation
    field, *see* SODE
self-duality equations, 220
Serret–Frenet equations, 37
sine-Gordon equation, 3, 18–21
    auto-Bäcklund transformation,
        21–29
    breather solution, 26–29
    breather surface, figures, 29
    multi-soliton solutions, 24–26
    single soliton solution, 25
SODE, 179–182, 189
soliton, 16–47, 60–65, 84–96
stability group, linear, *see* isotropy
        group, linear
stabilizer, *see* isotropy group
steady state boundary layer equa-
        tions, 142
stereographic projection, 203
symmetry, 5, 116, 119, 122, 140, 142,
        178, 185–187
    of defining system, 122–123
    of first order o.d.e., 140

tautological bundle, 211
Taylor series, 135, 137
Toda lattice, 36
torsion, 157–162
    absorption of, 157
    intrinsic, 157, 164
    normalisation of, 158
torus knot, 75
total length functional, 59
transfer matrix, 74, 90
transition functions, 210
transverse field, 186
trivial bundle, 210
twistor, 13, 201–221
    correspondence, 206, 215, 218
    space, 216
    transform, 216
two-sphere, 203
    tangent bundle to, 211
Tzitzeica equation, 36
    breather surface, 37
    one-soliton surface, 36

variational bicomplex, 172
vector field along f, 184

vector Poisson equation, 56
vertical endomorphism, 180
vertical subbundle, 179
vertical subspace, 180
vortex filament, 56–58, 63
vortex filament equation, 60, 64

wave equation, 205, 219
Weierstrass representation, 219, 220
Weingarten system, 22, 31
    Bäcklund transformation, 33
    in asymptotic coordinates, 32

Yang–Mills–Higgs energy, 220

zero curvature formulation, 62

DATE DUE

NOV 0 5 2008			
			Printed in USA

HIGHSMITH #45230